Nonlinear Systems and Complexity

For further volumes:
http://www.springer.com/series/11433

Valentin Afraimovich • Albert C. J. Luo
Xilin Fu
Editors

Nonlinear Dynamics and Complexity

Springer

Editors
Valentin Afraimovich
San Luis Potosi University
San Luis Potosi, Mexico

Xilin Fu
Shandong Normal University
Jínan, China, People's Republic

Albert C. J. Luo
School of Engineering Mechanical
and Industrial Engineering
Southern Illinois University Edwardsville
Edwardsville, Illinois, USA

ISSN 2195-9994 ISSN 2196-0003 (electronic)
ISBN 978-3-319-02352-6 ISBN 978-3-319-02353-3 (eBook)
DOI 10.1007/978-3-319-02353-3
Springer Cham Heidelberg New York Dordrecht London

Library of Congress Control Number: 2013955553

Printed on acid-free paper

Springer is part of Springer Science+Business Media (www.springer.com)

Preface

The chapters in this edited book are selected from the invited lectures in *Conference on Nonlinear Dynamics and Complexity*, held at Jinan, Shandong, China, on July 23–29, 2012. The aims of this edited book are to collect recent advances in nonlinear dynamics, including analytical solutions, chaos in Hamiltonian systems, time-delay, uncertainty, and bio-network dynamics, and to stimulate further research in nonlinear dynamics and complexity. The topics in this book cover:

(i) Nonlinear dynamics in complex networks
(ii) Fractional maps with memory
(iii) Accurate analytical solutions for complex motions
(iv) Spiking neural network dynamics and biological processes
(v) Scaling laws of human virtual motions and human actions in uncertainty
(vi) Time-delayed nonlinear systems and noise effects

During a workshop-extended conference, the comprehensive discussions on the above topics were made, which were led by the invited recognized scientists. From such discussions, the young scientists and students learned new methods, ideas, and results. This workshop-extended conference was supported by China National Foundation of Science (Funding Number 111171192), Doctoral Program of China Ministry of Education (Funding Number 20123741110001), and Shandong Provincial Key Program of Basic Mathematics at Shandong Normal University.

Herein, editors would like to thank all the financial supports from China, the authors and reviewers, for supporting the conference and collection. We hope the results presented in this edited book will be useful for other specialists in nonlinear dynamics and complexity.

San Luis Potosi, Mexico — Valentin Afraimovich
Edwardsville, IL, USA — Albert C. J. Luo
Jínan, China, People's Republic — Xilin Fu

Contents

From Long-Range Order to Complex Networks, an Hamiltonian Dynamics Perspective

Sarah de Nigris and Xavier Leoncini

Abstract In this chapter we discuss the influence of a nontrivial network topology on the thermodynamic behavior of an Hamiltonian model defined on it, the XY-rotors model. We first focus on network topology analysis, considering the regular chain and a Small World network, created with the Watt–Strogatz model. We parametrize these topologies via γ, giving the vertex degree $k \propto N^{\gamma-1}$ and p, the probability of rewiring. We then define two topological parameters, the average path length ℓ and the clustering coefficient C and we analyze their dependence on γ and p. We conclude this part presenting an algorithm, the tree algorithm, which enhances the calculation speed of ℓ and C. In the second part, we consider the behavior of the XY model on the regular chain and we find two regimes: one for $\gamma < 1.5$, which does not display any long-range order and one for $\gamma > 1.5$ in which a second order phase transition of the magnetization arises. Moreover we observe the existence of a metastable state appearing for $\gamma_c = 1.5$. Finally we illustrate in what conditions we retrieve the phase transition on Small World networks and how its critical energy $\varepsilon_c(\gamma, p)$ depends on the topological parameters γ and p.

1 Introduction

The concept of *network* can be found in systems which spread from sociology and information science to biology and physics. We can take as fundamental and well-known examples, among others, the Web or networks of cells, like the neurons.

S. de Nigris (✉) • X. Leoncini
Aix-Marseille Université, Centre de Physique Théorique, Campus de Luminy,

Aix-Marseille Université, CNRS, DGA and, if possible, she would like to acknowledge, at the end of the chapter, financial support from DGA/MRIS.

Case 907 - 13288 Marseille cedex 9, France
e-mail: denigris.sarah@gmail.com; Xavier.Leoncini@cpt.univ-mrs.fr

V. Afraimovich et al. (eds.), *Nonlinear Dynamics and Complexity*, Nonlinear Systems and Complexity 8, DOI 10.1007/978-3-319-02353-3_1,

Despite the difference between those systems, some common features arise in these real world networks as a high level of clustering and the "Small World" effect itself, which we will detail in the following. These shared properties are, looking at the heterogeneity of the examples, quite independent from the punctual nature of the agents interacting on networks and thus a lot of work has been done to point out how common *topological* features stem from simple assumptions which can be taken regardless of the particular model considered. We would like to stress here the importance of understanding and characterizing real world networks: actually the dynamical processes taking place on the top of them are profoundly influenced by the underlying structure and an enhanced knowledge of the latter could shed light on the rise of phenomena like synchronization or phase transitions (for a more complete overview, see [1]). We are thus dealing with two levels of analysis: the microscopical scale, which takes into account the very specific nature of the interaction between the agents on the network and the macroscopic one which is connected to the collective behaviors as the phase transitions. Thus a useful frame to understand the interplay between these two scales can be nonlinear dynamics, whose approach naturally deals with describing emergent collective phenomena. Indeed this approach could have the potential to turn the *qualitative* matching between network structure and dynamical processes into *quantitative* and the present work aims to inscribe itself on this line. To do so we chose a model for the network, the Watts and Strogatz model for Small World networks [2], and a model for the interaction, the XY model for rotators (see, for instance, [3, 4]). Our choice for the network has essentially two motivations: first the Watts and Strogatz model catches the two over mentioned topological features of many real world networks: high clustering and small average distance between two nodes of the network. This latter characteristic is "Small World" effect but, in fact, small world networks often show also a high clustering and this fact justifies our choice for the network model. Second, the operative algorithm provided by the model gives a natural way to *tune the topology* of the network: as we will see in Sect. 2.3.1, we have a continuous parameter with which it is possible to switch between two opposite topologies of networks, the regular and the random one, passing through the proper Small World topology. On the other hand, dealing with the interaction, the XY model for spins is paradigmatic for the study of phase transitions: this is a Hamiltonian model for two-dimensional vectors, described by an angle θ_i, interacting via their scalar product: $V(\theta_i - \theta_j) \sim 1 - (\cos(\theta_i - \theta_j))$. This model has been, over the years, the object of several numerical studies, especially on 2 and 3-D lattices [5–10]; moreover, this model displays, in the 2-D configuration with nearest neighbors interactions, *quasi* long-range in the form of a phase transition affecting the decaying of the correlation function, the Berezinskii–Kosterlitz–Thouless phase transition [4,11,12]. Moreover, recently, it has been shown that the XY model can host true long-range order even in the 1-D case, provided that the spin has a minimal number of interactions between each other [13]. In the case of full coupling of the spins, we recover the *Hamiltonian Mean Field* model (XY-HMF) which, as well as the XY model, has become a test bench for theories on system with long-range interactions. In spite of the simplicity of its definition, that we will detail in Sect. 3.2, both the equilibrium and the out-of-equilibrium regimes have a rich phenomenology: we mention here, as an example among others, an equilibrium phase transition of a global order parameter, the magnetization [3, 15]; while, concerning the out-of-equilibrium behavior, the emergence of out of equilibrium stationary states (*QSS*) has attracted a lot of attention especially since their length diverges with the system size [16–20]. The purpose of this work is to establish a connection between the *topological variations* of the network and the *response* of the equilibrium behavior of the Hamiltonian XY model we defined on it. Before starting with our analysis we would like to recall that this issue has also been tackled in [21, 22] but, differently from our case, they considered the XY model in the canonical ensemble via Monte Carlo simulations. We sketch here, as a guide for the reader, the steps of our analysis: in Sects. 2 and 3, we describe, respectively, some concepts defining networks in general and the XY model. Focusing on the latter, in Sect. 3.2, we make a digression on the XY-HMF model since we will use it as a benchmark to understand the behavior of the XY model on a network. In Sect. 2.3, we present in detail the characteristics of a Small World network, in particular its topological parameters, and an algorithm we developed to perform extensive and efficient calculations of these parameters. Finally, Sect. 4 consists in the synthesis of the previous ones: using the tools we introduced for the networks and numerical simulations, we will show symmetries and differences between the equilibrium properties of XY-HMF and the XY model on a regular chain first (Sect. 4.1) and successively on complex Small World network (Sect. 4.2) trying to specially highlight the role of *topological complexity*.

2 Introducing a Network

As explained in the introduction, we aim in this work to enlighten the influence of *network topology* in the thermodynamical behavior of an Hamiltonian system. There exist many famous examples which can illustrate, even with very simple models, this influence and thus justify our interest. We can consider, for instance, the XY model for spins, which we will introduce in the following section, on a 1-D linear chain with a nearest neighbors interaction. Because of the low dimensionality of the system, as an application of the Mermin–Wagner Theorem, it does not exhibit a phase transition of the global magnetization. On the contrary, if we take in account the same model on a 2-D network with a nearest neighbors interaction this one will show a Kosterlitz–Thouless phase transition, i.e. a transition from a order phase to a disordered one [23, 24]. Furthermore, considering the XY model in the full coupling limit, we recover another well-known case, the XY-HMF model, which displays a second order phase transition of the magnetization [25]. Those antipodal behaviors are hence intimately correlated with the shape and the dimensionality of the lattice connecting the particles and we shall introduce in the following some basic definitions in order to specify it.

2.1 Definitions

We encode the network connecting the particles in an *adjacency matrix* $\epsilon_{i,j}$ whose elements are 1 if the particles *i* and *j* are connected, 0 if they are not:

$$\epsilon_{i,j} = \begin{cases} 1 & \Rightarrow i, j \ connected \\ 0 & \Rightarrow i, j \ not \ connected \end{cases}. \tag{1}$$

We have, thus, a total number of links $N_L$connecting the N particles of the system and these particles are located at the nodes, or *vertices,* of the network. We introduce also a parameter γ which quantifies the total number of links N_L:

$$\gamma = \frac{log(N_L)}{log(N)}, \tag{2}$$

where $\gamma \in [1, 2]$ and for the two limit cases we have:

$$\gamma = \begin{cases} 1 & \Longrightarrow linear\ chain \\ 2 & \Longrightarrow fully\ coupled \end{cases}. \tag{3}$$

Finally we define the average number of links per particle, the *degree k*:

$$k = \frac{2^{2-\gamma} N^{\gamma}}{N} = 2^{2-\gamma} N^{\gamma-1}. \tag{4}$$

The prefactor $2^{2-\gamma}$ gives the "correct" value of *k* in the case of a linear chain ($\gamma = 1$) with closest neighbors interaction in which we have 2 links per vertex.[1]

2.2 The Regular Chain

Considering the example at the beginning of Sect. 2 and the definitions of Sect. 2.1, it comes naturally to the mind to first consider the regular one-dimensional chain as network topology (Fig. 1). At a first glance, this topology would seem trivial but we already discussed that, in the two extremal cases of Eq. (3), the degree is responsible for opposite thermodynamical behaviors. It is hence of interest to precisely understand in what conditions the pure one-dimensional topology is broken and the long-range order arises. Using γ in Eq. (2) is then possible to tune the range of interaction and investigate the parameter zone in between those two

[1] In the present discussion, the words "vertex" and "particle" are confounded since the particles are located at the vertices of links.

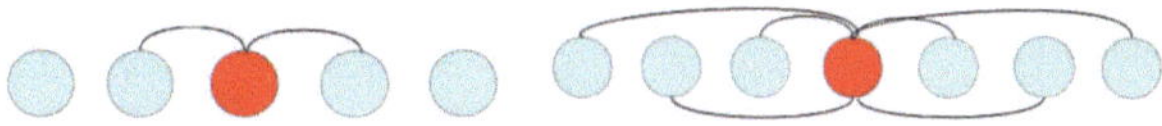

Fig. 1 Linear chain with (*left*) $k = 2$ and (*right*) $k = 6$

cases, the nearest neighbors and the full coupling interaction. In order to ensure the rotational invariance and avoid border effects we imposed periodic boundary conditions so that each spin possesses a symmetrical neighborhood of interaction (Fig. 1) and the chain is effectively a ring. As a final practical observation, we remark that, for fixed N and γ, Eq. (4) gives a non-integer result; so to construct our ring we take the integer part of Eq. (4).

2.3 *The Small World Networks*

2.3.1 General Characteristics

We introduced, in Sects. 2 and 2.2, some basic definitions of networks and the case of the regular chain with the spin neighborhood is controlled via γ. Moreover, at the beginning of this section, we suggested that a complex topology could be the key for interesting changes in the thermodynamical behavior[2]: at this point we shall introduce some tools to quantify this complexity and we will focus our analysis on a particular type of complex networks, the Small World Networks. We define as "Small World effect" the property of some networks to have a logarithmic growth of the average distance between two vertices $< \ell >$ with the system size:

$$< \ell > \sim \log(N). \tag{5}$$

In Eq. (5) we have that the average distance grows slower than linear with the system size as it happens in regular networks. This is the signature of the presence of *shortcuts* and, from the point of view of the statistical physics, these shortcuts can imply the emergence of *global coherence.* The Small World effect has first been detected in social networks [27] and it is often accompanied by a high clustering coefficient of the network, as we will see in the following. The algorithm chosen to produce such networks is issued from the seminal paper of Watts and Strogatz [2] and it acts on a regular network rewiring randomly the links. In practice we start from a regular network in which each vertex is connected to its k neighbors and according to a fixed probability p each link is either left untouched or rewired. Hence we have a parameter, the *probability* p, to tune the level of rewiring: for low p values, the network is almost regular; on the other hand, for high p values, almost

[2]We recall, for instance, that it has been shown how the thermodynamics of the XY model on *random* networks, hence topologically trivial, recovers the XY-HMF behavior [26].

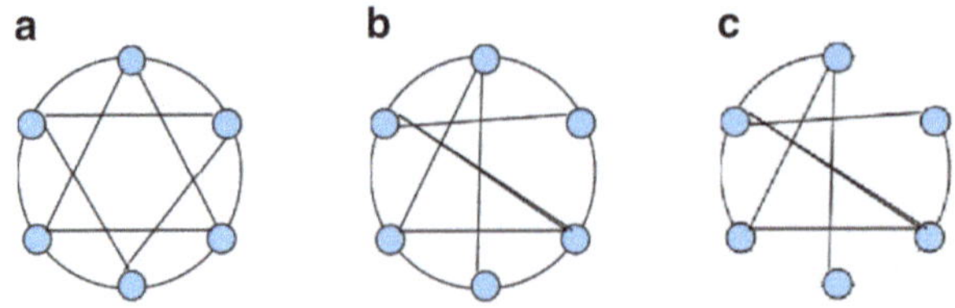

Fig. 2 Transition from regular network (**a**) to a random one (**c**) increasing the probability p. (**b**) corresponds to a Small World network

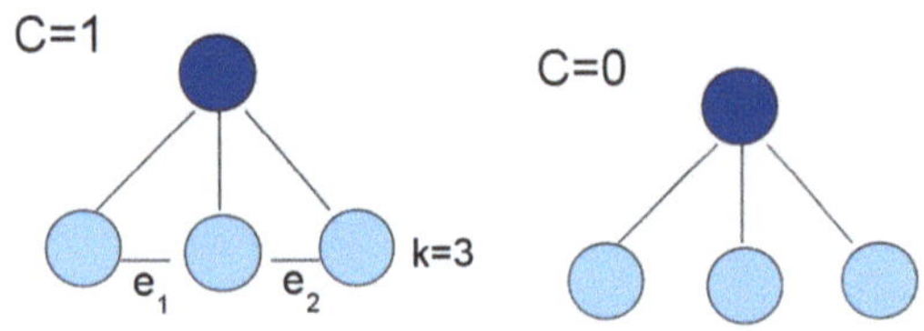

Fig. 3 (*left*) Configuration of maximal clustering coefficient (*clique*). (*right*) Configuration of minimal clustering coefficient

all the links are rewired and the network is random. With this parameter p, we can pass continuously between these two limit cases as shown in Fig. 2. Depending on the system size, a parameter region can be found in which the network has the two over mentioned features, high clustering coefficient and little average distance, being then what is now called a Small World network.

2.3.2 Important Parameters

Before proceeding further, we shall recall the definition of these two quantities, the clustering coefficient and average path length [27]. For the clustering coefficient we have:

$$C = \frac{1}{N}\sum_i c_i ,$$

where:

$$c_i = \frac{e_i}{\frac{1}{2}k_i(k_i - 1)}. \tag{6}$$

In Eq. (6), k_i stands for the number of links connected to the vertex i, the *degree* of the vertex. e_i is the number of links existing between the k_i neighbors of the vertex (Fig. 3): the maximal number of couples between the neighbors is $\frac{1}{2}k_i(k_i - 1)$ and we count how many of this links effectively exist, e_i. The fully coupled configuration will have then $c = 1$ (Fig. 3) and it is called a *clique*. In practice the clustering coefficient quantifies the average amount of clustering per

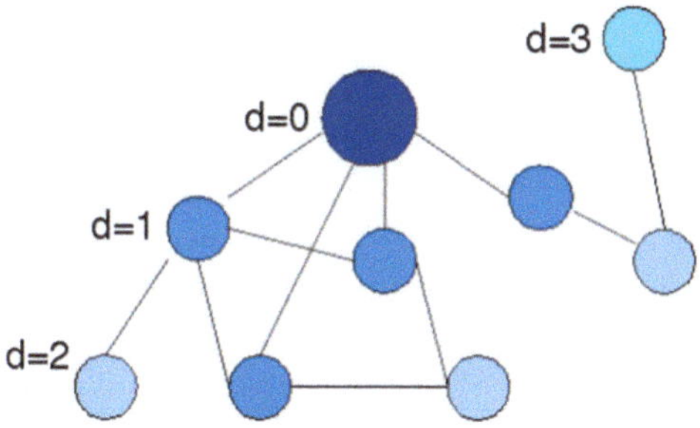

Fig. 4 Path lengths starting from the blue vertex

vertex and it is, by definition, a local parameter. For a regular network, the clustering coefficient has an analytical expression:

$$C_0 = \frac{3(k-1)}{2(2k-1)}.$$

where k is the degree of each vertex. Considering now the network globally, we define the average shortest path length as:

$$\ell = \frac{1}{N}\sum_i \ell_i, \tag{7}$$

with ℓ_i being the longest path attached to the vertex i. To quantify these paths, since the network lacks a metric, we count the number of edges between two vertices. Starting, for instance, from the i vertex, we have that its neighbors are at distance $d = 1$, the neighbors of the neighbors are at $d = 2$, and so on (Fig. 4). Finally, to ensure that the path taken is the shortest possible, we impose to consider each vertex only once avoiding this way to come back on links already explored. For this parameter too we can give an analytical expression in the case of a regular network:

$$\ell_0 = \frac{N(N+2k-2)}{4k(N-1)} \sim \frac{N}{4k}.$$

For big N values and k finite, which is the typical configuration, ℓ_0 is linear with the system size, while from Eq. (5), we have a logarithmic scaling for Small World networks. We mentioned above the centrality of the topological characterization: this centrality shall be justified now since the nature of the network (random, regular or Small World) (cfr Fig. 2) can be related to the values of the clustering coefficient c and the average path length $\tilde{\ell}$. As shown in Fig. 5 it is possible to identify three main zones: for low p values the two topological parameters are both high and we are thus in the regular network region. On the other side, for high p values, the network has low clustering coefficient and low average path length since it

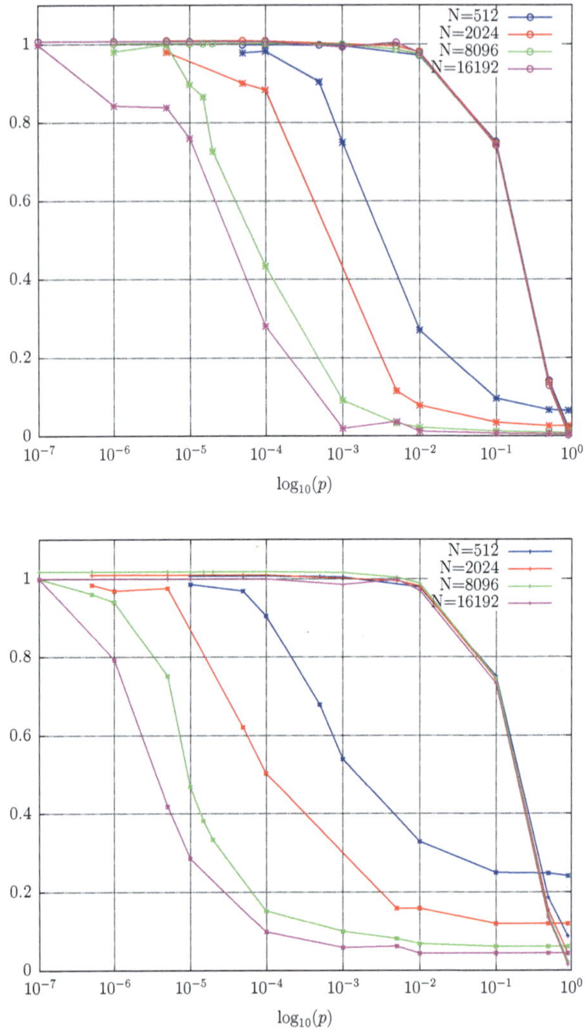

Fig. 5 Connectivity and average path length versus rewiring probability for $\gamma = 1.25$ (*up*) and for $\gamma = 1.5$ (*bottom*). Stars refer to length while circles to the clustering coefficient

is completely random. The intermediate zone is the one of our interests: the network is characterized by high clustering coefficient and low average path length having thus the features of a Small World network. This region varies with the system size since it is delimited by the fall of the average path length which is a global parameter. So far we focused our attention on the effects of the addition of randomness, via the probability p, in the regular networks. To complete now our analysis it is worth recalling that, in Sect. 2.2, we introduced a parameter γ whose role is controlling the interaction range in the regular chain case. Consequently, the Small World networks

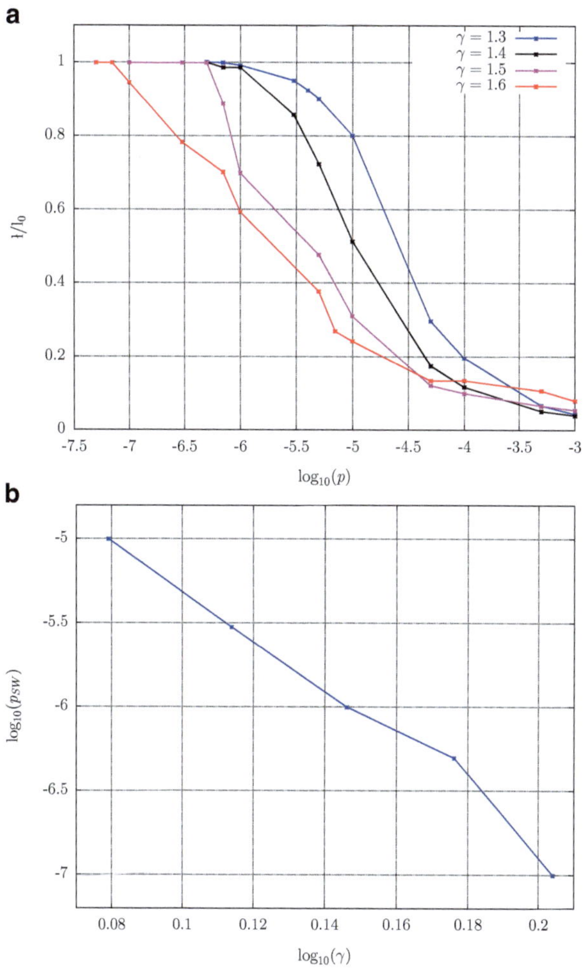

Fig. 6 (**a**) Average path length for different γ values and $N = 2^{14}$; (**b**) Scaling of the critical probability varying γ

we consider are parametrized by both p and γ and even by an heuristic inspection of Fig. 5, it is clear that the extensive parameter ℓ is heavily affected by γ, while the clustering coefficient which is intensive, does not seem to be much influenced by γ. Therefore to have a better insight on this effect we calculated numerically the average path length $\ell(N, \gamma, p)$ for several γ values (Fig. 6a). It is hence possible to define a threshold probability, p_{SW}, at which $\ell(N, \gamma, p)$ drops at the small values typical of random networks: this drop, as explained previously, signals the passage to the "Small World regime." This probability p_{SW} is expected to have the following expression [28]:

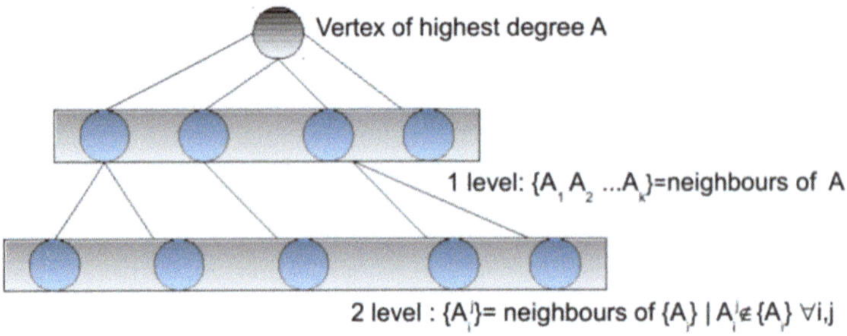

Fig. 7 Example of tree: starting from the vertex with the maximal degree k, we set at the first level its neighbors $\{A_1, \ldots, A_k\}$, at the second level the neighbors $\{A_1^j, .., A_k^j\}$ of the $\{A_i\}$ with the condition $A_i^j \notin \{A_i\}$ $\forall i, j$ and so on until having structured all the cluster

$$p_{SW} \sim \frac{1}{N^D k D} \propto \left(\frac{1}{N}\right)^{\gamma}, \tag{8}$$

in our case $D = 1$ since we deal with one-dimensional chains. In Fig. 6 we show the scaling with γ of the estimated p_{SW} from our simulations which is coherent with Eq. (8). The condition in Eq. (8) comes from a simple statistical argument: since, on average, the rewired links are $N_R = N^D k D p$, the expression in Eq. (8) ensures that, once the Watt–Strogatz algorithm has been performed on the regular network, at least one link would have been rewired, i.e. $N_R = N^D k D p_{SW} = 1$. We can hence conclude, from the evidence in Fig. 6, that the degree quantitatively affects the passage to the Small-World regime, enhancing the creation of shortcuts.

2.4 *The Tree Structure*

In Sect. 2.3.1, we introduced two parameters, the clustering coefficient and the average shortest path length, which imply a high degree of knowledge of the network and are thus computationally very expensive. More, to encode the information about the links between particles, we used the adjacency matrix which contains N^2 elements, many of them irrelevant since zero. Because of this need of efficiency and enhanced knowledge of the network, we thus developed a more organized structure, the *tree*, to rearrange the data concerning the links. The idea is to start from the vertex with the highest degree and to set at the first level its neighbors, at the second one the neighbors of the neighbors, and so on for the next levels. Finally we have at the nth level the nth generation of neighbors with the constraint to put in the next level only vertices which have not previously been visited (Fig. 7). The search for the new level stops when we have already explored all the cluster and the remaining vertices are connected to another one. This algorithm somehow recalls the one described in Sect. 2.3.1 for the average path length and we will see how they are related.

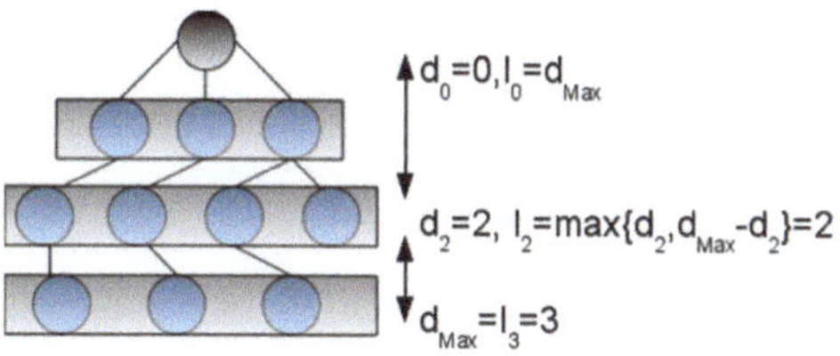

Fig. 8 Application of Eq. (9): the head vertex is at $d_0 = 0$; hence, its ℓ_0 is maximal, d_{max}. For the second level, it is at $d_2 = 2$ and, since there exist tree levels ($d_{max} = 3$), $\ell_2 = d_2 = 2$

Constructing the tree implies to access the detailed structure of the network: we can deduce the degree and cluster distributions and know how many particles are on the tree levels, characterizing completely the network (see Sect. 2.3.1). The second aim we pointed was to improve the calculation efficiency of the over mentioned topological parameters, the clustering coefficient and the average shortest path length. For the clustering coefficient we have, from Sect. 2.3.1, that its calculation is about counting triangles of links between a chosen vertex and its neighbors (Fig. 3). Hence the number of checks we are supposed to perform looking for these triangles will be a priori $Nk(k-1) \sim O(N^3)$ because of the Eq. (4) of the degree k. On the other hand, with the tree structure we have the additional information that these links can exist only between particles on the same level or in levels at a distance one: building the tree, we kept track of the links which affect the average path length; hence, for the clustering coefficient, we cannot have a triangle of links on two levels spaced by more than one step since, in this case, we would be looking for links which are either already considered in the tree or inexistent (Fig. 13). The restriction to a couple of levels instead of searching the whole adjacency matrix implies a remarkable improvement in the calculation of the clustering coefficient, allowing us to explore the behavior of this parameter for several sizes of the system, values of p and γ (see Sect. 2.3.1). For the average shortest path length, we decided to take an approximation instead of considering for each vertex all the possible paths which start from it. We considered the distribution of the level population as a statistical weight for the average path and we chose to associate at each level a distance defined as:

$$\tilde{\ell}_i = max(d_i; d_{max} - d_i),$$

where d_i is the distance of the i-level from the head of the tree and d_{max} is the total length of the tree (Fig. 8). The average shortest path length is:

$$\tilde{\ell} = \frac{1}{N} \sum_i \tilde{\ell}_i n_i, \tag{9}$$

n_i being the population of the i-level. We observe that $\tilde{\ell}$ calculated with Eq. (9) remains smaller than the one calculated with the classical definition in Eq. (7): if we re-write Eq. (9)

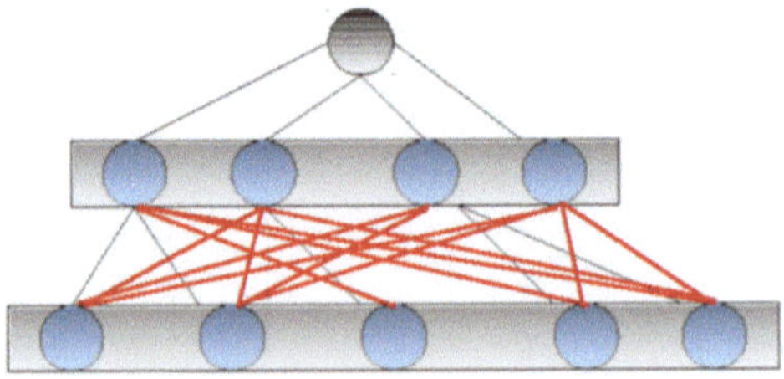

Fig. 9 Tree corresponding to the calculation of the length in Eq. (9). The *red links* are the ones "added" by the expression of $\tilde{\ell}$

$$\tilde{\ell} = \frac{1}{N} \sum_j \sum_{i=1}^{n_j} \tilde{\ell}_j,$$

it appears in a clearer way that $\tilde{\ell}$ is a sum of N terms like ℓ in Eq. (7) and, in fact, it corresponds to calculate ℓ on a network in which we added links to connect completely the levels (Fig. 9). So, because of the increased number of links which "constructs" new paths, $\tilde{\ell} < \ell$. In other words, if we calculated ℓ on the completely connected tree of Fig. 9, we would obtain the same value than $\tilde{\ell}$; while, since normally ℓ is calculated on a less connected network, the paths ℓ_i of Eq. (7) are longer. We discussed earlier in Sect. 2.3.1 that ℓ is important since its drop, occurring at p_{SW}, signals the crossover to the Small World regime. Hence, to validate the approximation $\tilde{\ell}$, it is of interest to compare the two threshold probabilities p_{SW} and $\tilde{p}_{SW}$, estimated, respectively, from the drop of ℓ and $\tilde{\ell}$. Just for comparison, we show in Fig. 10 that $\tilde{\ell}$ and ℓ [2] for $\gamma = 1.5$ have the same qualitative behavior so that the estimate p_{SW} and $\tilde{p}_{SW}$ correspond. To be more quantitative, we calculated $\tilde{p}_{SW}(N, \gamma)$ in the N range considered in Fig. 11b for $\gamma = 1.5$; in Fig. 11a we show that $\tilde{p}_{SW}(N, \gamma)$ displays the power law scaling of Eq. (8): consequently, we conclude that $\tilde{\ell}$ can be taken as a good indicator for ℓ to estimate the interesting parameter regions for the networks.

2.4.1 Algorithm Performance

As sketched in Sect. 2.4, the introduction of the tree structure achieves two aims: speeding up the calculation of the topological parameters and deepening the knowledge of the network. The speed of calculation is crucial since to enlighten clearly the influence of the network properties on the thermodynamical behavior we need their complete characterization. First, focusing on the clustering coefficient, we have that the improvement is given only by the reduced number of checks and it is a consequence of the tree structure. In Fig. 12 we show the increasing time of computation for different system sizes and values of gamma with the two algorithms. In this case, the clustering coefficient is calculated as in Eq. (6)

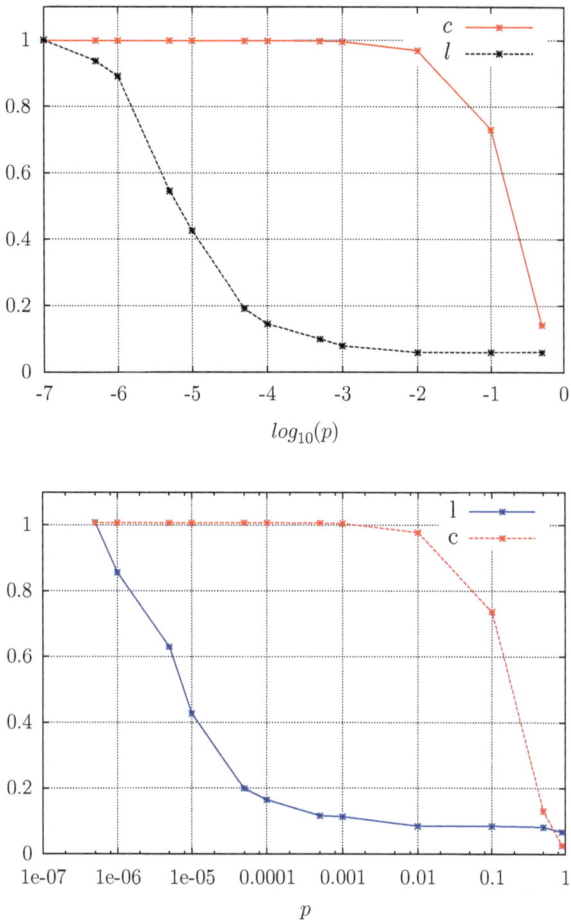

Fig. 10 (*up*) Plot of topological parameters for different p values for $N = 10,000$, $\gamma = 1.5$ using the standard algorithms. (*bottom*) Plot of topological parameters for different p values for N=10,000, $\gamma = 1.5$ using the tree algorithms

and we obtain the gain in speed because, calculating the clustering coefficient of the ith-vertex, we have a restriction in the choice of suitable connected vertices (Fig. 13). On the other hand, for the average path length, we took the approximation $\tilde{\ell}$ which is justified by the fact that its scaling reproduces correctly the theoretical one ℓ (Fig. 11). This approximation $\tilde{\ell}$ implies, however, a remarkable change in the times of calculation: from Eq. (7) we should explore, when considering a vertex, every path attached to it, scrolling hence a good part of the network for each of the neighbors. With the approximation $\tilde{\ell}$, building the tree gives in an automatic way the populations of the levels and their distance, allowing to calculate directly $\tilde{\ell}$ via Eq. (9). In Fig. 14 we show the curves of the computational time of the two

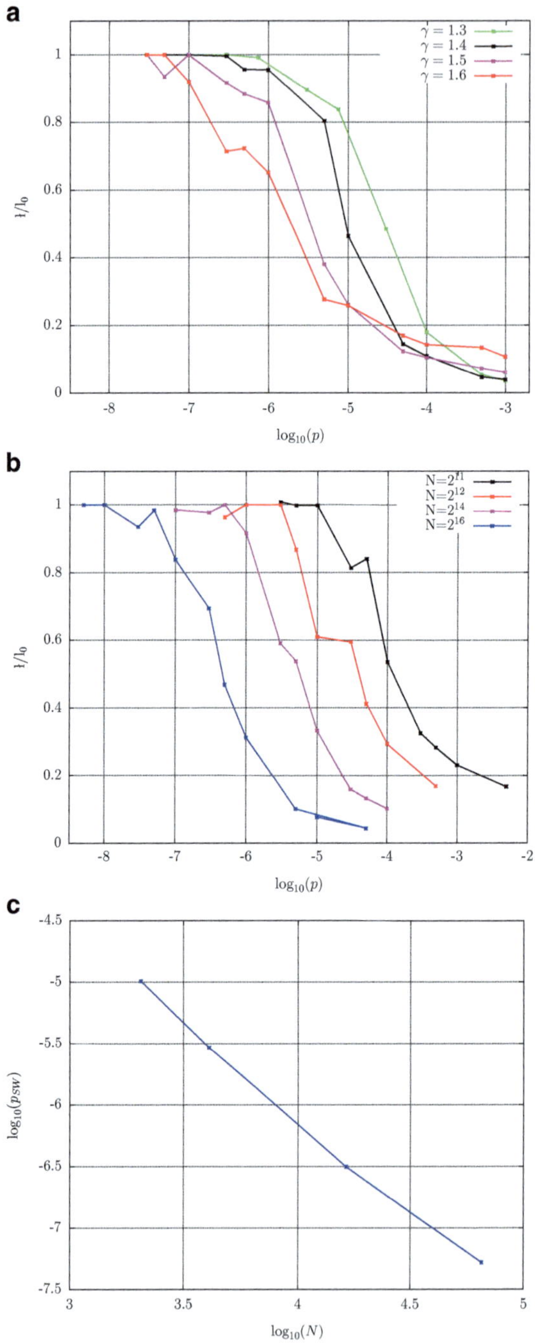

Fig. 11 Average path length $\tilde{\ell}$ for: (**a**) $N = 2^{14}$; (**b**) $\gamma = 1.5$; (**c**) Scaling of the critical probability $\tilde{p}_{SW}(N, \gamma)$ for $\gamma = 1.5$

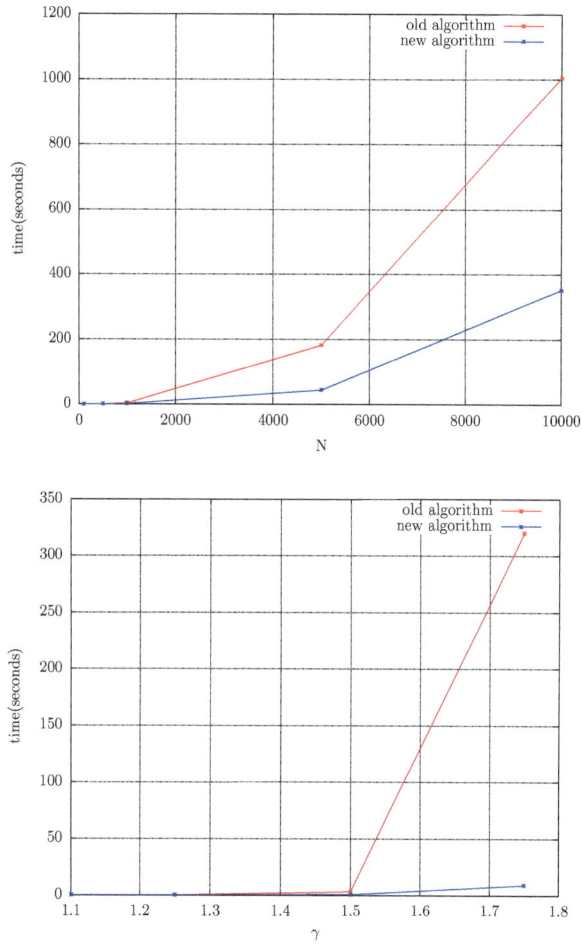

Fig. 12 Computational time for the two clustering coefficient algorithms: (*up*) for increasing system size ($\gamma = 1.5$; $p = 0.0001$); (*bottom*) for increasing γ ($N = 1000$; $p = 0.0001$)

algorithms: the difference between the two is of orders of magnitude when dealing with big system sizes. This result along with the good scaling of $\tilde{\ell}$ confirms the validity of its choice in order to explore the network properties.

3 The Model for the Interaction: The XY Model

We have considered in Sects. 2.2 and 2.3, respectively, the regular chain topology and the Small World one and investigated the passage from one to the other via p and γ. Now, as we sketched in the Introduction, the natural following step is to define a form of interaction between the vertices and tackle the issue of the

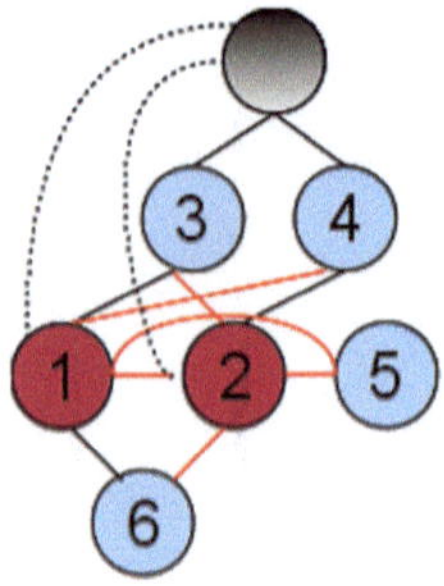

Fig. 13 Calculating the clustering coefficient of 1, we first consider the link between 1 and 2; then, we try to complete the triangle looking for the possible links with {3,4,5,6} (the *red* ones in figure). These links are the only ones which can exist (encoded in the adjacency matrix) leaving untouched the structure of the tree. For instance, the *dashed* links cannot exist since it would imply that the vertices 1 and 2 should have been placed at the first level instead of the second when building the tree

interface between the relative thermodynamic behavior of the dynamic model and the network properties we introduced in the aforementioned sections. To achieve this aim, we will study the XY model for rotors: in Sect. 3.1 we present the basic equations of the XY model, while Sect. 3.2 is devoted to a digression concerning the XY-HMF model, which is the full coupling limit of the XY model and which has become over the years a paradigm for theories on long-range systems.

3.1 Basic Definitions

We consider an interaction between spins of the form:

$$V(\theta) = \frac{J}{2\langle k\rangle}\sum_{i,j}\epsilon_{i,j}(1-\cos(\theta_i-\theta_j)) \tag{10}$$

with

$$\langle k\rangle = \sum_{i,j}\epsilon_{i,j}/N \tag{11}$$

and the θ_i $i \in [1, N]$ are angle variables associated with the spins. The coupling constant J is chosen positive to have a ferromagnetic behavior and, without losing generality, it can be set at 1. The phase space for the system is:

$$\begin{cases} p \in]-\infty, +\infty[\\ \theta \in [-\pi, \pi[\end{cases}$$

$$\Gamma = [-\pi, \pi[^N \times]-\infty, +\infty[^N,$$

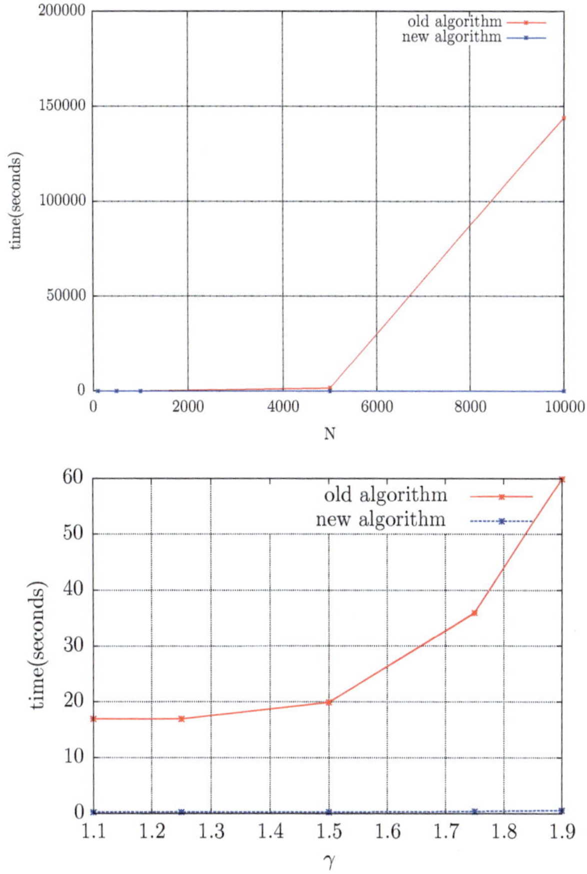

Fig. 14 Computational time for the two average path length algorithms: (*up*) for increasing system size ($\gamma = 1.5$; $p = 0.0001$); (*bottom*) for increasing γ ($N = 1000$; $p = 0.0001$)

and the total Hamiltonian reads:

$$H = \sum_i \frac{p_i^2}{2} + \frac{J}{2\langle k \rangle} \sum_{i,j} \epsilon_{i,j}(1 - \cos(\theta_i - \theta_j)). \tag{12}$$

Finally, for the Hamilton equations we have:

$$\begin{cases} \dot{\theta}_i = \frac{\partial H}{\partial p_i} = p_i \\ \dot{p}_i = -\frac{\partial H}{\partial \theta_i} = -\frac{J}{\langle k \rangle}(\sin\theta_i \sum \epsilon_{i,j} \cos\theta_j - \cos\theta_i \sum \epsilon_{i,j} \sin\theta_j) \end{cases}. \tag{13}$$

Starting from the definitions in Sect. 2 we recall that, in [26], the authors already took in account the case of a random network in which the degree $k \propto N^{\gamma-1}$ links per particle are randomly distributed and these particles interact via the potential (Eq. 10). It has been found that the thermodynamical behavior of the system on these networks recovers the case of the fully coupled model, i.e. the Hamiltonian Mean Field model (XY-HMF), for all values of γ, $1 < \gamma \leq 2$ [26]. We shall focus on the thermodynamical behavior of the XY-HMF in Sect. 3.2; but we want to stress here the implication of the result in [26]: actually introducing a "dilution," given by γ, in the strength of the interaction does not affect the thermodynamical properties in the TD limit when the network lacks any structure, being random. Hence we can hypothesize that the influential characteristic on dynamics should be a *nontrivial topology* of the network and, in the following, we will aim to shed light on this influence.

3.2 The Mean Field Model (Globally Coupled)

Since the XY-HMF model properties are the paradigm to which we compare the XY model on a network, we briefly recall in this section some of the principal features of the XY-HMF. We recover this model from Eq. (10) in the case of global coupling of the system, i.e. $\epsilon_{i,j} = 1 \ \forall i, j \in [1, N], \ i \neq j$:

$$V(\theta) = \frac{J}{2N} \sum_{i,j} (1 - \cos(\theta_i - \theta_j)), \tag{14}$$

so the Hamiltonian of the system reads:

$$H = \sum_i \frac{p_i^2}{2} + \frac{J}{2N} \sum_{i,j} (1 - \cos(\theta_i - \theta_j)). \tag{15}$$

From (15) we obtain the equations for the dynamics of a particle:

$$\begin{cases} \dot{\theta}_i = \frac{\partial H}{\partial p_i} = p_i \\ \dot{p}_i = -\frac{\partial H}{\partial \theta_i} = -\frac{J}{N}(\sin\theta_i \sum_j \cos\theta_j - \cos\theta_i \sum_j \sin\theta_j) \end{cases} \tag{16}$$

and, defining a global order parameter, the magnetization as:

$$\vec{M} = \begin{cases} M_x = \frac{1}{N} \sum_i \cos\theta_i \\ M_y = \frac{1}{N} \sum_i \sin\theta_i \end{cases} = \mathbf{M} \begin{pmatrix} \cos\phi \\ \sin\phi \end{pmatrix}, \tag{17}$$

where $\mathbf{M} = \left| \vec{M} \right|$, we can rewrite the system in Eq. (16) in terms of M:

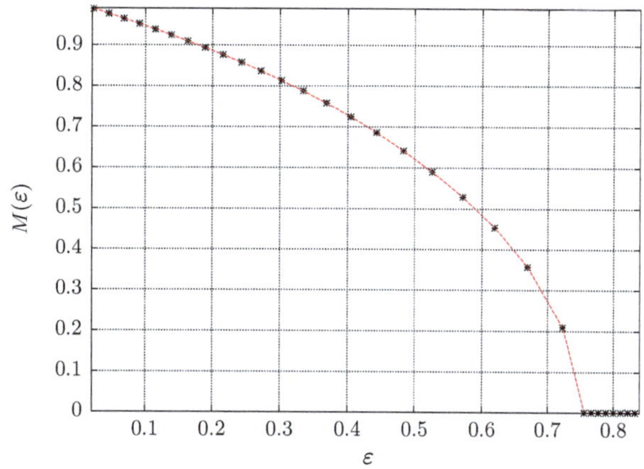

Fig. 15 Equilibrium magnetization versus energy per particle

$$\begin{cases} \dot{\theta}_i = \frac{\partial H}{\partial p_i} = p_i \\ \dot{p}_i = -\frac{\partial H}{\partial \theta_i} = -J\mathbf{M}\sin(\boldsymbol{\theta}_\mathbf{i} - \boldsymbol{\phi}) \end{cases}. \tag{18}$$

We observe that the potential in Eq. (14) belongs to the class of *long-range* ones.[3] Consequently, the system lacks the property of additivity and this implies the emergence of phenomena like phase transitions even in the 1-D case. For the XY-HMF model it is possible to find the analytical expression of the equilibrium magnetization [25]:

$$M = \frac{I_1(\beta M)}{I_0(\beta M)} \quad (M = |\mathbf{M}|),$$

where I_i are the Bessel functions of zeroth and first order and β is the inverse temperature. Varying the energy ε per particle, the magnetization shows a second order phase transition (Fig. 15) at $\varepsilon = 0.75$, which is also confirmed by numerical simulations. This transition is of capital importance since, in Sect. 4, we consider how it varies in correspondence with the variation of the network topology. If we consider a stationary state in which the magnetization is a constant, like the equilibrium, Eq. (18) show an interesting feature of the model: in the thermodynamic limit, the system behaves as uncoupled pendula. This implies that the dynamics remarkably regularize with the increasing size, instead of becoming more chaotic [29, 30]. The lack of additivity entails also the breaking of ergodicity: depending on the initial conditions, the system reaches a stationary state (*Quasi Stationary State*), different from the equilibrium, whose length τ_{QSS} varies with the size:

[3] Actually it is the first harmonic of the gravitational potential.

$$\tau_{QSS} \sim N^{\alpha}, \tag{19}$$

where the exponent α has been estimated around 1.7 see, for instance, [20, 26] for details and references. Hence the two limits $t \to \infty$ and $N \to \infty$ do not commute and, eventually, the magnetization gets stuck on values different from the equilibrium before relaxing for times diverging with the system size. This behavior thus can be of great interest since, as a perspective, we could eventually observe an influence on the non-equilibrium dynamics due to an initial inhomogeneity as the presence of clusters in the network. On the other hand, the existence of the QSS implies also that there is not a straight convergence to the equilibrium and this can be a limit when performing simulations: considering the equilibrium properties, we are forced to simulate the dynamics for times which are increasing with the system size (cfr. Sect. 4) and then simulations can eventually require days of calculation.

4 Thermodynamical Properties on Networks

In Sect. 3.1 we introduced the dynamic equations of the XY model and, in Sect. 3.2, we recalled that its full coupling limit, the XY-HMF model, undergoes a second order phase transition, signalled by the order parameter of Eq. (17), the magnetization. Now, since our purpose is to observe the influence of the network structure on the XY model thermodynamics, we introduce a generalization of the definition of the XY-HMF magnetization (Eq. 17) in order to embed the information on the network. We consider as order parameter a "new" magnetization $\mathbf{M}$:

$$\mathbf{M} = |\vec{M}| = \sqrt{M_x^2 + M_y^2}$$

$$\vec{M} = \begin{cases} M_x = \frac{1}{N}\sum_i M_x^i \\ M_y = \frac{1}{N}\sum_i M_y^i \end{cases}, \tag{20}$$

where:

$$\begin{cases} M_x^i = \frac{1}{k_i}\sum_j \epsilon_{i,j}\cos\theta_j \\ M_y^i = \frac{1}{k_i}\sum_j \epsilon_{i,j}\sin\theta_j \end{cases} = \mathbf{M^i}\begin{pmatrix}\cos\phi_i \\ \sin\phi_i\end{pmatrix}. \tag{21}$$

Equation (20) indeed is the generalization of the definition of the XY-HMF magnetization (Eq. 17): now, since each particle has a degree k_i, its *local* magnetization will be the sum of the contributions of its k_i neighbors encoded in the adjacency matrix $\epsilon_{i,j}$ (Eq. 21). We can rewrite the Hamilton equations in Eq. (13) using Eq. (21):

$$\begin{cases} \dot{\theta}_i = \frac{\partial H}{\partial p_i} = p_i \\ \dot{p}_i = -\frac{\partial H}{\partial \theta_i} = -J\mathbf{M}^{\mathbf{i}}\sin(\theta_i - \phi_i) \end{cases} . \tag{22}$$

We remark that, already at this early stage of discussion, a profound difference arises with the XY-HMF case from the definition of a local magnetization: we see, from Eq. (22), that the dynamics of a particle is driven by its k neighbors, differing hence from one particle to another according to their degree. This can be also the case for short and long range systems (see, for instance, [4, 31, 32]), but this is not the case of the XY-HMF, in which the fully coupling guarantees that, in the thermodynamic limit $N \longrightarrow \infty$, every particle feels the same mean field generated by the other $N - 1 \approx N$ particles. Interesting behaviors can stem from this inhomogeneity due to the network structure: in the frame of this work, we addressed its influence on the evolution of the total magnetization $\mathbf{M}$; but we would like to mention that a study relating the inhomogeneity to non-equilibrium properties is a perspective for further investigations due to the peculiarities of the model, as the *QSS* (see Sect. 3.2).

4.1 XY Model on Regular Chain

The first network topology we will take into account is the regular chain described in Sect. 2.2. As explained, the unidimensional spin chains can generate antipodal behaviors according to the spin degree k: a phase transition for the full coupling configuration (XY-HMF) while, on the other hand, no long-range order exists with the nearest neighbors coupling. The gist on this section is hence to interpolate between these two extrema, varying the coupling via the γ parameter (Eq. 2) and quantifying the emergence of the long-range order.

4.1.1 Simulations Procedure

We perform simulations integrating numerically the dynamic equations (Eq. 22) using a fifth order optimal symplectic integrator [33]. Then, having integrated the dynamics, we study the evolution of the total equilibrium magnetization $\overline{\mathbf{M}}$ (Eq. 20) where the bar denotes the temporal mean for $1 < \gamma < 2$. More in detail, we tune the energy $\varepsilon = H/N$ in the interval $\varepsilon \in [0.1; 0.9]$ and we take Gaussian initial conditions for the (p_i, θ_i). Then, having set the energy and the initial (p_i, θ_i), we let the dynamics evolve and we wait for the magnetization to reach a stationary state; finally, we perform the average to have the equilibrium value. Actually the relaxation to the equilibrium is a delicate passage of our analysis: as explained in Sect. 3.2, the system can reach a stationary state in which the magnetization has a different value from the equilibrium and the length τ_{QSS} of these states (*QSS*) increases with the system size as in Eq. (19). Therefore it is of crucial importance to check the evolution of the magnetization before taking the temporal mean, when these states can arise. Another point a bit technical that is worth to detail before proceeding is the numerical integration. We mentioned before the use of an optimal symplectic integrator of fifth order [33]: this integrator minimizes the introduction of numerical error. Indeed since we deal with an Hamiltonian system the energy is a conserved quantity along with the total momentum, because of the translational invariance. Hence, when integrating the dynamic equations with the symplectic integrator, we check the conservation of these momenta and, thus, we control the correctness of our simulation. Coming to the simulations results, as we explained at the beginning, we address here the issue of the long-range emergence increasing with γ the spin coupling. We hence tune γ in the interval $(1, 2)$ and in Fig. 16a we sum up in some sense our findings.

4.1.2 Numerical Results

We observe the presence of three different behaviors of the equilibrium magnetization: first, for $\gamma = 1.25$ in the figure, the magnetization seems to vanish for all the energies in the physical range; on the other hand, for $\gamma = 1.75$ in Fig. 16a, the magnetization undergoes a second order phase transition like the one of the XY-HMF model. In the middle between the two, for $\gamma = 1.5$, the magnetization for each energy is lower than the $\gamma = 1.75$ value but nevertheless remains finite and bigger than the $\gamma = 1.25$ case. This particular value of γ, $\gamma = 1.5$, signals in fact the passage between two parameter zones $\gamma < 1.5$ and $\gamma > 1.5$, represented, respectively, by $\gamma = 1.25$ and $\gamma = 1.75$ in the previous discussion. In the first interval $\gamma < 1.5$ the system behaves like a one-dimensional chain with nearest neighbors coupling in which long-range order is not possible. The other case, $\gamma > 1.5$, is thermodynamically like the XY-HMF case displaying a second order phase transition of the magnetization at $\varepsilon_c = 0.75$. Let us focus primarily on this latter case, $\gamma > 1.5$. Heuristically our result points out that in order to achieve a global alignment and give a nonzero magnetization, each spin does not need to be under the influence of all the others, like in the XY-HMF frame, but just a bunch of connections, for instance $\gamma = 1.6$, provides enough information to be representative of the whole system state and ensure the global alignment. More from the technical side, in Fig. 16a, the averages are taken when the magnetization is in a stationary state whose fluctuations, monitored via the variance

$$\sigma^2 = \overline{(\mathbf{M} - \overline{\mathbf{M}})^2}, \tag{23}$$

show the scaling canonically associated with equilibrium:

$$\sigma^2 \propto 1/N. \tag{24}$$

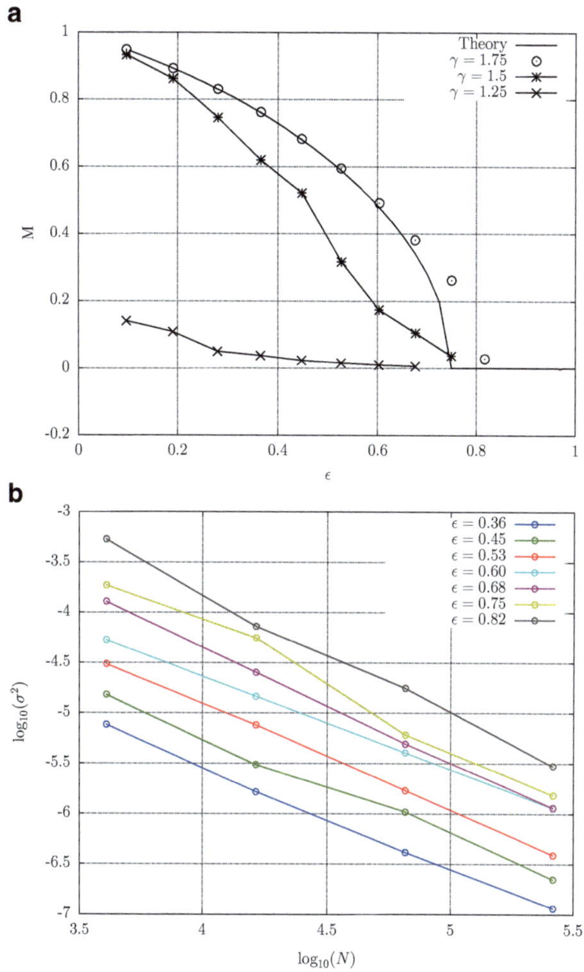

Fig. 16 (**a**) Equilibrium magnetization for $N = 2^{16}$ and different γ. For $\gamma \neq 1.5$ the error bars are of the size of the dots; (**b**) Scaling of the magnetization variance $\langle \sigma^2 \rangle$ with the size for $\gamma = 1.75$

In Fig. 16b, we control that this statement holds for each energy in the physical range considered so to justify our previous claim of an equilibrium second order phase transition. As a last remark on this part, we observe that, in Fig. 16a the phase transition seems to take place at a different energy than the XY-HMF one, $\varepsilon_c = 0.75$ but it is an artifact issued from the finite size of the simulations. Passing to the analysis of the other interval $\gamma < 1.5$, Fig. 17a confirms what anticipated before, i.e. that the magnetization vanishes for each energy in the thermodynamic (TD) limit since the residual values due to the finite size simulations decrease smoothly with the increasing size. To further support this thesis, in Fig. 17b, we monitored the magnetization for the lowest energy $\varepsilon = 0.1$, which is the most likely to give a finite

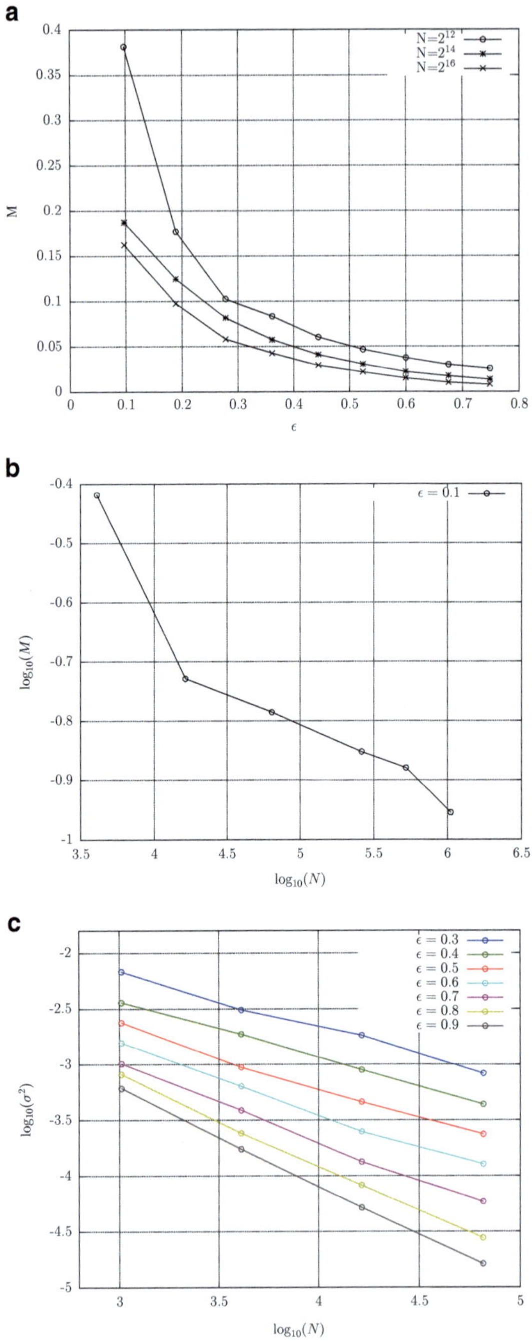

Fig. 17 (continued)

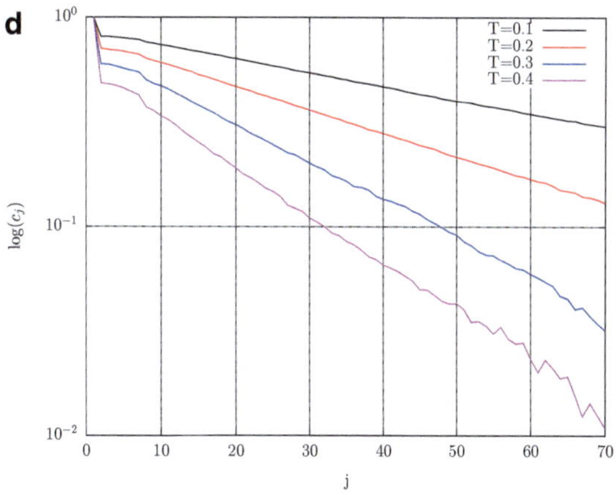

Fig. 17 (continued) (**a**) Equilibrium magnetization versus energy density for $\gamma = 1.25$ and different sizes. The error bars are of the size of the dots; (**b**) Residual magnetization for $\gamma = 1.25$ at $\epsilon = 0.1$ versus the system size; (**c**) Scaling of the magnetization variance $\langle\sigma^2\rangle$ with the size for $\gamma = 1.25$; (**d**) Correlation function c_j for $\gamma = 1.25$ and $N = 2^{14}$

value in the TD limit, and it is consistent with the decreasing trend shown in Fig. 17a. As for the $\gamma > 1.5$ case, the variance scaling in Fig. 17c shows the expected behavior of Eq. (24), suggesting that we are indeed observing the equilibrium state. Then in the $\gamma < 1.5$ regime the system behaves like if just the nearest neighbors coupling was at play and thus no long-range order is possible. In practice, interpreting this result in terms of information like before, the spins interact with a neighborhood which could be non negligible i.e. for $\gamma = 1.4$ but it does not bring enough information on the whole system state to allow alignment coherence and a finite magnetization. On the contrary our results point out that, in the TD limit, the spins behave like if coherence was destroyed by spin waves and the magnetization thus vanishes. To be more precise, this last statement should be supported by a further analysis: we show that true long-range order is absent but still a certain degree of coherence could eventually be retrieved in the spin correlation function

$$c(j) = \frac{1}{N}\sum_{i=1}^{N}\cos(\theta_i - \theta_{i+j[N]}) \ . \tag{25}$$

If, as claimed, the spin were uncorrelated, we would expect an exponential decay of $c(j)$ with the spin distance; but we recall that the XY model in two dimensions displays a power law decay of $c(j)$ for low temperatures. This slow decay turns into exponential for high temperatures, giving rise to the Berezinskii, Kosterlitz, and Thouless (BKT) phase transition [23]. This peculiar phase transition is the signature of *quasi* long-range order so, to eventually detect it in our frame, we monitored the correlation function for low temperatures: in Fig. 17d it has the exponential decay expected. This concludes our analysis, demonstrating that in the

$\gamma < 1.5$ range global coherence is not possible even in the slighter form of a BKT transition. The last but not less interesting case that remains to be considered is the one stemming from $\gamma = 1.5$. We already discussed that the magnetization gets stuck, for low energies $0.3 \lesssim \varepsilon \leq 0.75$, in a "middle" value which is inferior to the mean field value but still nonvanishing in the TD limit (Fig. 16a). But, in addition to this quantitative difference, the magnetization temporal behavior itself is intrinsically different for $\gamma = 1.5$: the "middle" value resulting from the temporal mean originates from important oscillations of the magnetization whose amplitude is orders of magnitude larger than in the other γ regimes, as illustrated by Fig. 18a. Naturally one would expect those oscillations to relax either in the $t \to \infty$ or in the $N \to \infty$ limit. We hence performed simulations increasing the simulation time and the size: in Fig. 18c the simulation time is ten times longer than what was required in the $\gamma \lessgtr 1.5$ regimes and still the oscillating behavior persists. The increasing size limit fails as well in smoothing down the fluctuations: in Fig. 18c we compare two time series for $N = 2^{12}$ and $N = 2^{18}$ and, in spite of the six order of magnitude difference, the fluctuation amplitude remains unchanged. This latter statement can be refined using the magnetization variance σ^2 (Eq. 23) as before: indeed, as displayed in Fig. 18d, the size influences very poorly the fluctuations amplitude giving an almost flat behavior to σ^2 in contrast to the scaling $\sigma^2 \propto 1/N$ which characterized the $\gamma \lessgtr 1.5$ configurations. We can hence interpretate this state as a kind of bistable regime: the spin degree for $\gamma = 1.5$, $k \propto \sqrt{N}$, is too high to let the magnetization vanish, but at the same time the spins are not "entangled" enough to restore the mean field picture and display the phase transition. Therefore the system gets trapped in this oscillating regime which seems persistent in time. Moreover, if we consider the definition of the susceptibility χ:

$$\chi \sim \lim_{N \to \infty} N\sigma^2,$$

we can deduce from the flat behavior of the variance that this fluctuating regime is characterized by an infinite susceptibility. As a final observation, which could give rise to interesting perspectives, we note that in Fig. 18c the oscillation period bears the information on the increasing size: for the biggest case is evidently longer than for the smaller system. This feature, mixing both the system size and an intrinsic time scale, sounds reminiscent with the *QSS* duration of Eq. (19) for the XY-HMF model so that it could be the trademark of a match between the out-of-equilibrium behavior of these two models.

4.1.3 Analytical Calculation

In the previous section we showed strong numerical evidence that changing the degree, via γ, creates the conditions for establishing the mean field behavior without, however, imposing the full coupling of the spins. A trace of such an effect should be then encoded somehow in the system dynamics and, in this section, we

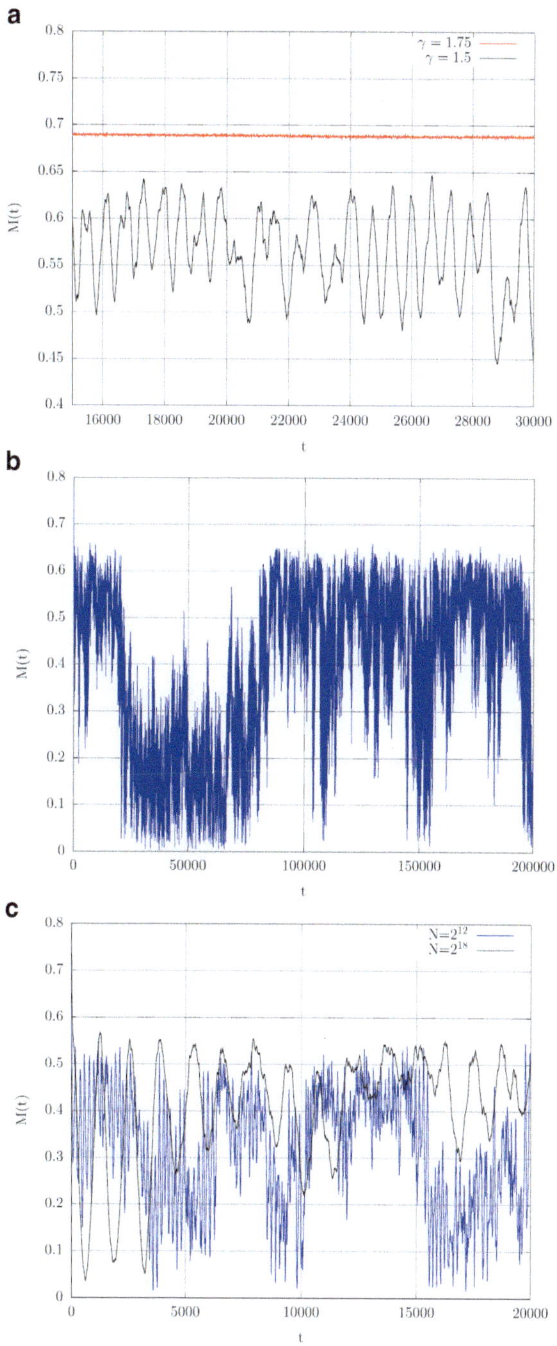

Fig. 18 (continued)

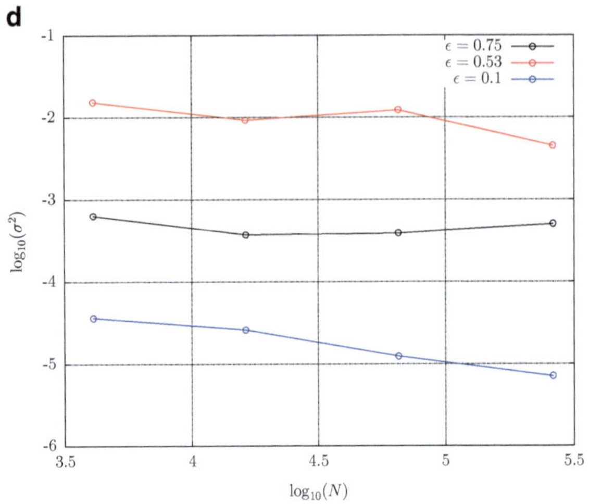

Fig. 18 (continued) (color online) Time series for the magnetization with (**a**) $N = 2^{18}$, $\epsilon = 0.60$; (**b**) $N = 2^{12}$; $\epsilon = 0.44$; (**c**) $\epsilon = 0.44$; (**d**) Scaling of the magnetization variance $\langle \sigma^2 \rangle$ with the size for $\gamma = 1.5$

present a calculation aiming to bridge the topological information about γ in the thermodynamic frame, calculating analytically the order parameter **M**. The first crucial observation is that the low energies regime is the appropriate one to compute the magnetization: when $\gamma < 1.5$ it vanishes in the TD limit while, once crossed $\gamma = 1.5$, the magnetization tends to 1 as $\varepsilon \to 0$. Hence in the low energies regime the shift between the two phases is at its clearest and, moreover, this regime offers a calculation advantage: we can assume the differences $\theta_i - \theta_j$ are small so that the free energy is maximized by the spin alignment. We can hence take the leading order, developing the Hamiltonian:

$$H = \sum_i \frac{p_i^2}{2} + \frac{J}{4k} \sum_{i,j} \epsilon_{i,j} (\theta_i - \theta_j)^2. \tag{26}$$

In practice, we deal via Eq. (26), with a collection of harmonic oscillators connected by the adjacency matrix. Then to compute the dynamics we make use of a procedure illustrated in [4, 34], starting from the following observation: since at equilibrium the momenta p_i are Gaussian variables we can choose represent them as a sum of random variables. The choice for this representation is a superposition of N modes, where the random variable is carried by the phase:

$$\begin{aligned} \theta_i &= \textstyle\sum_{l=0}^{N-1} \alpha_l(t) \cos(\frac{2\pi l i}{N} + \phi_l) \\ p_i &= \textstyle\sum_{l=0}^{N-1} \dot{\alpha}_l(t) \cos(\frac{2\pi l i}{N} + \phi_l) \end{aligned} . \tag{27}$$

We recall that we have periodic boundary conditions for the chain so that Eq. (27) in practice is the discrete Fourier transform. Moreover the linearity of the transformation is ensured by summing over N modes and the two equations for the angles and the momenta are connected by the first Hamilton equation $p_i = \dot{\theta}_i$. In [4, 34], another interpretation of the random phases is given, which allows to make the link between microscopic dynamics and the thermodynamics: a set of phases $\{\phi_l\}$ can be considered as a trajectory in the phase space so that averaging on random phases, hence on different sets $\{\phi_l\}_m$, could correspond to ensemble averaging. The aim of this procedure is then to obtain *averaged equations of motion* which govern the microscopic dynamics and at the same time carry some information about the *thermodynamic state*. In this purpose, we inject Eq. (27) in the Hamiltonian (26) and we obtain for the kinetic part K:

$$\frac{\langle K \rangle}{N} = \frac{1}{N}\left\langle \sum_i \frac{p_i^2}{2} \right\rangle = \frac{1}{4}\sum_l \dot{\alpha}_l^2, \tag{28}$$

where $\langle \dots \rangle$ stands for the average over random phases. In Eq. (28) we used the relation:

$$\langle \cos(k_i + \phi_i) cos(k_j + \phi_j) \rangle = \frac{1}{2}\delta_{i,j}.$$

The definition of the regular chain given in Sect. 2.2 gives a peculiar adjacency matrix: $\epsilon_{i,j}$ is a circulant so that it is diagonalizable and its spectrum $\{\lambda_l\}$ has an analytical expression:

$$\lambda_l = \frac{2}{k}\sum_{j=1}^{k/2} \cos\left(\frac{2\pi l j}{N}\right) = \frac{1}{k}\left[\frac{\sin[(k+1)l\pi/N]}{\sin(l\pi/N)} - 1\right], \tag{29}$$

where k is the spin degree of Eq. (4). Making use of this fact and of the identity

$$\frac{1}{kN}\sum_{i,j} \epsilon_{i,j}\theta_i\theta_j = \sum_l \lambda_l \left|\hat{\theta}_l\right|^2, \tag{30}$$

the potential will hence take the form:

$$\frac{V}{N} = \frac{1}{4kN}\sum_{i,j} \epsilon_{i,j}(\theta_i - \theta_j)^2 = \frac{1}{2}\sum_l (1-\lambda_l)\left|\hat{\theta}_l\right|^2. \tag{31}$$

Equation (30) derives from the fact that the eigenvectors of a circulant matrix of size N are the columns of the unitary discrete Fourier transform matrix of the same size. Using the linear waves representation and averaging over the phases in Eq. (31), we obtain:

$$\frac{\langle V \rangle}{N} = \left\langle \frac{1}{2} \sum_l (1 - \lambda_l) \left| \hat{\theta}_l \right|^2 \right\rangle = \frac{1}{4} \sum_l (1 - \lambda_l) \alpha_l^2.$$

At this point, we can finally achieve our aim: the *averaged equation of motion* comes from the second Hamilton equation applied on the averaged Hamiltonian $\langle H \rangle = \langle K \rangle + \langle V \rangle$:

$$\frac{d}{dt} \left(\frac{\partial \langle H \rangle}{\partial \dot{\alpha}_l} \right) = - \frac{\partial \langle H \rangle}{\partial \alpha_l} ,$$

which gives

$$\ddot{\alpha}_l = -(1 - \lambda_l) \alpha_l = -\omega_l^2 \alpha_l. \tag{32}$$

Equation (32) is then an equation for an harmonic oscillator where the frequency is driven by adjacency matrix spectrum. From the thermodynamics we obtain another relation to link the wave amplitudes $\{\alpha_l\}$ and frequencies $\{\omega_l\}$: actually we have, at equilibrium, the equipartition of the modes (p_i's are Gaussian):

$$T = \frac{1}{N} \sum_i \langle p_i^2 \rangle = \frac{1}{2} \sum_l \alpha_l^2 \omega_l^2 \Rightarrow \alpha_l^2 = \frac{2T}{N(1 - \lambda_l)}.$$

At this point we are able to compute M, given by Eq. (20): we substitute the representation Eq. (27) and we average on the phases as before, obtaining [4]:

$$\langle \mathbf{M} \rangle = \prod_l J_0(\alpha_l)(\cos \theta_0, \sin \theta_0), \tag{33}$$

where J_0 is the zeroth order Bessel function and θ_0 is the average of the angles $\{\theta_i\}$. This latter is conserved since

$$P = \frac{d\theta_0}{dt} = 0,$$

where P is the total momentum, which is a constant of motion because of translational invariance. P is set at $P = 0$ by our choice of initial conditions. As the final step to evaluate Eq. (33), we recall that we are dealing with a low temperatures approximation so we can consider that the amplitudes α_l^2 to be small at equilibrium and in the large system size limit [34]. This consideration allows to develop at leading order the product of the Bessel functions and, taking the logarithm of Eq. (33), we finally obtain:

$$\ln (\langle M \rangle) = - \sum_l \frac{\alpha_l^2}{4} = - \frac{T}{2N} \sum_l \frac{1}{1 - \lambda_l}. \tag{34}$$

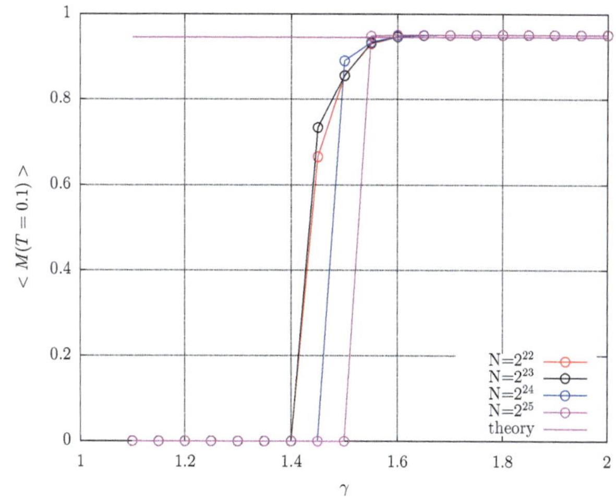

Fig. 19 (color online) Analytical magnetization $\langle M \rangle$ from Eq. (34) for $T = 0.1$ versus γ. Theory refers to the exact analytical solution of the XY-HMF model

In Fig. 19 we show, by evaluating numerically Eq. (34), that this approximated expression retrieves the correct asymptotic behavior: the mean field one in the $\gamma > 1.5$ regime while it vanishes for $\gamma < 1.5$. The transition becomes sharper at $\gamma_c = 1.5$ by increasing the size: as we argued from the numerical simulations, this analytical approach confirms too the critical signification of $\gamma_c = 1.5$. In Eq. (34) we then realized our project of bridging the topological information in the thermodynamics of the system via the matrix spectrum. This latter (Eq. 29) encodes the network complexity, so that it actually realizes the purpose of matching these two levels of description but at the same time it implies Eq. (34) to be evaluated numerically.

4.2 XY Model on Small World Networks

In the analysis of Fig. 5, we showed in Sect. 2.3 how it is possible, with our model, to interpolate continuously from a regular lattice to a totally random network and we defined a "working zone," i.e. an intermediate region of parameters in which the network has the Small World properties of our interest. We now focus our attention on the phase transition of the XY model on these Small World networks. Technically, to perform our simulations, we follow the same procedure devised in Sect. 4.1: we integrate the dynamic equations (Eq. 22) via the symplectic integrator and we take the temporal mean of the magnetization $\mathbf{M}$ (Eq. 20) when it reaches the equilibrium. In practice all the difference between the previous case of regular chains and the present one of Small World networks is encoded in the adjacency

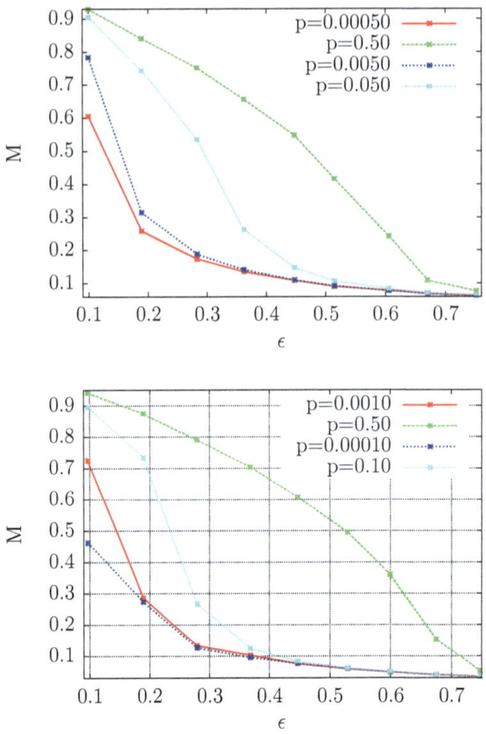

Fig. 20 (*up*) Magnetization for $N = 512$, $\gamma = 1.25$; (*bottom*) $N = 2024$, $\gamma = 1.25$

matrix $\epsilon_{i,j}$ which, by the way, loses here the property of circularity. Nevertheless a subtlety comes from the randomness introduced in the networks via p, implying that networks with the same parameters can be shaped differently. We hence take as order parameter $\left\langle \overline{\mathbf{M}} \right\rangle$ where the bar indicates, as usual, the temporal mean and $\langle \ldots \rangle$ stands for the mean over different network realizations with the same parameters. We recall that our model for Small World networks is parametrized by γ and p (Sect. 2.3.1): we hence first focus on how the transition of the magnetization answers to the tuning of the rewiring probability p for fixed γ. We show in Fig. 20 the two cases $N = 512$ and $N = 2024$ for $\gamma = 1.25$, the system does not display a transition for low values of p, which implies that the network still has not enough shortcuts to achieve global coherence and it behaves as 1-D network. Increasing p, the magnetization curve approaches the one of XY-HMF (see Fig. 15): for $p = 0.5$ the two systems undergo a phase transition confirming that the degree of randomness of the network introduces a long-range order via the shortcuts. Furthermore, refining the analysis, we observe that for $p = 0.5$ the two cases have the same clustering coefficient, which is an intensive parameter, but different average path length as it is extensive (see Fig. 5). The transition, at its turn, seems to take place at two different energies: $\epsilon_c \sim 0.67$ for $N = 512$ and $\epsilon_c \sim 0.75$ for $N = 2024$ like for the XY-HMF. This observation leaded us to investigate the case of fixed p and different sizes N: what we aimed was thus to separate the contribution of the average path length, which is affected by the change in size, and the clustering coefficient which is not, to the shape of the transition. In Fig. 21, we chose four representative values of the rewiring probability and we observe that the transition is at the same temperature for all the sizes at which it occurs as summarized in Table 1. We set the three lowest p values in order to have the same clustering coefficient, being in the plateau region, since we aimed to check if the clustering coefficient influences the value of ε_c quantitatively. This hypothesis was first motivated by the two simulations at $p_1 = 0.005$ and $p_4 = 0.05$: for those probabilities, the system shows different connectivities and different energies of transition. The results for $p_2 = 0.001$ and $p_3 = 0.0001$ show, on the contrary, that the transition shifts independently from the clustering coefficient, hence it is not the proper choice for an intensive parameter leading quantitatively the energy of transition. Furthermore, we observe that we do not have a phase transition for $N = 2024$ for and for $N = 16192$ with p_3 (Fig. 21). This difference according to the system size confirms the role of the extensive parameter, the average path length: the big sizes undergo the transition since their average path length ℓ is low enough because of the shortcuts. On the contrary, for sizes at which ℓ is still too high as for $N = 2024,\ 16192$, the transition cannot take place since the topology is still near to the one of a regular 1-D network. To conclude with the analysis of Fig. 21, we can resume our results in the following way: we confirmed that the average path length determines the rise of the transition while the clustering coefficient seems not to influence it decisively. The first is connected to the intrinsic character of the network: the value of ℓ is related to the amount of shortcuts that we introduced rewiring and, thus, it drives the passage from a 1-D network, which does not show phase transitions, to a mean field topology. This guess is confirmed by the fact that the rise of a phase transition appears to be connected to the system size so ℓ, because of its extensive nature is the most suitable candidate to lead this *qualitative* changein the network. On the other hand, the clustering coefficient does not *quantitatively* affect the temperature of the phase transition: being an intensive parameter, it was the first natural choice; moreover, since it quantifies the local amount of clustering, we could heuristically foresee that a high clustering coefficient entailed a lowering of the ε_c. But, interpreting our numerical results, we observe that both the regular network and the Small World one share the same high clustering coefficient while their thermodynamic behavior is totally different. Hence the importance of the clustering coefficient is more related to its fall for high p values which points out the topological transition between the Small World regime and the random network case; on the other hand, in the plateau region, the clustering coefficient does not *see* the passage from regular to Small World and, as a consequence, it is not a complete topological descriptor. We recall that in Sect. 2.3 we discussed how γ influences the average path length ℓ and the probability of crossover p_{SW} (Fig. 4a–b). Now, the natural extension of our previous analysis is hence to investigate the thermodynamic response of the phase transition to the variations of this parameter. In addition to this, we observe that Table 1 points out a strong dependence of the critical energy on p. We then tackled this issue of

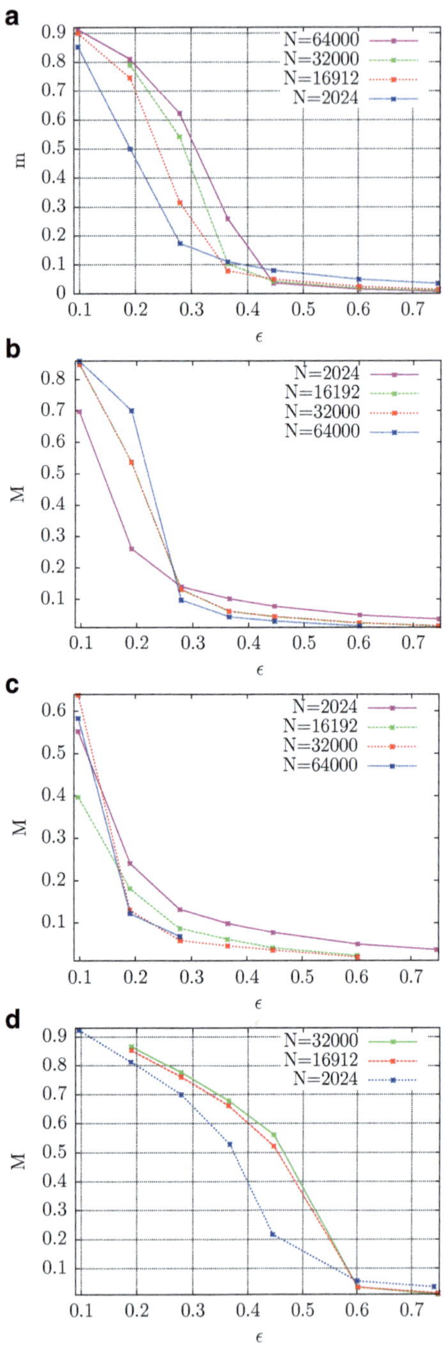

Fig. 21 (**a**) Magnetization for $\gamma = 1.25$ and $p_1 = 0.005$; (**b**) $p_2 = 0.001$; (**c**) $p_3 = 0.0001$; (**d**) $p_4 = 0.05$

Table 1 Critical energies ε_c for different p values

p	ε_c
0.0001	0.2
0.001	0.27
0.005	0.45
0.05	0.6

the γ and p dependence of ε_c performing simulations for $\gamma < 1.5$ and different p values. Our choice of the γ interval is motivated by the results on the regular chain topology of Sect. 4.1: we demonstrated that, when γ overcomes the $\gamma_c = 1.5$ threshold, the mean field behavior is at play and the second order phase transition takes place at $\varepsilon_c = 0.75$. As displayed in Fig. 22a, the critical energy seems to have the following scaling on p, γ:

$$\varepsilon_c = \log(g(\gamma)p^c), \tag{35}$$

with $C \sim 0.1$. This result indeed confirms what before argued: the topology of the Small World network influences quantitatively the equilibrium properties of the XY model, shifting the critical energy, and both parameters γ and p are involved in this interplay. Furthermore, in Fig. 22a there is a clear saturation effect: for each γ it exists a probability above which the critical energy acquires the mean field value $\varepsilon_c = \varepsilon_{MF} = 0.75$. Hence it naturally appears another intrinsic threshold probability p_{MF} signalling the transition to the mean field frame: in practice tuning appropriately p and γ, we obtain a Small World network entailing a mean field behavior of the XY model, so that it is thermodynamically equivalent to the regular chain with $\gamma > 1.5$. This new threshold p_{MF} seems to have a power law dependence on γ (Fig. 22b) but it is necessary to observe that near $\gamma_c = 1.5$ the determination of p_{MF} becomes particularly delicate: for the regular chain, we demonstrated that it separates the "1-D zone" from the mean field one; so we would expect $p_{MF}(\gamma_c = 1.5) \to 0$. In other words, at $\gamma_c = 1.5$ the system requires a very small, morally vanishing, amount of randomness to achieve the mean field behavior; nevertheless, since we deal with finite size simulations and our averages are performed on a limited number of network realizations, p_{MF} remains, in his turn finite. In addition to this, the determination of p_{MF} proves very delicate in general way since it relies on the determination of the critical energy, which is as well a tricky passage.

To conclude this part, we think that, as a further development, more refined parameters should be considered to understand quantitatively the shown shift in the phase transition: here, as an hypothesis, we could mention that *heterogeneity* is one of the crucial characteristics of complex networks and hence parameters like the degree and the clusters distributions could be more appropriate to pursue our analysis.

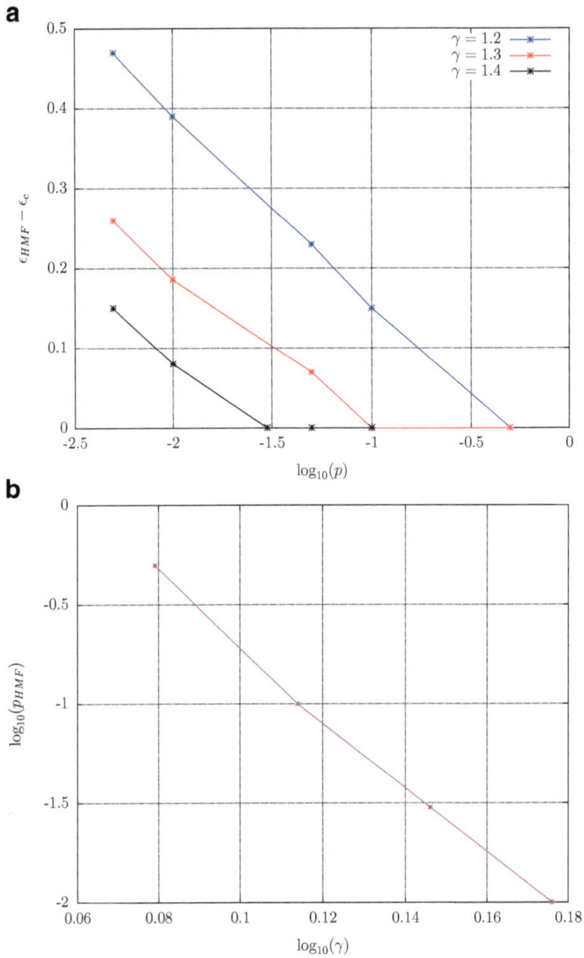

Fig. 22 (color online) (**a**) Logarithmic dependence of the critical energy ε_c versus the rewiring probability p for different γ values. (**b**) Power law scaling of p_{MF} versus γ

5 Conclusions

In this last section, we would like to resume the logical steps of our work and to give a perspective of further developments. In Sect. 2 and in Sect. 3, we first introduced some basic tools to define networks in general and our model for the interaction, the XY model. More in detail we presented, in Sect. 3.2, a particular case of the XY model, the XY-HMF model: we specially stressed the presence of a second order phase transition which we retrieve, with some differences, in the more general XY model on networks. Section 2.3 is devoted to Small World networks with particular attention to their topological parameters, the clustering coefficient and the average path length. In this section we described a model, the Watts–Strogatz model, which by the means of the rewiring probability p, allowed us to explore the topologies of the three main configurations taken in account: random, regular, and Small World networks. We hence developed an algorithm, described in Sect. 2.4, in order to enhance efficiency calculating topological parameters. Actually this algorithm structures in the *tree* the data encoded in the adjacency matrix and, in addition to the increasing of the calculation speed, it gives a complete network characterization as, for instance, details like the degree and clusters distributions. We would like to stress here the importance of this development effort: understanding networks complexity requires a profound knowledge of their structure; moreover, to achieve making connections to the dynamical behavior, we should be able to access this knowledge efficiently for a huge interval of parameters. Hence the numerical results presented in Sect. 4.2, concerning the Small World network, are a consequence of this computational improvement and, furthermore, the successive refined analysis we could perform would stem from it. Finally, in Sect. 4, we presented the numerical results of simulations. We first focused, in Sect. 4.1, on the regular chain topology: using the γ parameter, we increased the spin degree and we identified the threshold above which the mean field behavior can be reestablished, $\gamma_c = 1.5$, from the previous 1-D topology. Then, in the $\gamma > 1.5$, we obtained the same phase transition of the XY-HMF model, while for $\gamma < 1.5$ no long-range order was displayed. Furthermore, for $\gamma_c = 1.5$, we discovered a metastable state in which the order parameter is affected by important fluctuations and which does not seem to relax to equilibrium. Those numerical results were also analytically confirmed at the end of Sect. 4.1. We then passed, in Sect. 4.2, to the Small network topology. We highlighted the effect of the average path length l in giving the system *global coherence:* we observe the rise of the phase transition only in regimes of low l which imply the fundamental presence of *shortcuts*. These shortcuts are responsible for the efficiency of information transmission throughout the network and they allow the emergence of a collective behavior in a 1-D network. On the other hand, our guess for the clustering coefficient to be the guide of *quantitative* changes in the energy of transition is contradicted by our numerical results. As we explained, we think it is not a complete topological descriptor: it is certainly useful, along with l, to determine the Small World parameter zone but it lacks in precision when we deal with a trial of *quantification* of the observed transition shifts. The final part of Sect. 4.2 was dedicated to the interplay between the γ and p parameters in influencing critical energy $\varepsilon_c(p, \gamma)$ of the phase transition. By extensive numerical simulations we identified a logarithmic dependence of $\varepsilon_c(p, \gamma)$ on p while γ affects it linearly. Finally our investigations showed a saturation phenomenon for which $\varepsilon_c \to \varepsilon_{MF} = 0.75$ increasing the topological parameters γ and p. Therefore, at the end of Sect. 4.2, we defined the threshold probability p_{MF} as the minimal randomness to inject in our network to obtain the mean field regime at the same energy of the XY-HMF model and we showed that it depends as a power law on γ. From these results, a widespread of successive investigations can originate and, as a conclusion, we would like to sketch some of these research perspectives. First the above discussion proved the necessity of a deeper understanding of the topological tools we already have and, probably, the definition of new ones which should be able to encode properly the network complexity. Furthermore a refining of the network model itself, the Watts–Strogatz model, could be an interesting path to follow. This model, even if it is paradigmatic for its simplicity, has strong limits: a Watts–Strogatz network is a Small World network, but a *real* Small World network is not a Watt–Strogatz one in the sense that the latter has many peculiarities which are not mirrored by the simple algorithm we utilized in this work; a more complex model could hence be required. Concluding, we remark once more the interest in pursuing these researches: networks and dynamical processes on them are object of growing attention because of their widespread diffusion in real systems dealing with different fields and, thus, understanding topological effects on these processes, even in simple models like the ones presented here, might give a clue to interpret real world phenomena as, for instance, disease spreading [27].

References

1. Dorogovtsev SN, Goltsev AV, Mendes JFF (2008) Rev Mod Phys 80(4):1275
2. Watts DJ, Strogatz SH (1998) Nature 393:440
3. Antoni M, Ruffo S (1995) Phys Rev E 52:3261
4. Leoncini X, Verga A, Ruffo S (1998) Phys Rev E 57(6):6377
5. Jain S, Young AP (1986) J Phys C Solid State Phys 19(20):3913
6. Janke W, Nather K (1991) Phys Lett A 157(1):11
7. Kim JK (1994) Europhys Lett 28(3):211
8. Lee DH, Joannopoulos JD, Negele JW, Landau DP (1984) Phys Rev Lett 52(6):433
9. Loft R, DeGrand TA (1987) Phys Rev B 35(16):8528
10. McCarthy JF (1986) Nucl Phys B 275(3):421
11. Kosterlitz JM, Thouless DJ (1973) J Phys C Solid State Phys 6(7):1181
12. Zheng B, Schulz M, Trimper S (1999) Phys Rev Lett 82:1891. doi:10.1103/PhysRevLett.82.1891. http://link.aps.org/doi/10.1103/PhysRevLett.82.1891
13. De Nigris S, Leoncini X (2013) Europhys Lett 101(1):10002
14. De Nigris S., Leoncini X. (2013) Phys Rev E 88:012131
15. Campa A, Dauxois T, Ruffo S (2009) Phys Rep 480(3–6):57
16. Antoniazzi A, Fanelli D, Barré J, Chavanis PH, Dauxois T, Ruffo S (2007) Phys Rev E 75(1):011112
17. Ettoumi W, Firpo MC (2011) J Phys A Math Theor 44(17):175002
18. Latora V, Rapisarda A, Tsallis C (2002) Phys A 305(1–2):129
19. Chavanis PH, Vatteville J, Bouchet F (2005) Eur Phys J B 46(1):6
20. Chavanis PH, Ninno GD, Fanelli D, Ruffo S (2008) In: Chandre C, Leoncini X, Zaslavsky G (eds) Chaos, Complexity and Transport. World Scientific, Singapore, 2008, pp 3–26
21. Kim BJ, Hong H, Holme P, Jeon GS, Minnhagen P, Choi MY (2001) Phys Rev E 64:056135
22. Medvedyeva K, Holme P, Minnhagen P, Kim BJ (2003) Phys Rev E 67(3):036118
23. Kosterlitz JM, Thouless DJ (1973) J Phys C Solid State Phys 6:1181
24. Fröhlich J, Spencer T (1981) Comm Math Phys 81(4):527
25. Campa A, Dauxois T, Ruffo S (2009) Phys Rep 480:57
26. Ciani A, Ruffo S, Fanelli D (2010) Long-range interaction, stochasticity and fractional dynamics: dedication to George M. Zaslavsky (1935–2008). HEP, Beijing and Springer, Berlin
27. Barrat A, Barthélemy M, Vespignani A (2008) Dynamical processes on complex networks. Cambridge University Press, New York

28. Newman MEJ, Watts DJ (1999) Phys Rev E 60(6):7332
29. Bachelard R, Chandre C, Fanelli D, Leoncini X, Ruffo S (2008) Phys Rev Lett 101(26):260603
30. Leoncini X, Van den Berg TL, Fanelli D (2009) Europhys Lett 86:20002
31. Van den Berg TL, Fanelli D, Leoncini X (2010) Europhys Lett 89:50010
32. Turchi A, Fanelli D, Leoncini X (2011) Comm Nonlinear Sci Numer Simulat 16(12):4718–4724. doi:10.1016/j.cnsns.2011.03.013. http://arxiv.org/abs/1007.2065v1;http://arxiv.org/pdf/1007.2065v1
33. McLachlan RI, Atela P (1992) Nonlinearity 5:541
34. Leoncini X, Verga A (2001) Phys Rev E 64(6):066101

Lyapunov Exponent Sign Reversal: Stability and Instability by the First Approximation

G.A. Leonov and N.V. Kuznetsov

1 Introduction

This chapter is a concise and updated version of authors' survey *Time-Varying Linearization and the Perron effects* [52], devoted to the rigorous mathematical justification of the use of Lyapunov exponents to investigate the stability, instability, and chaos. In his thesis A.M. Lyapunov [57] proved that if the first approximation system is *regular* and its largest Lyapunov exponent is negative, then the solution of the original system is asymptotically stable. Then it was stated by O. Perron [62] that the requirement of regularity is substantial: he constructed an example of second-order system such that a solution of the first approximation system has negative largest Lyapunov exponent while the solution of the original system with the same initial data has positive largest Lyapunov exponent. The effect of Lyapunov exponent sign reversal of solutions of the first approximation system and of the original system under the same initial data, we shall call the Perron effect.

Later, [14, 58, 60, 63] there were obtained sufficient conditions of stability by the first approximation for nonregular linearizations generalizing the Lyapunov theorem. At the same time, according to [58]: "... The counterexample of Perron shows that the negativeness of Lyapunov exponents is not a sufficient condition of stability by the first approximation. In the general case necessary and sufficient conditions of stability by the first approximation are not obtained."

Recently, it was also shown [47, 52] that, in general, the positiveness of the largest Lyapunov exponent is not a sufficient condition of instability by the first approximation and chaos.

In the 1940s N.G. Chetaev [15] published the criterion of instability by the first approximation for regular linearizations. However, in the proof of these criteria a

G.A. Leonov • N.V. Kuznetsov (✉)
Saint Petersburg State University, Russia; University of Jyväskylä, Finland
e-mail: leonov@math.spbu.ru; nkuznetsov239@gmail.com

V. Afraimovich et al. (eds.), *Nonlinear Dynamics and Complexity*, Nonlinear Systems and Complexity 8, DOI 10.1007/978-3-319-02353-3_2,

flaw was discovered [48, 52] and, at present, a complete proof of Chetaev theorems is given for a more weak condition in comparison with that for instability in the sense of Lyapunov, namely, for instability in the sense of Krasovsky.

The discovery of strange attractors and chaos in the investigation of complex nonlinear dynamical systems led to the use and study of instability by the first approximation. At present, many specialists in chaotic dynamics use various numerical methods for computation of Lyapunov exponent (see, e.g., [3, 8, 10, 12, 13, 16, 17, 24, 26–28, 30, 55, 56, 64, 65, 68, 70–72, 75–77, 79], and others) and believe that the positiveness of the largest Lyapunov exponent of linear first approximation system implies the instability of solutions of the original system.

As a rule, the authors ignore the justification of linearization procedure and use numerical values of exponents so obtained to construct various numerical characteristics of attractors of the original nonlinear systems (Lyapunov dimensions, metric entropies, and so on). Sometimes, computer experiments serve as arguments for the partial justification of the linearization procedure. For example, some computer experiments [61, 67] show the coincidence of the Lyapunov and Hausdorff dimensions of the attractors of Henon, Kaplan–Yorke, and Zaslavskii. But for B-attractors of Henon and Lorenz, such a coincidence does not hold [46, 48].

So, the approach, based on linearizations along the nonstationary trajectories on the strange attractors, requires justification. This motivates the development of nonstationary theory of instability by the first approximation.

In this work for the discrete and continuous systems the results of stability by the first approximation for regular and nonregular linearizations are given, the Perron effects are considered, the criteria of stability and instability of flow and cascade of solutions, and the criteria of instability in the sense of Lyapunov and Krasovsky are demonstrated. Some recent consideration of Lyapunov exponents, stability, and chaos can be found, e.g., [7, 18, 22, 31–33, 39, 53, 54, 59, 66, 69, 78]

2 Classical Definitions of Stability

Consider a continuous system

$$
\begin{gathered}
\frac{dx}{dt} = F(x,t), \quad x \in \mathbb{R}^n,\ t \in \mathbb{R} \\
F(\cdot,\cdot)\colon \mathbb{R}^n \times \mathbb{R} \to \mathbb{R}^n
\end{gathered} \tag{1}
$$

and its discrete analog

$$
\begin{gathered}
x(t+1) = F\big(x(t),t\big), \qquad x \in \mathbb{R}^n,\ t \in \mathbb{Z}, \\
F(\cdot,\cdot)\colon \mathbb{R}^n \times \mathbb{Z} \to \mathbb{R}^n.
\end{gathered} \tag{1'}
$$

Consider the solution $x(t)$ of system (1) or (1′), given on the interval $a < t < +\infty$.

Definition 1. The solution $x(t)$ is said to be stable in the sense of Lyapunov (Lyapunov stable) if for any $\varepsilon > 0$ and $t_0 > a$ there exists a number $\delta = \delta(\varepsilon, t_0)$ such that

1. all solutions $y(t)$, satisfying the condition

$$|y(t_0) - x(t_0)| < \delta,$$

   are defined in the interval $t_0 \le t < +\infty$;
2. for these solutions the inequality

$$|x(t) - y(t)| < \varepsilon, \quad \forall t \ge t_0$$

is valid. If $\delta(\varepsilon, t_0)$ is independent of t_0, then Lyapunov stability is called uniform.

Definition 2. The solution $x(t)$ is said to be asymptotically Lyapunov stable if it is Lyapunov stable and for any $t_0 > a$ there exists a positive number $\Delta = \Delta(t_0)$ such that all solutions $y(t)$, defined in the interval $t_0 \le t < +\infty$ and satisfying the condition

$$|y(t_0) - x(t_0)| < \Delta,$$

have the following property:

$$\lim_{t \to +\infty} |y(t) - x(t)| = 0.$$

In other words, for any $\varepsilon' > 0$ there exists a positive number $T = T(\varepsilon', y(t_0), t_0)$ such that the inequality $|x(t) - y(t)| < \varepsilon', \forall t \ge t_0 + T$ is valid. If $x(t)$ is uniformly stable and $\Delta(t_0)$ and $T(\varepsilon', y(t_0), t_0)$ is independent of t_0, then Lyapunov asymptotic stability is called uniform.

Definition 3. The solution $x(t)$ is said to be exponentially stable if for any $t_0 > a$ there exist positive numbers $\delta = \delta(t_0)$, $R = R(t_0)$, and $\alpha = \alpha(t_0)$ such that

1. all solutions $y(t)$, satisfying the condition

$$|y(t_0) - x(t_0)| < \delta,$$

   are defined in the interval $t_0 \le t < +\infty$;
2. the inequality

$$|y(t) - x(t)| \le R \exp\big(-\alpha\,(t - t_0)\big)|y(t_0) - x(t_0)|, \quad \forall t \ge t_0$$

is satisfied. If δ, R, and α are independent of t_0, then exponential stability is called uniform.

Assuming $\alpha = 0$, one obtains the following definition.

Definition 4. The solution $x(t)$ is said to be stable in the sense of Krasovsky (Krasovsky stable) if for any $t_0 > a$ there exist positive numbers $\delta = \delta(t_0)$ and $R = R(t_0)$ such that

1. all solutions $y(t)$, satisfying the condition
$$|y(t_0) - x(t_0)| < \delta,$$
are defined in the interval $t_0 \le t < +\infty$;
2. the following inequality
$$|x(t) - y(t)| \le R|y(t_0) - x(t_0)|, \quad \forall t \ge t_0$$

is valid. If δ and R are independent of t_0, then stability in the sense of Krasovsky is called uniform.

Hence, it follows that the stability of solution in the sense of Krasovsky yields its stability in the sense of Lyapunov. Relations with uniform stability can be found in [Willems, 1970].

Further without loss of generality, consider solutions with $t_0 = 0$. Denote by $x(t, x_0)$ a solution of either system (1) or system (1′) with the initial data $x(0, x_0) = x_0$, and suppose that all solutions $x(t, x_0)$ of continuous system are defined on the interval $[0, +\infty)$ and the solutions of discrete system are defined on the set $\mathbb{N}_0 = 0, 1, 2 \ldots$.

3 Characteristic Exponents, Regular Systems, Lyapunov Exponents

The problem of the investigation of the solution $x(t, x_0)$ can be reduced to the problem of the stability of the trivial solution $y(t) \equiv 0$ by transformation $x = y + x(t, x_0)$. Then one can consider systems (1) and (1′) with a marked linear part. In the continuous case one has

$$\frac{dx}{dt} = A(t)x + f(t, x), \ x \in \mathbb{R}^n, \quad t \in [0, +\infty), \tag{2}$$

where $A(t)$ is a continuous $(n \times n)$-matrix, $f(\cdot, \cdot) : [0, +\infty) \times \mathbb{R}^n \to \mathbb{R}^n$ is a continuous vector-function.

In the discrete case, one has

$$x(t + 1) = A(t)x(t) + f\big(t, x(t)\big), \ x(t) \in \mathbb{R}^n, \quad t \in \mathbb{N}_0, \tag{2′}$$

where $A(t)$ is an $(n \times n)$-matrix, $f(\cdot, \cdot)\colon \mathbb{N}_0 \times \mathbb{R}^n \to \mathbb{R}^n$.

Suppose, in a certain neighborhood $\Omega(0)$ of the point $x = 0$ the nonlinear parts of systems (2) and (2′) satisfy the following condition

$$|f(t,x)| \leq \kappa |x|^{\nu} \ \ \forall t \geq 0, \quad \forall x \in \Omega(0), \quad \kappa > 0, \nu > 1. \tag{3}$$

We shall say that the first approximation system for (2) *is the following linear system*

$$\frac{dx}{dt} = A(t)x \tag{4}$$

and that for discrete system (2′) *is the linear system*

$$x(t+1) = A(t)x(t). \tag{4'}$$

Consider a fundamental matrix $X(t) = \big(x_1(t), \ldots, x_n(t)\big)$, consisting of the linear-independent solutions $\{x_i(t)\}_1^n$ of the first approximation system. For the determinant of the fundamental matrix one has the Ostrogradsky–Liouville formula, which in the continuous case is as follows

$$\det X(t) = \det X(0) \exp\left(\int_0^t \mathrm{Tr}A(\tau)d\tau\right), \tag{5}$$

and in the discrete one takes the form

$$\det X(t) = \det X(0) \prod_{j=0}^{t-1} \det A(j). \tag{5'}$$

The fundamental matrices are often considered to satisfy the following condition

$$X(0) = I_n,$$

where I_n is a unit $(n \times n)$-matrix.

The following definitions and results are valid for continuous system as well as for the discrete one. The proofs will be given, if necessary, for each situation separately.

Consider the vector-function $f(t)$ such that $\lim\limits_{t\to+\infty} \sup |f(t)| \neq 0$.

Definition 5. The value (or the symbol $+\infty, or -\infty$), defined by formula

$$\mathscr{X}[f(t)] = \lim_{t\to+\infty} \sup \frac{1}{t} \ln |f(t)|,$$

is called a characteristic exponent (or upper characteristic exponent) of the vector-function $f(t)$. The value

$$\lim_{t \to +\infty} \inf \frac{1}{t} \ln |f(t)|,$$

is called lower characteristic exponent.

The characteristic exponent is equal to that taken with inverse sign characteristic number, introduced by [57].

Definition 6. The characteristic exponent of the vector-function $f(t)$ is said to be exact if the finite limit

$$\lim_{t \to +\infty} \frac{1}{t} \ln |f(t)|$$

exists.

Consider the characteristic exponents of solutions of linear system (4) or (4′).

Definition 7 ([19]). A set of distinctive characteristic exponents of all solutions (except a zero solution), being different from $\pm\infty$, of linear system is called its spectrum.

Note that the number of different characteristic exponents is bounded by the dimension of the considered space of system states. Imposing conditions on $A(t)$ one can get boundedness of characteristic exponents (see, e.g., [19, 20, 43]).

3.1 Regular Systems

Consider the normal fundamental systems of solutions [57].

Definition 8. A fundamental matrix is said to be normal if the sum of characteristic exponents of its columns is minimal in comparison with other fundamental matrices.

For continuous systems [19] and discrete [20] the following result is well known.

Lemma 1. *In all normal fundamental systems of solutions, the number of solutions with equal characteristic exponents is the same. Each normal fundamental system realizes a spectrum of linear system.*

Thus, one can introduce the following definition.

Definition 9 ([19]). The set of characteristic exponents

$$\lambda_1, \ldots, \lambda_n$$

of a certain normal fundamental system of solutions is called a complete spectrum and the number $\sigma = \sum_1^n \lambda_i$ is a sum of characteristic exponents of linear system.

Note that any fundamental system of solutions has a solution with the largest characteristic exponent $\max\limits_{1\le j\le n} \lambda_j$.

Consider a class of regular systems, introduced by Lyapunov.

Definition 10. A linear system is said to be regular if for the sum of its characteristic exponents σ the following relation holds

$$\sigma = \liminf_{t\to+\infty} \frac{1}{t} \ln|\det X(t)|.$$

Taking into account formula (5), in the continuous case one obtains a classical definition [1, 19] of the regularity of system

$$\sigma = \liminf_{t\to+\infty} \frac{1}{t} \int\limits_0^t \mathrm{Tr}A(\tau)\,\mathrm{d}\tau.$$

Similarly, formula (5′) gives a definition of regularity [20] in the discrete case

$$\sigma = \liminf_{t\to+\infty} \frac{1}{t} \ln \prod_{j=0}^{t-1} |\det A(j)|.$$

Definition 11. The number

$$\Gamma = \sigma - \liminf_{t\to+\infty} \frac{1}{t} \ln|\det X(t)|$$

is called an irregularity coefficient of linear system.

As was shown in [19], the systems with constant and periodic coefficients are regular.

For continuous [19] and discrete systems [20, 25] the following is well known

Lemma 2 (Lyapunov inequality). *Let all characteristic exponents of solutions of linear system be $< +\infty$ (or all characteristic exponents be $> -\infty$.) Then, for any fundamental system of solutions $X(t)$ the following inequality*

$$\limsup_{t\to+\infty} \frac{1}{t} \ln|\det X(t)| \le \sigma_X, \tag{6}$$

where σ_X is a sum of characteristic exponents of the system of solutions $X(t)$, is satisfied.

Thus, for regular systems there exists the limit $\lim\limits_{t\to+\infty} \frac{1}{t} \ln|\det X(t)|$.

Note that also from the condition of regularity of linear system it follows [19] that for its solutions $x(t) \neq 0$ there exist the limits

$$\lim_{t\to+\infty} \frac{1}{t} \ln |x(t)|.$$

Example 1 (Nonregular system with exact characteristic exponents). As was shown in [11], the opposite, generally speaking, is not valid. Consider an example of nonregular system, all characteristic exponents of which are exact [11]. Consider system (4) with the matrix

$$A(t) = \begin{pmatrix} 0 & 1 \\ 0 & (\cos\ln t - \sin\ln t - 1) \end{pmatrix}, \quad t \geq 1 \tag{7}$$

and its fundamental matrix $X(t)$

$$X(t) = \big(x_1(t), x_2(t)\big) = \begin{pmatrix} 1 & \int\limits_1^t e^{\gamma(\tau)}\,\mathrm{d}\tau \\ 0 & e^{\gamma(t)} \end{pmatrix},$$

where $\gamma(t) = t(\cos\ln t - 1)$. In this case for the determinant of fundamental matrix the following relation

$$\lim_{t\to+\infty} \inf \frac{1}{t} \ln |\det X(t)| = -2 \tag{8}$$

is satisfied. Consider characteristic exponents of solutions. For $x_1(t)$ one has

$$\lim_{t\to+\infty} \sup \frac{1}{t} \ln |x_1(t)| = \lim_{t\to+\infty} \inf \frac{1}{t} \ln |x_1(t)| = 0. \tag{9}$$

Since $e^{\gamma(t)} \leq 1$ for $t \geq 1$, one concludes that the characteristic exponent $x_2(t)$ is less than or equal to zero

$$\lim_{t\to+\infty} \sup \frac{1}{t} \ln |x_2(t)| \leq 0.$$

On the other hand, since the integral of $e^{\gamma(\tau)}$ is divergent, namely

$$\int\limits_1^{+\infty} e^{\gamma(\tau)}\,\mathrm{d}\tau = +\infty, \tag{10}$$

for $x_2(t)$ one has the following estimate

$$\lim_{t\to+\infty} \inf \frac{1}{t} \ln |x_2(t)| \geq 0.$$

This implies that

$$\lim_{t\to+\infty}\sup\frac{1}{t}\ln|x_2(t)| = \lim_{t\to+\infty}\inf\frac{1}{t}\ln|x_2(t)| = 0. \tag{11}$$

Thus, by (8), (9), and (11) the linear system with matrix (7) has exact characteristic exponents but it is nonregular:

$$\Gamma = 2.$$

Let us prove that the integral of $e^{\gamma(\tau)}$ is divergent.

Suppose, $t^u(k) = e^{2k\pi+\delta(k)}$ and $t^l(k) = e^{2k\pi-\delta(k)}$, where $\delta(k) = e^{-k\pi}$, $k = 1,2\ldots$. From the definition of $t^l(k)$ and $t^u(k)$ one obtains

$$t^u(k) - t^l(k) \geq e^{2k\pi-\delta(k)}(e^{2\delta(k)} - 1) \geq e^{2k\pi-\delta(k)}2\delta(k) \geq 2e^{k\pi-1}. \tag{12}$$

In the case $\tau \in \left[t^l(k), t^u(k)\right]$ for $\gamma(\tau)$ the estimate

$$-\gamma(\tau) \leq \tau\big(1-\cos(\delta(k))\big) \leq t^u(k)\frac{\delta^2(k)}{2} \leq \frac{1}{2}e^{2k\pi+\delta(k)}e^{-2\pi k} \leq \frac{e^{\delta(k)}}{2} \leq \frac{e}{2} \tag{13}$$

is valid. Then one has

$$\int_1^{t^u(k)} e^{\gamma(\tau)}d\tau \geq \big(t^u(k) - t^l(k)\big)\,e^{-e/2} \geq 2e^{k\pi-1-e/2} \to +\infty$$

as $k \to +\infty$. ∎

3.2 Lyapunov Exponents and Singular Values

Consider singular values (see, e.g., [9]) of the matrix $X(t)$.

Definition 12. The singular values $\{\alpha_j\big(X(t)\big)\}_1^n$ of the matrix $X(t)$ are the square roots of the eigenvalues of the matrix $X(t)^*X(t)$.

The following geometric interpretation of singular values is known: the numbers $\alpha_j\big(X(t)\big)$ coincide with a principal semiaxis of the ellipsoid $X(t)B$, where B is a ball of unit radius.

Definition 13 ([73]). The Lyapunov exponent μ_j is as follows

$$\mu_j = \lim_{t\to+\infty}\sup\frac{1}{t}\ln\alpha_j\big(X(t)\big). \tag{14}$$

In the case (14) the terms *upper singular exponent* are also used [5].

Let μ_1 and λ_1 be the largest Lyapunov exponent and the largest characteristic exponent, respectively.

Lemma 3. *For the linear systems the largest characteristic exponent is equal to the largest Lyapunov exponent.*

Proof. Recall that a geometric interpretation of singular values implies the relation $|X(t)| = \alpha_1(X(t))$. Here $|X|$ is a norm of the matrix X, defined by formula $|X| = \max\limits_{|x|=1} |Xx|, \quad x \in \mathbb{R}^n$. Then the relation $\lim\limits_{t\to+\infty} \sup \frac{1}{t} \ln |X(t)| = \lambda_1$ yields the relation $\lambda_1 = \mu_1$. ∎

Example 2 (Characteristic exponents do not coincide with Lyapunov exponents). Consider [51] system (4) with the matrix

$$A(t) = \begin{pmatrix} 0 & \sin(\ln t) + \cos(\ln t) \\ \sin(\ln t) + \cos(\ln t) & 0 \end{pmatrix} \qquad t > 1$$

and with the fundamental normal matrix

$$X(t) = \begin{pmatrix} e^{\gamma(t)} & e^{-\gamma(t)} \\ e^{\gamma(t)} & -e^{-\gamma(t)} \end{pmatrix},$$

where $\gamma(t) = t\sin(\ln t)$. It is obvious that $\lambda_1 = \lambda_2 = 1$ and

$$\alpha_1(X(t)) = \sqrt{2}\max(e^{\gamma(t)}, e^{-\gamma(t)}), \quad \alpha_2(X(t)) = \sqrt{2}\min(e^{\gamma(t)}, e^{-\gamma(t)}).$$

This implies the following relations $\mu_1 = 1$, $\mu_2 = 0$. Thus, one has $\lambda_2 \neq \mu_2$. ∎

Example 3 (Nemytskii–Vinograd counterexample). Consider [11] a continuous system

$$\frac{dx}{dt} = A(t)x$$

with the matrix

$$A(t) = \begin{pmatrix} 1 - 4(\cos 2t)^2 & 2 + 2\sin 4t \\ -2 + 2\sin 4t & 1 - 4(\sin 2t)^2 \end{pmatrix}.$$

In this case, its solution is the vector-function

$$x(t) = \begin{pmatrix} e^t \sin 2t \\ e^t \cos 2t \end{pmatrix}. \tag{15}$$

It follows that

$$\det(A(t) - pI_n) = p^2 + 2p + 1.$$

Therefore for the eigenvalues $\nu_1(t)$ and $\nu_2(t)$ of the matrix $A(t)$ one has

$$\nu_1(t) = \nu_2(t) = -1.$$

On the other hand, the characteristic exponent λ of solution (15) is equal to 1. ∎

This counterexample shows that all eigenvalues of the matrix $A(t)$ can have negative real parts even if the corresponding linear system has positive characteristic exponents.

It also shows that the formulas, obtained in the book [2], namely

$$\lambda_j = \limsup_{t \to +\infty} \frac{1}{t} \int_0^t \operatorname{Re} \nu_j(\tau)\, d\tau$$

are untrue.

4 The Perron Effects

In 1930, O. Perron [62] showed that the negativeness of the largest Lyapunov exponent of the first approximation system does not always result in the stability of zero solution of the original system. Furthermore, in an arbitrary small neighborhood of zero, the solutions of the original system with positive Lyapunov exponent can be found.

We now present the outstanding result of Perron [1930] and its discrete analog [25,41] (see, also, [4,6,21,29,34–38]).

Consider the following system

$$\begin{aligned} \frac{dx_1}{dt} &= -ax_1 \\ \frac{dx_2}{dt} &= \big(\sin(\ln(t+1)) + \cos(\ln(t+1)) - 2a\big)x_2 + x_1^2 \end{aligned} \tag{16}$$

and its discrete analog

$$\begin{aligned} x_1(t+1) &= \exp(-a)x_1(t) \\ x_2(t+1) &= \frac{\exp\big((t+2)\sin\ln(t+2) - 2a(t+1)\big)}{\exp\big((t+1)\sin\ln(t+1) - 2at\big)} x_2(t) + x_1(t)^2. \end{aligned} \tag{16'}$$

Here a is a number satisfying the following inequalities

$$1 < 2a < 1 + \frac{1}{2}\exp(-\pi). \tag{17}$$

The solution of the first approximation system for systems (16) and (16′) takes the form

$$\begin{aligned} x_1(t) &= \exp(-at)x_1(0) \\ x_2(t) &= \exp\big((t+1)\sin(\ln(t+1)) - 2at\big)x_2(0). \end{aligned}$$

It is obvious that by condition (17) for the solution of the first approximation system for $x_1(0) \neq 0, x_2(0) \neq 0$ one has

$$\mathscr{X}[x_1(t)] = -a, \ \mathscr{X}[x_2(t)] = 1 - 2a < 0.$$

This implies that a zero solution of linear system of the first approximation is Lyapunov stable.

Consider the solution of system (16)

$$\begin{aligned} x_1(t) &= \exp(-at)x_1(0), \\ x_2(t) &= \exp\big((t+1)\sin(\ln(t+1)) - 2at\big)\times \\ &\times\left(x_2(0) + x_1(0)^2 \int\limits_0^t \exp\big(-(\tau+1)\sin(\ln(\tau+1))\big)d\tau\right). \end{aligned} \tag{18}$$

Assuming $t = t_k = \exp\left((2k + \frac{1}{2})\pi\right) - 1$, where k is an integer, one obtains

$$\exp\big((t+1)\sin(\ln(t+1)) - 2at\big) = \exp\big((1-2a)t + 1\big), \quad (1+t)e^{-\pi} - 1 > 0,$$

$$\int\limits_0^t \exp\big(-(\tau+1)\sin(\ln(\tau+1))\big)d\tau >$$

$$> \int\limits_{f(k)}^{g(k)} \exp\big(-(\tau+1)\sin(\ln(\tau+1))\big)\,d\tau >$$

$$> \int\limits_{f(k)}^{g(k)} \exp\left(\frac{1}{2}(\tau+1)\right)\,d\tau > \int\limits_{f(k)}^{g(k)} \exp\left(\frac{1}{2}(\tau+1)\exp(-\pi)\right)\,d\tau =$$

$$= \exp\left(\frac{1}{2}(t+1)\exp(-\pi)\right)(t+1)\left(\exp\big(-\frac{2\pi}{3}\big) - \exp(-\pi)\right),$$

where

$$f(k) = (1+t)\exp(-\pi) - 1,$$

$$g(k) = (1+t)\exp\left(-\frac{2\pi}{3}\right) - 1.$$

Hence one has the following estimate

$$\begin{aligned} &\exp\big((t+1)\sin(\ln(t+1)) - 2at\big)\int_0^t \exp\big(-(\tau+1)\sin(\ln(\tau+1))\big)d\tau > \\ &\quad > \exp\left(\frac{1}{2}(2+\exp(-\pi)\right)\left(\exp\big(-\frac{2\pi}{3}\big) - \exp(-\pi)\right)(t+1)\times \\ &\quad\quad \times\exp\left(\left(1-2a+\frac{1}{2}\exp(-\pi)\right)t\right). \end{aligned} \tag{19}$$

From the last inequality and condition (17) it follows that for $x_1(0) \neq 0$ one of the characteristic exponents of solutions of system (16) is positive:

$$\mathcal{X}[x_1(t)] = -a,\ \mathcal{X}[x_2(t)] \geq 1 - 2a + e^{-\pi}/2 > 0. \tag{20}$$

Thus, one obtains that all characteristic exponents of the first approximation system are negative but almost all solutions of the original system (16) tend exponentially to infinity as $t_k \to +\infty$.

Consider now the solution of discrete system (16′)

$$\begin{aligned} x_1(t) &= x_1(0)e^{-at} \\ x_2(t) &= \exp\big((t+1)\sin\ln(t+1) - 2at\big)\times \\ &\quad\times\left(x_2(0) + x_1(0)^2\sum_{k=0}^{t-1}\exp\big(-(k+2)\sin\ln(k+2) + 2a\big)\right), \end{aligned} \tag{21}$$

and show that for this system inequalities (20) are also satisfied. For this purpose one obtains the estimate similar to estimate (19) in the discrete case.

Obviously, for any $N > 0$ and $\delta > 0$ there exists a natural number ($t' = t'(N,\delta)$, $t' > N$) such that

$$\sin\ln(t'+1) > 1 - \delta.$$

Then

$$\exp\big((t'+1)\sin\ln(t'+1) - 2at'\big) \geq \exp\big((1-\delta-2a)t' + 1 - \delta\big). \tag{22}$$

Estimate from below the second multiplier in the expression for $x_2(t)$. For sufficiently large t' there exists a natural number m

$$m \in \left(\frac{t'+1}{e^{\pi}} - 2, t'\right)$$

such that

$$\sin \ln(m+2) \le -\frac{1}{2}.$$

Then one has

$$-(m+2)\sin\ln(m+2) + 2a \ge \frac{t'+1}{2e^{\pi}}.$$

This implies the following estimate

$$\sum_{k=0}^{t'-1} \exp\big(-(k+2)\sin\ln(k+2) + 2a\big) \ge \exp\left((t'+1)\frac{1}{2}e^{-\pi}\right). \tag{23}$$

From (22), (23), and condition (17) it follows that for $x_1(0) \neq 0$ one of characteristic exponents of solutions (21) of system (16′) is positive and inequalities (20) are satisfied.

Consider an example, which shows the possibility of the sign reversal of characteristic exponents "on the contrary," namely the solution of the first approximation system has a positive characteristic exponent while the solution of the original system with the same initial data has a negative exponent [47].

Consider the following continuous system [49]

$$\begin{aligned} \dot{x}_1 &= -ax_1 \\ \dot{x}_2 &= -2ax_2 \\ \dot{x}_3 &= \big(\sin(\ln(t+1)) + \cos(\ln(t+1)) - 2a\big)x_3 + x_2 - x_1^2 \end{aligned} \tag{24}$$

and its discrete analog

$$\begin{aligned} x_1(t+1) &= e^{-a}x_1(t) \\ x_2(t+1) &= e^{-2a}x_2(t) \end{aligned}$$

$$x_3(t+1) = \frac{\exp\big((t+2)\sin\ln(t+2) - 2a(t+1)\big)}{\exp\big((t+1)\sin\ln(t+1) - 2at\big)} x_3(t) + x_2(t) - x_1(t)^2 \tag{24′}$$

on the invariant manifold

$$M = \{x_3 \in \mathbb{R}^1, \; x_2 = x_1^2\}.$$

Here the value a satisfies condition (17).

The solutions of (24) and (24′) on the manifold M take the form

$$\begin{aligned} x_1(t) &= \exp\big(-at\big)x_1(0)\\ x_2(t) &= \exp\big(-2at\big)x_2(0)\\ x_3(t) &= \exp\big((t+1)\sin(\ln(t+1)) - 2at\big)x_3(0),\\ x_1(0)^2 &= x_2(0). \end{aligned} \tag{25}$$

Obviously, these solutions have negative characteristic exponents.

For system (24) in the neighborhood of its zero solution, consider the first approximation system

$$\begin{aligned} \dot{x}_1 &= -ax_1\\ \dot{x}_2 &= -2ax_2\\ \dot{x}_3 &= \big(\sin(\ln(t+1)) + \cos(\ln(t+1)) - 2a\big)x_3 + x_2. \end{aligned} \tag{26}$$

The solutions of this system are the following

$$\begin{aligned} x_1(t) &= \exp\big(-at\big)x_1(0)\\ x_2(t) &= \exp\big(-2at\big)x_2(0)\\ x_3(t) &= \exp\big((t+1)\sin(\ln(t+1)) - 2at\big)\times\\ &\times\left(x_3(0) + x_2(0)\int_0^t \exp\big(-(\tau+1)\sin(\ln(\tau+1))\big)d\tau\right). \end{aligned} \tag{27}$$

For system (24′) in the neighborhood of its zero solution, the first approximation system is as follows

$$\begin{aligned} x_1(t+1) &= \exp(-a)x_1(t)\\ x_2(t+1) &= \exp(-2a)x_2(t)\\ x_3(t+1) &= \frac{\exp\big((t+2)\sin\ln(t+2) - 2a(t+1)\big)}{\exp\big((t+1)\sin\ln(t+1) - 2at\big)}x_3(t) + x_2(t). \end{aligned} \tag{26′}$$

Then the solutions of system (26′) take the form

$$\begin{aligned} x_1(t) &= \exp\big(-at\big)x_1(0)\\ x_2(t) &= \exp\big(-2at\big)x_2(0)\\ x_3(t) &= \exp\big((t+1)\sin\ln(t+1) - 2at\big)\times\\ &\times\left(x_3(0) + x_2(0)^2\sum_{k=0}^{t-1}\exp\big(-(k+2)\sin\ln(k+2) + 2a\big)\right). \end{aligned} \tag{27′}$$

By estimates (19) and (23) for solutions (27) and (27′) for $x_2(0) \neq 0$ one obtains

$$\mathscr{X}[x_3(t)] > 0.$$

It is easily shown that for the solutions of systems (24) and (26) the following relations

$$\big(x_1(t)^2 - x_2(t)\big)^{\bullet} = -2a\big(x_1(t)^2 - x_2(t)\big)$$

are valid. Similarly, for system (26′) one has

$$x_1(t+1)^2 - x_2(t+1) = \exp(-2a)\big(x_1(t)^2 - x_2(t)\big).$$

Then

$$x_1(t)^2 - x_2(t) = \exp\big(-2at\big)\big(x_1(0)^2 - x_2(0)\big).$$

It follows that the manifold M is an invariant exponentially attractive manifold for solutions of continuous systems (24) and (26), and for solutions of discrete systems (24′) and (26′).

This means that the relation $x_1(0)^2 = x_2(0)$ yields the relation $x_1(t)^2 = x_2(t)$ for all $t \in \mathbb{R}^1$ and for any initial data one has

$$\big|x_1(t)^2 - x_2(t)\big| \leq \exp\big(-2at\big)\big|x_1(0)^2 - x_2(0)\big|.$$

Thus, systems (24) and (26) have the same invariant exponentially attractive manifold M on which almost all solutions of the first approximation system (26) have a positive characteristic exponent and all solutions of the original system (24) have negative characteristic exponents. The same result can be obtained for discrete systems (24′) and (26′).

The Perron effect occurs here on the whole manifold

$$\{x_3 \in \mathbb{R}^1,\ x_2 = x_1^2 \neq 0\}.$$

To construct exponentially stable system, the first approximation of which has a positive characteristic exponent we represent system (24) in the following way

$$\begin{aligned}
\dot{x}_1 &= F(x_1, x_2) \\
\dot{x}_2 &= G(x_1, x_2) \\
\dot{x}_3 &= \big(\sin\ln(t+1) + \cos\ln(t+1) - 2a\big)x_3 + x_2 - x_1^2.
\end{aligned} \tag{28}$$

Here the functions $F(x_1, x_2)$ and $G(x_1, x_2)$ have the form

$$F(x_1,x_2) = \pm 2x_2 - ax_1, \quad G(x_1,x_2) = \mp x_1 - \varphi(x_1,x_2),$$

in which case the upper sign is taken for $x_1 > 0$, $x_2 > x_1^2$ and for $x_1 < 0$, $x_2 < x_1^2$, the lower one for $x_1 > 0$, $x_2 < x_1^2$ and for $x_1 < 0$, $x_2 > x_1^2$.

The function $\varphi(x_1,x_2)$ is defined as

$$\varphi(x_1,x_2) = \begin{cases} 4ax_2 \text{ for } |x_2| > 2x_1^2 \\ 2ax_2 \text{ for } |x_2| < 2x_1^2. \end{cases}$$

The solutions of system (28) are regarded in the sense of Filippov [23]. By definition of $\varphi(x_1,x_2)$ the following system

$$\begin{aligned} \dot{x}_1 &= F(x_1,x_2) \\ \dot{x}_2 &= G(x_1,x_2) \end{aligned} \tag{29}$$

on the lines of discontinuity $\{x_1 = 0\}$ and $\{x_2 = x_1^2\}$ has sliding solutions, which are given by the equations

$$x_1(t) \equiv 0, \quad \dot{x}_2(t) = -4ax_2(t)$$

and

$$\dot{x}_1(t) = -ax_1(t), \quad \dot{x}_2(t) = -2ax_2(t), \quad x_2(t) \equiv x_1(t)^2.$$

In this case the solutions of system (29) with the initial data $x_1(0) \neq 0$, $x_2(0) \in \mathbb{R}^1$ attain the curve $\{x_2 = x_1^2\}$ in a finite time, which is less than or equal to 2π.

This implies that for the solutions of system (28) with the initial data $x_1(0) \neq 0$, $x_2(0) \in \mathbb{R}^1$, $x_3(0) \in \mathbb{R}^1$, for $t \geq 2\pi$ one obtains the relations $F(x_1(t),x_2(t)) = -ax_1(t)$, $G(x_1(t),x_2(t)) = -2ax_2(t)$. Therefore, based on these solutions for $t \geq 2\pi$ system (26) is a system of the first approximation for system (28).

System (26), as was shown above, has a positive characteristic exponent. At the same time, all solutions of system (28) tend exponentially to zero. ■

The considered technique permits us to construct the different classes of nonlinear continuous and discrete systems for which the Perron effects occur.

5 Stability Criteria by the First Approximation

Consider a normal fundamental matrix $X(t)$ of the linear part of the system, and let

$$\Lambda = \max_j \lambda_j, \quad \lambda = \min_j \lambda_j.$$

Here $\{\lambda_j\}_1^n$ is a complete spectrum of linear system. We shall say that $X(t)X(\tau)^{-1}$ is a Cauchy matrix. Represent the solutions of systems (2) and (2′) in the Cauchy form. In the continuous case one has

$$x(t) = X(t)x(0) + \int_0^t X(t)X(\tau)^{-1} f\big(\tau, x(\tau)\big)\, d\tau, \tag{30}$$

and in the discrete one

$$x(t) = X(t)x(0) + \sum_{\tau=0}^{t-1} X(t)X(\tau+1)^{-1} f\big(\tau, x(\tau)\big),\ t = 1, 2 \ldots. \tag{31}$$

The following result is well known and is often used.

Theorem 1. *For any number $\varepsilon > 0$ there exists a number $C > 0$ such that the following inequalities*

$$|X(t)X(\tau)^{-1}| \le C \exp\big((\Lambda + \varepsilon)(t - \tau) + (\Gamma + \varepsilon)\tau\big), \quad \forall\, t \ge \tau \ge 0 \tag{32}$$

$$|X(t)X(\tau)^{-1}| \le C \exp\big(\lambda(t - \tau) + (\Gamma + \varepsilon)\tau\big), \quad \forall\, \tau \ge t \ge 0, \tag{33}$$

where Γ is the irregularity coefficient, are satisfied.

Recall that by condition (3) the nonlinear part $f(t, x)$ of systems (2) and (2′) in a certain neighborhood $\Omega(0)$ of the point $x = 0$ satisfies the following condition

$$|f(t, x)| \le \kappa |x|^{\nu}\ \forall t \ge 0, \quad \forall x \in \Omega(0), \quad \kappa > 0, \nu > 1.$$

Let us describe the most famous stability criteria by the first approximation.

Consider the continuous case. Assume that there exists a number $C > 0$ and a piecewise continuous function $p(t)$ such that for the Cauchy matrix $X(t)X(\tau)^{-1}$ the estimate

$$|X(t)X(\tau)^{-1}| \le C \exp \int_\tau^t p(s)\, ds, \quad \forall t \ge \tau \ge 0 \tag{34}$$

is valid.

Theorem 2 ([51]). *If condition* (3) *with $\nu = 1$ and the inequality*

$$\lim_{t \to +\infty} \sup \frac{1}{t} \int_0^t p(s)\, ds + C\kappa < 0$$

are satisfied, then the solution $x(t) \equiv 0$ of system (2) *is asymptotically Lyapunov stable.*

Proof. From (30) and the hypotheses of theorem one has

$$|x(t)| \le C \exp\left(\int_0^t p(s)\,ds\right)|x(0)| + C\int_0^t \exp\left(\int_\tau^t p(s)\,ds\right)\kappa|x(\tau)|\,d\tau.$$

This estimate can be rewritten as

$$\exp\left(-\int_0^t p(s)\,ds\right)|x(t)| \le C|x(0)| + C\kappa\int_0^t \exp\left(-\int_0^\tau p(s)\,ds\right)|x(\tau)|)\,d\tau.$$

By Bellman–Gronwall Lemma the following estimate

$$|x(t)| \le C|x(0)|\exp\left(\int_0^t p(s)\,ds + C\kappa t\right), \quad \forall t \ge 0$$

is satisfied. This completes the proof of theorem. ∎

Consider a discrete analog of this theorem. In the discrete case it is assumed that in place of inequality (34) one has

$$|X(t)X(\tau)^{-1}| \le C\prod_{s=\tau}^{t-1} p(s), \quad \forall t > \tau \ge 0, \tag{35}$$

where $p(s)$ is a positive function.

In the discrete case one has a similar theorem

Theorem 3 ([44,52]). *If condition* (3) *with* $\nu = 1$ *and the inequality*

$$\limsup_{t\to+\infty} \frac{1}{t}\ln\prod_{s=0}^{t-1}\big(p(s) + C\kappa\big) < 0 \tag{36}$$

are satisfied, then the solution $x(t) \equiv 0$ *of system* (2′) *is asymptotically Lyapunov stable.*

Corollary 1. *For the first-order system* (2) *or* (2′) *the negativeness of characteristic exponent implies the asymptotic stability of its zero solution.*

Assume that for the Cauchy matrix $X(t)X(\tau)^{-1}$ the following estimate

$$|X(t)X(\tau)^{-1}| \le C\exp\big(-\alpha(t-\tau) + \gamma\tau\big), \quad \forall t \ge \tau \ge 0, \tag{37}$$

where $\alpha > 0$, $\gamma \ge 0$, is satisfied.

Theorem 4 ([14,58,60]). *Let condition* (3) *with sufficiently small* κ *and condition* (37) *be valid. Then if the inequality*

$$(\nu - 1)\alpha - \gamma > 0 \tag{38}$$

holds, then the solution $x(t) \equiv 0$ *is asymptotically Lyapunov stable.*

Theorem 4 strengthens the well-known Lyapunov theorem [57] on stability by the first approximation for regular systems.

5.1 Stability Criteria for the Flow and Cascade of Solutions

Consider system (1) or (1′) where $F(\cdot,\cdot)$ is a twice continuously differentiable vector-function. Consider the linearizations of these systems along solutions with the initial data $y = x(0, y)$ from the open set Ω, which is bounded in $\mathbb{R}^n$

$$\frac{dz}{dt} = A_y(t)z, \tag{39}$$

$$z(t+1) = A_y(t)z(t). \tag{39'}$$

Here the matrix

$$A_y(t) = \frac{\partial F(x,t)}{\partial x}|_{x=x(t,y)}$$

is Jacobian matrix of the vector-function $F(x,t)$ on the solution $x(t, y)$. Let $X(t, y)$ be a fundamental matrix of linear system and $X(0, y) = I_n$.

Assume that for the largest singular value $\alpha_1(t, y)$ of systems (39) and (39′) for all t the following estimate

$$\alpha_1(t, y) < \alpha(t), \quad \forall y \in \Omega, \tag{40}$$

where $\alpha(t)$ is a scalar function, is valid.

Theorem 5 ([43, 50]). *Suppose the function* $\alpha(t)$ *is bounded on the interval* $(0, +\infty)$*. Then the flow (cascade) of solutions* $x(t, y)$*,* $y \in \Omega$*, of systems* (1) *and* (1′) *is Lyapunov stable. If, in addition,*

$$\lim_{t\to+\infty} \alpha(t) = 0,$$

then the flow (cascade) of solutions $x(t, y)$*,* $y \in \Omega$*, is asymptotically Lyapunov stable.*

Proof. It is well known that

$$\frac{\partial x(t,y)}{\partial y} = X(t,y), \quad \forall t \geq 0.$$

It is also known that for any vectors y, z, and a number $t \geq 0$ there exists a vector w such that the relations

$$|w - y| \leq |y - z|,$$

$$|x(t,y) - x(t,z)| \leq \left|\frac{\partial x(t,w)}{\partial w}\right| |y - z|$$

are satisfied. Therefore for any vector z from the ball centered at y and placed entirely in Ω the following estimate

$$|x(t,y) - x(t,z)| \leq |y - z| \sup \alpha_1(t,w) \leq \alpha(t)|y - z|, \quad \forall t \geq 0 \tag{41}$$

is valid. Here the supremum is taken over all w from the ball $\{w : |w - y| \leq |y - z|\}$.

Estimate (41) gives at once the assertions of theorem. ■

Corollary 2. *The Perron effects are possible on the boundary of the stable by the first approximation solutions flow (cascade) only.*

Consider the flow of solutions of system (16) with the initial data in a neighborhood of the point $x_1 = x_2 = 0$: $x_1(0, x_{10}, x_{20}) = x_{10}$, $x_2(0, x_{10}, x_{20}) = x_{20}$.

Hence it follows easily that

$$x_1(t, x_{10}, x_{20}) = \exp\big(-at\big)x_{10}.$$

Therefore for continuous system the matrix $A(t)$ of linear system takes the form

$$A(t) = \begin{pmatrix} -a & 0 \\ 2\exp\big(-at\big)x_{10} & r(t) \end{pmatrix}, \tag{42}$$

where

$$r(t) = \sin(\ln(t+1)) + \cos(\ln(t+1)) - 2a.$$

For the discrete system one has

$$A(t) = \begin{pmatrix} e^{-a} & 0 \\ 2\exp\big(-at\big)x_{10} & r(t) \end{pmatrix}, \tag{42'}$$

$$r(t) = \frac{\exp\big((t+2)\sin\ln(t+2) - 2a(t+1)\big)}{\exp\big((t+1)\sin\ln(t+1) - 2at\big)}.$$

The solutions of system (39) and (39′) with matrices (42) and (42′), respectively, are the following

$$
\begin{aligned}
z_1(t) &= \exp\big(-at\big)z_1(0),\\
z_2(t) &= p(t)(z_2(0) + 2x_{10}z_1(0))q(t)).
\end{aligned}
\tag{43}
$$

Here

$$
p(t) = \exp\big((t+1)\sin(\ln(t+1)) - 2at\big),
$$

$$
q(t) = \int_0^t \exp\big(-(\tau+1)\sin(\ln(\tau+1))\big)\,d\tau
$$

in the continuous case and

$$
p(t) = \exp\big((t+1)\sin(\ln(t+1)) - 2at\big),
$$

$$
q(t) = \sum_{k=0}^{t-1} \exp\big(-(k+2)\sin\ln(k+2) + 2a\big)
$$

in the discrete case.

As was shown above (20), if relations (17) are satisfied and

$$
z_1(0)x_{10} \neq 0,
$$

then the characteristic exponent of $z_2(t)$ is positive.

Hence in an arbitrary small neighborhood of the trivial solution $x_1(t) \equiv x_2(t) \equiv 0$ there exist the initial data x_{10}, x_{20} such that for $x_1(t, x_{10}, x_{20})$, $x_2(t, x_{10}, x_{20})$ the first approximation system has the positive largest characteristic exponent (and Lyapunov exponent μ_1).

Therefore in this case there does not exist a neighborhood Ω of the point $x_1 = x_2 = 0$ such that uniform estimates (40) are satisfied. Thus, for systems (16), (16′) the Perron effect occurs.

6 Instability Criteria by the First Approximation

6.1 *The Perron–Vinograd Triangulation Method*

One of the basic procedures for analysis of instability is a reduction of the linear part of the system to the triangular form. In this case the Perron–Vinograd triangulation method for a linear system [11, 19] turns out to be most effective.

Let $Z(t) = \big(z_1(t), \ldots, z_n(t)\big)$ be a fundamental system of solutions of linear continuous system (4) or discrete system (4′). Apply the Schmidt orthogonalization procedure to the solutions $z_j(t)$.

$$\begin{aligned} v_1(t) &= z_1(t) \\ v_2(t) &= z_2(t) - v_1(t)^* z_2(t) \frac{v_1(t)}{|v_1(t)|^2} \\ &\ldots\ldots\ldots\ldots\ldots \\ v_n(t) &= z_n(t) - v_1(t)^* z_n(t) \frac{v_1(t)}{|v_1(t)|^2} - \ldots - v_{n-1}(t)^* z_n(t) \frac{v_{n-1}(t)}{|v_{n-1}(t)|^2}. \end{aligned} \tag{44}$$

Relations (44) yield the following relations

$$v_i(t)^* v_j(t) = 0, \quad \forall j \neq i, \tag{45}$$

$$|v_j(t)|^2 = v_j(t)^* z_j(t). \tag{46}$$

If for the fundamental matrix $Z(t)$ the relation $Z(0) = I_n$ holds, one concludes that $V(0) = \big(v_1(0), .., v_n(0)\big) = I_n$.

Proceed now to the description of *the triangulation procedure of Perron–Vinograd.*

Consider the unitary matrix

$$U(t) = \left(\frac{v_1(t)}{|v_1(t)|}, \cdots, \frac{v_n(t)}{|v_n(t)|}\right),$$

and make the change of variable: $z = U(t)w$ in the linear system. In the continuous case one obtains the system

$$\frac{dw}{dt} = B(t)w, \tag{47}$$

where

$$B(t) = U(t)^{-1} A(t) U(t) - U(t)^{-1} \dot{U}(t), \tag{48}$$

and in the discrete case the system

$$w(t+1) = B(t)w(t), \tag{47'}$$

where

$$B(t) = U(t+1)^{-1}A(t)U(t). \tag{48'}$$

The unitarity of the matrix $U(t)$ implies that for the columns $w(t)$ of the fundamental matrix

$$W(t) = \big(w_1(t), \dots, w_n(t)\big) = U(t)^* Z(t), \tag{49}$$

the relations $|w_j(t)| = |z_j(t)|$ are satisfied.

By (44)–(46) one obtains that the matrix $W(t)$ has the upper triangular form with the diagonal elements $|v_1(t)|$,..,$|v_n(t)|$, namely

$$W(t) = \begin{pmatrix} |v_1(t)| & \cdots & \\ & \ddots & \vdots \\ 0 & & |v_n(t)| \end{pmatrix}. \tag{50}$$

From the fact that $W(t)$ is an upper triangular matrix it follows that $W(t)^{-1}$, $\dot{W}(t)$ are also upper triangular matrices. Hence $B(t)$ is an upper triangular matrix with the diagonal elements $b_1(t)$,..,$b_n(t)$:

$$B(t) = \begin{pmatrix} b_1(t) & \cdots & \\ & \ddots & \vdots \\ 0 & & b_n(t) \end{pmatrix}, \tag{51}$$

where in the continuous case $b_i(t) = (\ln|v_i(t)|)^\bullet$ and in the discrete one

$$b_i(t) = \frac{|v_i(t+1)|}{|v_i(t)|}.$$

Thus, it is proved the following

Theorem 6 (Perron triangulation). *By means of the unitary transformation $z = U(t)w$ the linear system can be reduced to the linear system with the upper triangular matrix $B(t)$.*

Note that if $|A(t)|$ is bounded for $t \geq 0$, then $|B(t)|$, $|U(t)|$, and $|\dot{U}(t)|$ are also bounded for $t \geq 0$. If in the discrete case, in addition, $|A(t)^{-1}|$ is bounded for $t \geq 0$, then $|B(t)^{-1}|$ is also bounded for $t \geq 0$.

Lemma 4. *The following estimate*

$$\frac{|v_n(t)|}{|v_n(\tau)|} \geq \frac{|\det Z(t)|}{|\det Z(\tau)|} \prod_{j=1}^{n-1} \frac{|v_j(\tau)|}{|z_j(t)|}, \qquad \forall t, \tau \geq 0 \tag{52}$$

is valid.

Define the vector $z_i' = z_i - v_i$. Then the vector z_i' is orthogonal to the vector v_i, where $i \geq 2$. Consider the angle included between the vectors z_i and z_i'. Note that from definition of the angle included between the vectors one has $\angle(z_i, z_i') \leq \pi$. In this case the following relation

$$|v_i| = |z_i| \sin\left(\angle(z_i, z_i')\right) \quad i \geq 2 \tag{53}$$

is valid.

By (53) from (49) and (50) one has

$$|\det Z(t)| = |\det U(t)| \prod_{i=1}^{n} |v_i| = \prod_{i=1}^{n} |z_i| \prod_{k=2}^{n} |\sin\left(\angle(z_k, z_k')\right)|.$$

With the help of this relation in [74] the following criterion of system regularity was obtained.

Theorem 7 ([74]). *Consider a linear system with bounded coefficients and its certain fundamental system of solutions* $Z(t) = \left(z_1(t), \ldots, z_n(t)\right)$. *Let there exist the exact characteristic exponents of* $|z_i(t)|$

$$\lim_{t\to+\infty} \frac{1}{t} \ln |z_i(t)| \quad i = 1, \ldots, n \tag{54}$$

and let there exist and be equal to zero the exact characteristic exponents of sines of the angles $\angle(z_i, z_i')$

$$\lim_{t\to+\infty} \frac{1}{t} \ln \left|\sin\left(\angle(z_i, z_i')\right)\right| = 0 \quad i = 2, \ldots, n. \tag{55}$$

Then the linear system is regular and $Z(t)$ *is a normal system of solutions.*

Conversely, if the linear system is regular and $Z(t)$ *is a normal system of solutions, then (54) and (55) are satisfied.*

6.2 Instability Criterion by Krasovsky

Consider instability in the sense of Krasovsky for the solution $x(t) \equiv 0$ of continuous system (2) and of discrete system (2′).

Theorem 8 ([42,45,49]). *If the relation*

$$\sup_{1\le k\le n}\ \lim_{t\to+\infty}\inf\left[\frac{1}{t}\left(\ln\left|\det Z(t)\right| - \sum_{j\ne k}\ln\left|z_j(t)\right|\right)\right] > 1 \tag{56}$$

is satisfied, then the solution $x(t)\equiv 0$ is unstable in the sense of Krasovsky.[1]

Proof. One can assume, without loss of generality, that in (56) the supremum, taken over k, is attained for $k = n$. Then by Lemma 4 with $\tau = 0$ from condition (56) one obtains that there exists a number $\mu > 1$ such that for sufficiently large t the following estimate

$$\ln|v_n(t)| \ge \mu t, \quad \mu > 1 \tag{57}$$

holds. Suppose now that the solution $x(t)\equiv 0$ is stable by Krasovsky. This means that in a certain neighborhood of the point $x = 0$ there exists a number $R > 0$ such that the estimate

$$|x(t,x_0)| \le R|x_0|, \quad \forall t \ge 0 \tag{58}$$

is valid. Make use of the Perron–Vinograd change of variable

$$x = U(t)y \tag{59}$$

to obtain a system with the upper triangular matrix $B(t)$ of the type (51).

1. Consider the continuous case. Using (59), from continuous system (2) one obtains

$$\frac{dy}{dt} = B(t)y + g(t,y), \quad g(t,y) = U(t)^{-1}f\big(t,U(t)y\big). \tag{60}$$

Thus, the last equation of system (60) takes the form

$$\frac{dy_n}{dt} = \big(\ln|v_n(t)|\big)^{\bullet}y_n + g_n(t,y). \tag{61}$$

Here y_n and g_n are the nth components of the vectors y and g, respectively. Conditions (3) and (58) yield the estimate

$$\big|g\big(t,y(t)\big)\big| \le \varkappa R^{\nu}\big|y(0)\big|^{\nu}. \tag{62}$$

Note that the solution $y_n(t)$ of (61) can be represented in the form

[1] Condition (56) can be weakened [40,42,52].

$$y_n(t) = \frac{|v_n(t)|}{|v_n(0)|}\left(y_n(0) + \int_0^t \frac{|v_n(0)|}{|v_n(s)|} g\big(s, y(s)\big) ds\right). \tag{63}$$

Estimate (57) implies that there exists a number $\rho > 0$ such that the following inequalities

$$\int_0^t \frac{|v_n(0)|}{|v_n(s)|} ds \le \rho, \quad \forall t \ge 0 \tag{64}$$

are valid. Take the initial condition $x_0 = U(0)y(0)$ in such a way that $y_n(0) = |y(0)| = \delta$, where the number δ satisfies the inequality

$$\delta > \rho \varkappa R^\nu \delta^\nu. \tag{65}$$

Then from (62)–(64) for sufficiently large $t \ge 0$ one obtains the following estimate

$$y_n(t) \ge t^\mu(\delta - \rho\varkappa R^\nu \delta^\nu), \quad \mu > 1.$$

By (65)

$$\lim_{t\to+\infty} \inf y_n(t) = +\infty.$$

The latter contradicts the assumption on stability in the sense of Krasovsky of a trivial solution of system (2).

2. Let us prove the theorem in the discrete case. By (59), from discrete system (2′) one has

$$y(t+1) = B(t)y(t) + g\big(t, y(t)\big), \tag{66}$$

where

$$g\big(t, y(t)\big) = U(t+1)^{-1} f\big(t, U(t)y(t)\big).$$

Then the last equation of system (66) takes the form

$$y_n(t+1) = \frac{|v_n(t+1)|}{|v_n(t)|} y_n(t) + g_n\big(t, y(t)\big), \tag{67}$$

where y_n and g_n are the nth components of the vectors y and g, respectively. Conditions (3) and (58) give the following estimate

$$\big|g\big(t, y(t)\big)\big| \le \varkappa R^\nu \big|y(0)\big|^\nu. \tag{68}$$

Note that the solution $y_n(t)$ of (67) can be represented as

$$y_n(t) = \frac{|v_n(t)|}{|v_n(0)|}\left(\sum_{j=0}^{t-1}\frac{|v_n(0)|}{|v_n(j+1)|}g_n\big(j, y(j)\big) + y_n(0)\right). \tag{69}$$

Estimate (57) implies that there exists a number $\rho > 0$ such that the following inequality

$$\sum_{j=0}^{t-1}\frac{|v_n(0)|}{|v_n(j+1)|} < \rho, \quad t \geq 1 \tag{70}$$

is satisfied. Taking the same initial data as in the continuous case (65), one obtains

$$\lim_{t\to+\infty}\inf y_n(t) = +\infty.$$

The latter contradicts the assumption on stability in the sense of Krasovsky of a trivial solution of system (2′).

This proves the theorem. ∎

Remark. Concerning the method for the proof of theorem.

Assuming that the zero solution of the considered system is stable in the sense of Lyapunov and using the same reasoning as in the case of stability in the sense of Krasovsky, one need to prove in the continuous case the following inequality

$$y_n(0) + \int_0^{+\infty}\frac{|v_n(0)|}{|v_n(s)|}g\big(s, y(s)\big)\,ds \neq 0. \tag{71}$$

While the above inequality is easily proved in the case of stability in the sense of Krasovsky, this becomes an intractable problem in the case of stability in the sense of Lyapunov.

A scheme similar to that, considered above for reducing the problem to one scalar equation of the type (61), was used by N.G. Chetaev [1990; 1948] to obtain instability criteria. In the scheme, suggested by N.G. Chetaev for proving inequality (71), a similar difficulty occurs. Therefore, at present, Chetaev's technique permits us to obtain the criteria of instability in the sense of Krasovsky only.

The method to obtain the criteria of instability in the sense of Lyapunov invites further development. Such development under certain additional restrictions will be presented in Theorem 10.

Corollary 3. *Condition (56) of Theorem 8 is satisfied if the following inequality*

$$\Lambda - \Gamma > 0 \tag{72}$$

is valid, where Λ is the largest Lyapunov exponent, Γ is the irregularity coefficient.

Note 1. An open problem is to prove Chetayev theorem on Lyapunov instability or refute the assertion, i.e. to build an example of a regular (non-regular) system with a positive Lyapunov exponent, which is stable in the sence of Lyapunov but unstable in the sense of Krasovsky.

Recall here stability condition (38) of Theorem 4, which by Theorem 1 can be represented as

$$(\nu - 1)\Lambda + \Gamma < 0. \tag{73}$$

Since Theorems 2–4 give, at the same time, the criteria of stability in the sense of Krasovsky, one can formulate the following

Theorem 9 ([49]). *If*

$$\Lambda < \frac{-\Gamma}{(\nu - 1)},$$

then the solution $x(t) \equiv 0$ is stable in the sense of Krasovsky and if

$$\Lambda > \Gamma,$$

then the solution $x(t) \equiv 0$ is unstable in the sense of Krasovsky.

For regular systems (the case $\Gamma = 0$), Theorem 9 gives a complete solution of the problem of stability in the sense of Krasovsky in the noncritical case ($\Lambda \neq 0$).

Note that for system (26) the relation $\Gamma = \Lambda + 2a + 1$ holds. Therefore for system (26) condition (72) is untrue.

Consider now Lyapunov instability of the solution $x(t) \equiv 0$ of multidimensional continuous system (2) and of discrete system (2′).

Theorem 10 ([42, 49, 52]). *Let for certain values $C > 0$, $\beta > 0$, $\alpha_1, \ldots, \alpha_{n-1}$ ($\alpha_j < \beta$ for $j = 1, \ldots, n-1$) the following conditions hold:*

1.

$$\begin{gathered} |z_j(t)| \leq C \exp(\alpha_j (t - \tau)) |z_j(\tau)|, \\ \forall t \geq \tau \geq 0, \quad j = 1, \ldots, n-1, \end{gathered} \tag{74}$$

2.

$$\frac{1}{(t - \tau)} \ln |\det Z(t)| > \beta + \sum_{j=1}^{n-1} \alpha_j, \quad \forall t \geq \tau \geq 0, \tag{75}$$

and, if $n > 2$,

3.

$$\prod_{j=1}^{n} |z_j(t)| \le C|\det Z(t)|, \quad \forall t \ge 0. \tag{76}$$

Then the zero solution of the system considered is Lyapunov unstable.

Corollary 4. *For the first-order system* (2) *or* (2′) *with bounded coefficients the positiveness of lower characteristic exponent of the first approximation system results in exponential instability of zero solution of the original system.*

The problem arises naturally as to the weakening of instability conditions, which are due to Theorems 8 and 10. However the Perron effects impose restrictions on such weakening.

Consider continuous and discrete systems (1) and (1′), respectively.

Suppose, for a certain vector-function $\xi(t)$ the following relations

$$|\xi(t)| = 1, \quad \inf_{y \in \Omega} \left| X(t, y)\xi(t) \right| \ge \alpha(t), \quad \forall t \ge t_0 \tag{77}$$

hold.

Theorem 11 ([42, 50]). *Let for the function $\alpha(t)$ the following condition*

$$\lim_{t \to +\infty} \sup \alpha(t) = +\infty \tag{78}$$

be satisfied.

Then the flow (cascade) of solutions $x(t, y)$, $y \in \Omega$ is Lyapunov unstable.

Proof. Holding a certain pair $x_0 \in \Omega$ and $t \ge t_0$ fixed, choose the vector y_0 in any δ-neighborhood of the point x_0 in such a way that

$$x_0 - y_0 = \delta\xi(t). \tag{79}$$

Let δ be so small that the ball of radius δ centered at x_0 is entirely placed in Ω.

For any fixed values t, j and for the vectors x_0, y_0 there exists a vector $w_j \in \mathbb{R}^n$ such that

$$|x_0 - w_j| \le |x_0 - y_0|,$$

$$x_j(t, x_0) - x_j(t, y_0) = X_j(t, w_j)(x_0 - y_0). \tag{80}$$

Here $x_j(t, x_0)$ is the jth component of the vector-function $x(t, x_0)$, $X_j(t, w)$ is the jth row of the matrix $X(t, w)$.

By (80) one has

$$|x(t, x_0) - x(t, y_0)| = \sqrt{\sum_j |X_j(t, w_j)(x_0 - y_0)|^2} \geq$$

$$\geq \delta \max\{|X_1(t, w_1)\xi(t)|, \dots, |X_n(t, w_n)\xi(t)|\} \geq$$

$$\geq \delta \max_j \inf_\Omega |X_j(t, x_0)\xi(t)| = \delta \inf_\Omega \max_j |X_j(t, x_0)\xi(t)| \geq$$

$$\geq \frac{\delta}{\sqrt{n}} \inf_\Omega |X(t, x_0)\xi(t)| \geq \frac{\alpha(t)\delta}{\sqrt{n}}.$$

This estimate and conditions (78) imply that for any positive numbers ε and δ there exist a number $t \geq t_0$ and a vector y_0 such that

$$|x_0 - y_0| = \delta, \quad |x(t, x_0) - x(t, y_0| > \varepsilon.$$

The latter means that the solution $x(t, x_0)$ is Lyapunov unstable. ∎

Consider the hypotheses of Theorem 11.

The hypotheses of Theorem 11 is, in essence, the requirement that, at least, one Lyapunov exponent of the linearizations of the flow of solutions with the initial data from Ω is positive under the condition that the "unstable directions $\xi(t)$" (or unstable manifolds) of these solutions depend continuously on the initial data x_0. Actually, if this property holds, then, regarding (if necessary) the domain Ω as the union of the domains Ω_i, of arbitrary small diameter, on which conditions (77) and (78) are valid, one obtains Lyapunov instability of the whole flow of solutions with the initial data from Ω.

Apply Theorem 11 to systems (24) and (24′).

For the solutions $x(t, t_0, x_0)$ with the initial data $t_0 = 0$,

$$x_1(0, x_{10}, x_{20}, x_{30}) = x_{10},$$

$$x_2(0, x_{10}, x_{20}, x_{30}) = x_{20},$$

$$x_3(0, x_{10}, x_{20}, x_{30}) = x_{30}$$

in the continuous case one has the following relations

$$x_1(t, x_{10}, x_{20}, x_{30}) = \exp(-at)x_{10},$$

$$\frac{\partial F(x,t)}{\partial x}|_{x=x(t,0,x_0)} = \begin{pmatrix} -a & 0 & 0 \\ 0 & -2a & 0 \\ -2\exp(-at)x_{10} & 1 & r(t) \end{pmatrix}, \tag{81}$$

where

$$r(t) = \sin(\ln(t+1)) + \cos(\ln(t+1)) - 2a.$$

For discrete system one obtains

$$\frac{\partial F(x,t)}{\partial x}|_{x=x(t,0,x_0)} = \begin{pmatrix} \exp(-a) & 0 & 0 \\ 0 & \exp(-2a) & 0 \\ -2\exp(-at)x_{10} & 1 & r(t) \end{pmatrix}, \tag{81'}$$

where

$$r(t) = \frac{\exp\big((t+2)\sin\ln(t+2) - 2a(t+1)\big)}{\exp\big((t+1)\sin\ln(t+1) - 2at\big)}.$$

Solutions (39) and (39′) with matrices (81) and (81′), respectively, have the form

$$\begin{aligned} z_1(t) &= \exp(-at)z_1(0), \\ z_2(t) &= \exp(-2at)z_2(0), \\ z_3(t) &= p(t)\big(z_3(0) + (z_2(0) - 2x_{10}z_1(0))\,q(t)\big). \end{aligned} \tag{82}$$

Here in the continuous case one has

$$p(t) = \exp\big((t+1)\sin(\ln(t+1)) - 2at\big),$$

$$q(t) = \int_0^t \exp\big(-(\tau+1)\sin(\ln(\tau+1))\big)\,d\tau.$$

and in the discrete case

$$p(t) = \exp\big((t+1)\sin(\ln(t+1)) - 2at\big),$$

$$q(t) = \sum_{k=0}^{t-1} \exp(-(k+2)\sin\ln(k+2) + 2a).$$

Relations (82) give

$$X(t,0,x_0) = \begin{pmatrix} \exp(-at) & 0 & 0 \\ 0 & \exp(-2at) & 0 \\ -2x_{10}p(t)q(t) & p(t)q(t) & p(t) \end{pmatrix}.$$

If it is assumed that

$$\xi(t) = \begin{pmatrix} 0 \\ 1 \\ 0 \end{pmatrix},$$

then for $\Omega = \mathbb{R}^n$ and

$$\alpha(t) = \sqrt{\exp(-4at) + (p(t)q(t))^2}$$

relations (77) and (78) are satisfied (see estimate (19)).

Thus, by Theorem 11 any solution of system (24) is Lyapunov unstable.

Restrict ourselves to the consideration of the manifold

$$M = \{x_3 \in \mathbb{R}^1, \quad x_2 = x_1^2\}.$$

In this case the initial data of the unperturbed solution x_0 and the perturbed solution y_0 belong to the manifold M:

$$x_0 \in M, \quad y_0 \in M. \tag{83}$$

The analysis of the proof of Theorem 11 (see (79)) implies that the vector-function $\xi(t)$ satisfies the following additional condition: if (79) and (83) hold, then the inequality $\xi_2(t) \neq 0$ yields the relation $\xi_1(t) \neq 0$.

In this case (77) and (78) are not valid since for either $2x_{10}\xi_1(t) = \xi_2(t) \neq 0$ or $\xi_2(t) = 0$ the value

$$|X(t, x_0)|$$

is bounded on $[0, +\infty)$.

Thus, since in conditions (77) and (78) the uniformity with respect to x_0 is violated, for system (24) on the set M the Perron effects are possible under certain additional restrictions on the vector-function $\xi(t)$. ■

7 Conclusion

We summarize the investigations of stability by the first approximation.

Theorems 5 and 11 give a complete solution for the problem on the flows and cascade of solutions in the noncritical case when for small variations of the initial data of the original system, the first approximation system preserves its stability (or instability in the certain "direction" $\xi(t)$). Thus, the classical problem of stability by the first approximation of nonstationary motions is completely proved in the general case [58].

The Perron effects of largest Lyapunov exponent sign reversal are possible only on the boundaries of the flows that are either stable or unstable by the first approximation. Thus, the difficulties, arising in studying the individual solutions, are connected to the fact that these solutions can be situated on the boundaries of the flows that are stable (or unstable) by the first approximation. In this case a special situation occurs which requires the development of more complicated tools for investigation. Such methods of investigation of the individual solutions are given in the present study.

It is shown that Perron effects may occur on the boundaries of a flow of solutions that is stable by the first approximation. Inside a flow, stability is completely determined by the negativeness of the characteristic exponents of linearized systems.

References

1. Andrianova L (1998) Introduction to linear systems of differential equations. American Mathematical Society, Providence, Rhode Island
2. Anishchenko V, Astakhov V, Neiman A, Vadivasova T, Schimansky-Geier L (2002) Nonlinear dynamics of chaotic and stochastic systems. Springer, New York
3. Aston P, Dellnitz M (2003) Computation of the dominant Lyapunov exponent via spatial integration using matrix norms. Proc R Soc A Math Phys Eng Sci 459(2040):2933–2955
4. Balibrea F, Caballero M (2013) Stability of orbits via Lyapunov exponents in autonomous and non-autonomous systems. Int J Bifurcat Chaos. doi:10.1142/S0218127413501277
5. Barabanov E (2005) Singular exponents and properness criteria for linear differential systems. Differ Equat 41:151–162
6. Barreira L, Pesin Y (2001) Lectures on Lyapunov exponents and smooth ergodic theory. In: Proc. of. Symposia in Pure Math., vol 69. AMS, Providence, RI, pp 3–89
7. Barreira L, Vails C (2008) Stability of nonautonomous differential equations. Lect Notes Math 1926:1–290
8. Benettin G, Galgani L, Giorgilli A, Strelcyn JM (1980) Lyapunov characteristic exponents for smooth dynamical systems and for hamiltonian systems. A method for computing all of them. Part 1: Theory. Meccanica 15(1):9–20
9. Boichenko VA, Leonov GA, Reitmann V (2005) Dimension theory for ordinary differential equations. Teubner, Stuttgart
10. Brunton S, Rowley C (2010) Fast computation of finite-time Lyapunov exponent fields for unsteady flows. Chaos 20(1):017503
11. Bylov BE, Vinograd RE, Grobman DM, Nemytskii VV (1966) Theory of characteristic exponents and its applications to problems of stability [in Russian]. Nauka, Moscow
12. Caponetto R, Fazzino S (2013) A semi-analytical method for the computation of the Lyapunov exponents of fractional-order systems. Comm Nonlinear Sci Numer Simulat 18(1):22–27
13. Carbonell F, Jimenez J, Biscay R (2002) A numerical method for the computation of the Lyapunov exponents of nonlinear ordinary differential equations. Appl Math Comput 131(1):21–37
14. Chetaev N (1955) Stability of motion [in Russian]. Gostekhizdat, Moscow
15. Chetaev NG (1948) Some problems of stability and instability for nonregular systems [in Russian]. Prikl Mat Mekh 12(6):639–642
16. Christiansen F, Rugh H (1997) Computing Lyapunov spectra with continuous gram-schmidt orthonormalization. Nonlinearity 10(5):1063–1072
17. Conti C, Rossinelli D, Koumoutsakos P (2012) Gpu and apu computations of finite time Lyapunov exponent fields. J Comput Phys 231(5):2229–2244

18. Czornik A, Nawrat A, Niezabitowski M (2013) Lyapunov exponents for discrete time-varying systems. Stud Comput Intell 440:29–44
19. Demidovich BP (1967) Lectures on mathematical theory of stability [in Russian]. Nauka, Moscow
20. Demidovich V (1969) Stability criterion for difference equations [in Russian]. Diff Uravneniya 5(7):1247–1255
21. Demir B, Koak S (2013) A note on positive Lyapunov exponent and sensitive dependence on initial conditions. Chaos Soliton Fract 12:2119–2121
22. Dieci L, Jolly M, Van Vleck E (2011) Numerical techniques for approximating Lyapunov exponents and their implementation. J Comput Nonlinear Dynam 6(1):011003
23. Filippov AF (1988) Differential equations with discontinuous right-hand sides. Kluwer, Dordrecht
24. Galvanetto U (2000) Numerical computation of Lyapunov exponents in discontinuous maps implicitly defined. Comput Phys Comm 131(1):1–9
25. Gayschun I (2001) Systems with discrete time [in Russian]. Inst. matemat. RAN Belarusi, Minsk
26. Greene J, Kim JS (1987) The calculation of Lyapunov spectra. Physica D: Nonlinear Phenomena 24(1–3):213–225
27. He DH, Xu JX, Chen YH (2000) Study on Lyapunov characteristic exponents of a nonlinear differential equation system. Wuli Xuebao Acta Phys Sin 49(5):836–837
28. Hegger R, Kantz H, Schreiber T (1999) Practical implementation of nonlinear time series methods: The TISEAN package. Chaos 9:413–435
29. Izobov N, Korovin S (2011) Multidimensional analog of the two-dimensional Perron effect of sign change of characteristic exponents for infinitely differentiable differential systems. Differ Equat 48(11):1444–1460
30. Janaki T, Rangarajan G, Habib S, Ryne R (1999) Computation of the Lyapunov spectrum for continuous-time dynamical systems and discrete maps. Phys Rev E 60(6 A):6614–6626
31. Jost J (2005) Dynamical systems: examples of complex behaviour. Universitext (1979), Springer, Heidelberg
32. Kloeden P, Rasmussen M (2011) Nonautonomous dynamical systems. Mathematical surveys and monographs, vol 176. American Mathematical Society, Providence, RI
33. Kocak H, Palmer K (2010) Lyapunov exponents and sensitive dependence. J Dynam Differ Equat 22(3):381–398
34. Korovin S, Izobov N (2010a) On the Perron sign change effect for Lyapunov characteristic exponents of solutions of differential systems. Differ Equat 46(10):1395–1408
35. Korovin S, Izobov N (2010b) The Perron effect of change of values of characteristic exponents for solutions of differential systems. Dokl Math 82(2):798–800
36. Korovin S, Izobov N (2010c) Realization of the Perron effect whereby the characteristic exponents of solutions of differential systems change their values. Differ Equat 46(11):1537–1551
37. Korovin S, Izobov N (2011) Generalization of the Perron effect whereby the characteristic exponents of all solutions of two differential systems change their sign from negative to positive. Differ Equat 47(7):942–954
38. Korovin S, Izobov N (2012) Multidimensional Perron sign change effect for all characteristic exponents of differential systems. Dokl Math 85(3):315–320
39. Kunze M (2000) Non-smooth dynamical systems. Lecture Notes in Mathematics, vol 1744. Springer, Berlin
40. Kuznetsov NV (2008) Stability and oscillations of dynamical systems: Theory and applications. Jyvaskyla University Printing House, Jyvaskyla
41. Kuznetsov NV, Leonov GA (2001) Counterexample of Perron in the discrete case. Izv RAEN, Diff Uravn 5:71
42. Kuznetsov NV, Leonov GA (2005a) Criteria of stability by the first approximation for discrete nonlinear systems. Vestnik Sankt-Peterburgskogo Universiteta Ser 1 Matematika Mekhanika Astronomiya (3):30–42

43. Kuznetsov NV, Leonov GA (2005b) Criterion of stability to first approximation of nonlinear discrete systems. Vestnik Sankt-Peterburgskogo Universiteta Ser 1 Matematika Mekhanika Astronomiya (2):55–63
44. Kuznetsov NV, Leonov GA (2005c) On stability by the first approximation for discrete systems. 2005 International Conference on Physics and Control, PhysCon 2005 Proceedings Volume 2005:596–599, doi:10.1109/PHYCON.2005.1514053
45. Leonov G (2002a) Instability in the first approximation for time–dependent linearization. J Appl Math Mech 66(2):323–325
46. Leonov G (2002b) Lyapunov dimension formulas for Henon and Lorenz attractors. St Petersburg Math J 13(3):453–464
47. Leonov G (2003a) A modification of Perron's counterexample. Differ Equat 39(11):1651–1652
48. Leonov G (2003b) Problems of the first approximation in the theory of stability of motion. Russ Mech Surv [in Russian] 2(3):3–35
49. Leonov G (2004) Criteria of instability in the first approximation for time–dependent linearizations. J Appl Math Mech 68(6):858–871
50. Leonov GA (1998) On stability in the first approximation. J Appl Math Mech 62(4):511–517
51. Leonov GA (2008) Strange attractors and classical stability theory. St.Petersburg University Press, St.Petersburg
52. Leonov GA, Kuznetsov NV (2007) Time-varying linearization and the Perron effects. Int J Bifurcat Chaos 17(4):1079–1107. doi:10.1142/S0218127407017732
53. Leonov GA, Kuznetsov NV (2013) Hidden attractors in dynamical systems. From hidden oscillations in Hilbert-Kolmogorov, Aizerman, and Kalman problems to hidden chaotic attractors in Chua circuits. Int J Bifurcat Chaos 23(1). doi:10.1142/S0218127413300024, art. no. 1330002
54. Lian Z, Lu K (2010) Lyapunov exponents and invariant manifolds for random dynamical systems in a Banach space. Memoirs of the American Mathematical Society, American Mathematical Society, Providence, RI
55. Liu JH, Feng J, Yang DS (2009) Improved method of computing Lyapunov-spectrum and its application. Xitong Fangzhen Xuebao/J Syst Simul 21(16):5124–5126
56. Luca D, Russell, RD Van Vleck E (1997) On the computation of Lyapunov exponents for continuous dynamical systems. SIAM J Numer Anal 34:402–423
57. Lyapunov AM (1892) The general problem of the stability of motion. Kharkov Mathematical Society, Kharkov
58. Malkin I (1966) Stability of Motion [in Russian]. Nauka, Moscow
59. Maruskin J (2012) Introduction to dynamical systems and geometric mechanics. Solar Crest Publishing, LLC
60. Massera J (1956) Contributions to stability theory. Ann Math 64:182–206
61. Neimark YI, Landa PS (1992) Stochastic and Chaotic Oscillations. Kluwer Academic Publishers, Dordrecht
62. Perron O (1930) Die Stabilitätsfrage differentialgleichungen. Math Z 32(5):702–728
63. Persidskii K (1947) Characteristic numbers of differential equations. Izv AN Kasakh SSR Ser Matem i Mekh [in Russian] 1
64. Ramasubramanian K, Sriram M (2000) A comparative study of computation of Lyapunov spectra with different algorithms. Phys Nonlinear Phenom 139(1–2):72–86
65. Rangarajan G, Habib S, Ryne R (1998) Lyapunov exponents without rescaling and reorthogonalization. Phys Rev Lett 80(17):3747–3750
66. Robinson R (2013) An Introduction to dynamical systems: continuous and discrete. Pure and Applied Undergraduate Texts. American Mathematical Society, Providence, RI
67. Russel D, Hanson J, Ott E (1980) Dimension of strange attractors. Phys Rev Lett 45(14):1175–1178
68. Shimada I, Nagashima T (1979) A numerical approach to ergodic problem of dissipative dynamical systems. Progr Theor Phys 61(6):1605–1616
69. Shorten R, Wirth F, Mason O, Wulff K, King C (2007) Stability criteria for switched and hybrid systems. SIAM Rev 49(4):545–592

70. Skokos C (2010) The Lyapunov characteristic exponents and their computation. Lect Notes Phys 790:63–135
71. Stachowiak T, Szydlowski M (2011) A differential algorithm for the Lyapunov spectrum. Phys Nonlinear Phenom 240(16):1221–1227
72. Tancredi G, Sanchez A (2001) A comparison between methods to compute Lyapunov exponents. Astron J 121:1171–1179
73. Temam R (1988) Infinite-dimensional dynamical systems in mechanics and physics. Springer, New York
74. Vinograd R (2012) A new proof of Perron's theorem and certain properties of regulating systems. Usp Mat Nauk [in Russian] 9(2):129–136
75. Von Bremen H, Udwadia F, Proskurowski W (1997) An efficient QR based method for the computation of Lyapunov exponents. Phys Nonlinear Phenom 101:1–16
76. Wolf A, Swift JB, Swinney HL, Vastano JA (1985) Determining Lyapunov exponents from a time series. Phys Nonlinear Phenom 16(D):285–317
77. Wu X, Huang TY (2003) Computation of Lyapunov exponents in general relativity. Phys Lett A 313(1–2):77–81
78. Xie W (2006) Dynamic stability of structures. Cambridge University Press, Cambridge
79. Zhang X, Li Z, Zhang L (2005) Method based on singular value decomposition for computation of Lyapunov exponent. Beijing Keji Daxue Xuebao/J Univ Sci Tech Beijing 27(3):371–374

Fractional Maps as Maps with Power-Law Memory

Mark Edelman

Abstract The study of systems with memory requires methods which are different from the methods used in regular dynamics. Systems with power-law memory in many cases can be described by fractional differential equations, which are integro-differential equations. To study the general properties of nonlinear fractional dynamical systems we use fractional maps, which are discrete nonlinear systems with power-law memory derived from fractional differential equations. To study fractional maps we use the notion of α-families of maps depending on a single parameter $\alpha > 0$ which is the order of the fractional derivative in a nonlinear fractional differential equation describing a system experiencing periodic kicks. α-families of maps represent a very general form of multi-dimensional nonlinear maps with power-law memory, in which the weight of the previous state at time t_i in defining the present state at time t is proportional to $(t - t_i)^{\alpha-1}$. They may be applicable to studying some systems with memory such as viscoelastic materials, electromagnetic fields in dielectric media, Hamiltonian systems, adaptation in biological systems, human memory, etc. Using the fractional logistic and standard α-families of maps as examples we demonstrate that the phase space of nonlinear fractional dynamical systems may contain periodic sinks, attracting slow diverging trajectories, attracting accelerator mode trajectories, chaotic attractors, and cascade of bifurcations type trajectories whose properties are different from properties of attractors in regular dynamical systems.

M. Edelman (✉)
Stern College at Yeshiva University, 245 Lexington Street, New York, NY 10016, USA

Courant Institute of Mathematical Sciences at NYU, New York, NY 10012, USA
e-mail: edelman@cims.nyu.edu

V. Afraimovich et al. (eds.), *Nonlinear Dynamics and Complexity*,
Nonlinear Systems and Complexity 8, DOI 10.1007/978-3-319-02353-3_3,

1 Introduction

Many natural and social systems are systems with memory. Their mathematical description requires solving integro-differential equations and is quite complicated. Maps with memory are used to model real systems with memory in order to derive their basic properties.

1.1 Systems with Memory

Writing this text I am recalling the content of my latest papers. It is easy to recall the content of my last paper but it becomes more and more difficult as I try to recall papers that are more and more distant in time. Memory is a significant property of human beings and is the subject of extensive biophysical and psychological research. As it has been demonstrated in experiments, forgetting—the accuracy on a memory task decays as a power law, $\sim t^{-\beta}$, with $0 < \beta < 1$ [37, 66, 87–89]. It is interesting that fractional maps corresponding to fractional differential equations of the order $0 < \alpha < 1$ are maps with the power-law decaying memory in which the power is $-\beta = \alpha - 1$ and $0 < \beta < 1$ [20]. Human learning is closely related to memory. It can also be described by a power law: the reduction in reaction times that comes with practice is a power function of the number of training trials [1]. There are multiple publications where power-law adaptation has been applied in describing the dynamics of biological systems at levels ranging from single ion channels up to human psychophysics [23, 43, 82, 83, 89, 96].

Power-law memory applies not only to the human being as a whole but also to the hierarchy of its building blocks, from individual neurons and proteins to the tissue of individual organs. It has been shown recently [47, 48] that processing of external stimuli by individual neurons can be described by fractional differentiation. The orders of fractional derivatives α derived for different types of neurons fall within the interval $[0, 1]$. For neocortical pyramidal neurons it is quite small: $\alpha \approx 0.15$. Fluctuations within single protein molecules demonstrate a power-law memory kernel with the exponent -0.51 ± 0.07 [57].

Viscoelastic properties of human tissues were demonstrated in many examples: the brain and the central nervous system in general [10, 44, 49], the breast [12], the liver [41, 81], the spleen [61], the prostate [36, 95], the arteries [13, 14], the muscles [31] (see also references for some other human and animal organ tissues [16, 50, 55, 60, 69]). Viscoelastic materials obey the following stress–strain relationship:

$$\sigma(t) = E(\gamma)\frac{d^{\alpha}\gamma(t)}{dt^{\alpha}}, \tag{1}$$

where σ is the stress, γ is the strain, α is the order of the fractional derivative, and t is time. In most of the cases for human tissues $0 < \alpha < 1$ and is close to zero.

In some cases, e.g. for modeling of the accurate placement of the needle tip into the target tissue during needle insertion treatments for liver tumors, nonlinearity of $E(\gamma)$ should be taken into account [41]. In the last example a simple quadratic nonlinearity and $\alpha = 0.1$ were used.

A Fourier transform of a fractional derivative is [40, 63, 67]

$$F\{D^{\alpha} g(t); \omega\} = (-i\omega)^{\alpha} \hat{g}(\omega), \tag{2}$$

where $\hat{g}(\omega) = F\{g(t); \omega\}$. As a result, whenever the term $(\omega)^{\alpha} \hat{g}(\omega)$ appears in the frequency domain, there is a good chance that function $g(t)$ is a solution of a fractional differential equation with a fractional derivative of the order α and the corresponding system is a system with power-law memory. Well-known examples of such systems are dielectrics. Electromagnetic fields in dielectric media are described by equations with time fractional derivatives due to the "universal" response—the power-law frequency dependence of the dielectric susceptibility in a wide range of frequencies [73, 74, 77, 78]. Similarly, elastic wave attenuation in biological tissue over a wide range of frequencies follows the power law $\alpha(\omega) \propto \omega^{\eta}$ with $\eta \in [0, 2]$ [17, 34, 59, 71] which implies a fractional wave equation. The establishment of accurate fractional wave-propagation models is important for many medical applications [59].

Above we concentrated on biological systems with memory in order to emphasize the importance of the study of nonlinear fractional dynamical systems described by fractional differential equations of the order $0 < \alpha < 2$ and especially α close to zero which is a major subject of the following sections (Sects. 3.2 and 3.3). Now we'll list some (not all) other examples of systems with power-law memory. As has been mentioned above, time fractional derivatives and correspondingly systems with power-law memory in many cases are used to describe viscoelasticity and rheology (for the original papers and reviews, see [4, 5, 8, 9, 51–54] and for nonlinear effects, see [64, 65, 85, 86]). Electromagnetic fields in dielectric media were also mentioned above. Hamiltonian systems and billiards are also systems with power-law memory, in which the fractal structure of the phase space and stickiness of trajectories in time imply description of transport by the fractional (fractional time and space derivatives) Fokker–Plank–Kolmogorov equation [90–93]. In some cases [38–40] fractional differential equations are equivalent to the Volterra integral equations of the second kind. Systems considered in population biology and epidemiology are systems with memory and Volterra integral equations are frequently used to describe such systems [6, 35]. Long-term memory provides more robust control in liner and nonlinear control theory (see [7, 62]).

1.2 Maps with Memory

As in the study of regular dynamics, in the study of systems with memory use of discrete maps significantly simplifies investigation of the general properties of the corresponding systems. In some cases of kicked systems maps are equivalent to the original differential equations. Historically, maps with memory were first considered as analogues of the integro-differential equations of non-equilibrium statistical physics [26,27,32], with regard to thermodynamic theory of systems with memory [30], and to model non-Markovian processes in general [28, 29]. The most general form of a map with memory is

$$\mathbf{x}_{n+1} = \mathbf{f}_{n+1}(\mathbf{x}_n, \mathbf{x}_{n-1}, \ldots, \mathbf{x}_0, P), \tag{3}$$

where $\mathbf{x}_k$ are N-dimensional vectors, $k, N \in \mathbb{Z}, k \geq 0$, and P is a set of parameters. It is almost impossible to derive the general properties of systems with memory from Eq. (3) and simplified forms of maps with memory are used. The most commonly used form is the one-dimensional map with long-term memory

$$x_{n+1} = \sum_{k=0}^{n} V_\alpha(n,k) G_K(x_k), \tag{4}$$

where $V_\alpha(n,k)$ and α characterize memory effects and K is a parameter. In many cases weights are taken as convolutions with $V_\alpha(n,k) = V_\alpha(n-k)$. The particular form of Eq. (4) with constant weights

$$x_{n+1} = c \sum_{k=0}^{n} G_K(x_k) \tag{5}$$

is called a full-memory map. It is easy to note that (5) is equivalent to

$$x_{n+1} = x_n + c G_K(x_n), \tag{6}$$

which means that maps with full memory are maps with one-step memory in which all memory is accumulated in the present state of a system and the next values of map variables are fully defined by their present values. We won't consider maps with short memory in which the number of terms in the sum in Eq. (4) is bounded (from $k = n - M + 1$ to $k = n$).

Initial investigations of long-term memory maps were done mostly on different modifications of the logistic map and exponential memory. The general applicability of their results to systems with memory in general is limited. Recently Stanislavsky [70] considered the maps Eq. (4) with $G_K(x) = Kx(1-x)$ (the logistic map) and the weights $V_\alpha(n,k)$ as a combination of power-law functions taken from one of the algorithms of numerical fractional integration. He came to the conclusion that increase in long-term memory effects leads to a less chaotic behavior.

First maps with power-law memory equivalent to fractional differential equations were derived in [75, 76, 78–80] by integrating fractional differential equations describing systems under periodic kicks. The method used is similar to the way in which the universal map is derived in regular dynamics.

1.3 Universal Map

In the following section (Sect. 2) we will modify the way presented in Sect. 1.3 to derive the universal map in regular dynamics (see [11], and Chap. 5 from [90]) in order to derive the universal fractional map.

The universal map can be derived from the differential equation

$$\ddot{x} + KG(x) \sum_{n=-\infty}^{\infty} \delta\Big(\frac{t}{T} - (n+\varepsilon)\Big) = 0, \tag{7}$$

where $0 < \varepsilon < 1$ and K is a parameter, with the initial conditions:

$$x(0) = x_0, \;\; p(0) = \dot{x}(0) = p_0. \tag{8}$$

This equation is equivalent to the Volterra integral equation of second kind

$$x(t) = x_0 + p_0 t - K \int_0^t d\tau G(x(\tau)) \sum_{n=-\infty}^{\infty} \delta\Big(\frac{\tau}{T} - (n+\varepsilon)\Big)(t-\tau). \tag{9}$$

Equation (9) for $(n+\varepsilon)T < t < (n+1+\varepsilon)T$ has a solution

$$x(t) = x_0 + p_0 t - KT \sum_{k=0}^{n} G(x(Tk+T\varepsilon))(t - Tk - T\varepsilon),$$

$$p(t) = \dot{x}(t) = p_0 - KT \sum_{k=0}^{n} G(x(Tk+T\varepsilon)). \tag{10}$$

After the introduction of the map variables

$$x_n = x(Tn), \;\; p_n = p(Tn) \tag{11}$$

Equation (10) considered for time instances t=(n+1)T gives

$$x_{n+1} = x_0 + p_0(n+1)T - KT^2 \sum_{k=0}^{n} G(x(Tk+T\varepsilon))(n-k+1-\varepsilon),$$

$$p_{n+1} = p_0 - KT \sum_{k=0}^{n} G(x(Tk+T\varepsilon)). \tag{12}$$

As it follows from Eq. (10), $\dot{x}(t) = p(t)$ is a bounded function with the discontinuities at the time instances of the kicks (at $t = Tk + T\varepsilon$) and $x(t)$ is a continuous function. This allows us to calculate $G(x)$ at the time instances of the kicks. In the limit $\varepsilon \to 0$ Eq. (12) gives

$$x_{n+1} = x_0 + p_0(n+1)T - KT^2 \sum_{k=0}^{n} G(x_k)(n-k+1),$$

$$p_{n+1} = p_0 - KT \sum_{k=0}^{n} G(x_k). \tag{13}$$

Equation (13) is a form of the universal map which allows further simplifications. It can be written in a symmetric form as a map with full memory (see Sect. 1.2):

$$x_{n+1} = x_0 + T \sum_{k=1}^{n+1} p_k,$$

$$p_{n+1} = p_0 - KT \sum_{k=0}^{n} G(x_k). \tag{14}$$

As we saw in Sect. 1.2, maps with full memory are equivalent to maps with one-step memory. Map Eq. (14) can be written as the iterative area preserving ($\partial(p_{n+1}, x_{n+1})/\partial(p_n, x_n) = 1$) process with one-step memory which is called the universal map:

$$p_{n+1} = p_n - KTG(x_n), \tag{15}$$

$$x_{n+1} = x_n + p_{n+1}T. \tag{16}$$

This map represents the relationship between the values of the physical variables in Eq. (7) on the left sides of the consecutive kicks. The standard map may be obtained from the universal map by assuming $G(x) = \sin(x)$:

$$p_{n+1} = p_n - K \sin x, \quad (\text{mod } 2\pi),$$

$$x_{n+1} = x_n + p_{n+1}, \quad (\text{mod } 2\pi). \tag{17}$$

Here we assumed $T = 1$ and consider this map on a torus (mod 2π).

Derivation of the fractional universal map in the next section (Sect. 2) follows [20] and the analysis of this map for $\alpha \in (0,1)$ and $\alpha \in (1,2)$ in Sects. 3.2 and 3.3 follows [19–22].

2 Fractional Universal Map

The one-dimensional logistic map

$$x_{n+1} = Kx_n(1 - x_n) \tag{18}$$

be presented in the 2D form

$$p_{n+1} = -G_{lK}(x_n),$$

$$x_{n+1} = x_n + p_{n+1}, \tag{19}$$

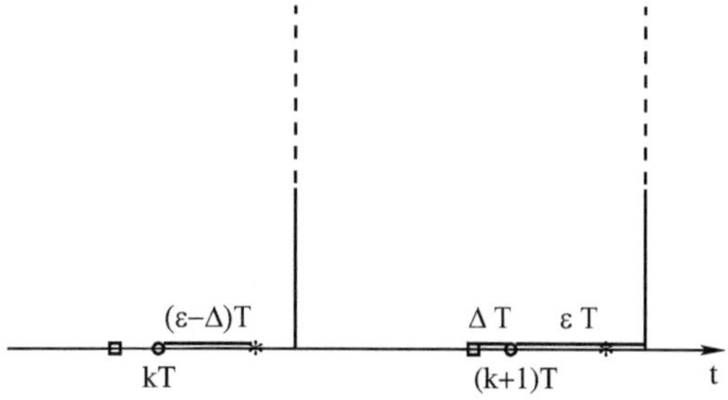

Fig. 1 The universal map is a relationship between values of $x(t)$ considered at the times kT (*small circles*). The kicks occur at the time instances $(k+\varepsilon)T$ (*vertical lines*). With the time delay ΔT (distance between the *squares* and the *circles*) the function $G_K(x(t))$ is calculated at the time instances $t = (k+\varepsilon-\Delta)T$ (*stars*)

where

$$G_{lK}(x) = x - Kx(1-x). \tag{20}$$

It can't be written as a particular form of the universal map Eqs. (15) and (16). In order to derive the logistic map from the universal map we'll introduce the notion of the n-dimensional universal map depending on a single parameter.

2.1 Universal Integer-Dimensional Maps

Solution of the one-dimensional analog of Eq. (7) would require calculations of the function $G(x)$ at the time instances of the kicks $T(n+\varepsilon)$ at which $x(t)$ is discontinuous. To enable us to introduce the universal fractional map we'll include a time delay ΔT into the argument of the function $G(x(t))$ (see Fig. 1). In order to extend the class of maps which are particular forms of the universal map we'll also consider K not as a factor but as a parameter. Let's consider the following generating equation:

$$\dot{x} + G_K(x(t-\Delta T)) \sum_{n=-\infty}^{\infty} \delta\Big(\frac{t}{T} - (n+\varepsilon)\Big) = 0, \tag{21}$$

where $0 < \varepsilon < 1$ and $0 < \Delta < 1$ with the initial condition:

$$x(0) = x_0. \tag{22}$$

1D analog of Eq. (10) (for $(n+\varepsilon)T < t < (n+1+\varepsilon)T$) can be written as

$$x(t) = x_0 - T\sum_{k=0}^{n} G_K(x[T(k+\varepsilon-\Delta)]). \tag{23}$$

From the fact that $\dot{x} = 0$ for $t \in (T(k+\varepsilon-1), T(k+\varepsilon))$ it follows that $x[T(k+\varepsilon-\Delta)] = x(Tk)$ and the corresponding 1D map can be written as a map with full memory

$$x_{n+1} = x_0 - T\sum_{k=0}^{n} G_K(x_k). \tag{24}$$

From Sect. 1.2 it follows that this map can be written as the 1D form of the universal map with one-step memory

$$x_{n+1} = x_n - TG_K(x_n). \tag{25}$$

It would be impossible to derive the logistic map from Eq. (25) if K were a factor, but from the present form the logistic map can be obtained by assuming

$$G_K(x) = G_{lK}(x) = \frac{1}{T}[x - Kx(1-x)]. \tag{26}$$

In [6, 72] Eq. (21) with no time delay, no delta functions, and $G_K(x)$ defined by Eq. (26) is used as one of the most general models in population biology and epidemiology. Three terms in $G_K(x)$ represent a growth rate proportional to the current population, restrictions due to the limited resources, and the death rate. The logistic map appears and plays an important role not only in population biology but also in economics, condensed matter physics, and other areas of science [3, 72]. In population biology and epidemiology time delays can be related to the time of the development of an infection in a body until a person becomes infectious, or to the time of the development of an embryo. For the importance of time delay in many scientific applications of the logistic map, see, e.g., Chap. 3 from [72] and Chap. 3 from [3]. Changes which occur as periodically following discrete events can be modeled by the delta function.

The n-dimensional universal map can be derived from the following generating equation:

$$\frac{d^n x}{dt^n} + G_K(x(t-\Delta)) \sum_{k=-\infty}^{\infty} \delta\Big(t - (k+\varepsilon)\Big) = 0, \tag{27}$$

where $n \geq 0$, $n \in \mathbb{Z}$, and $\varepsilon > \Delta > 0$ in the limit $\varepsilon \to 0$. This means that in the general case time delay is not essential. Without losing the generality, in Eq. (27) we assumed $T = 1$. The case $T \neq 1$ is considered in [20] and can be reduced to this case by rescaling the time variable and the map generating function $G_K(x)$. In Sect. 3 T denotes periods of trajectories. The 2D universal map Eqs. (15) and (16) corresponds to $n = 2$ and the 1D universal map (25) corresponds to $n = 1$. In the consistent introduction of fractional derivatives integer derivatives appear as the limits when the order of a fractional derivative assumes an integer value. Correspondingly, the general form of the n-dimensional universal map appears if we assume an integer value of α in the general form of the fractional universal map. In the following sections we'll consider the general forms of the fractional universal map which will be derived from Eq. (27) with integer n replaced by $\alpha \in \mathbb{R}$ ($\alpha \geq 0$). The Riemann–Liouville universal map will be derived in Sect. 2.2 and the Caputo universal map will be derived in Sect. 2.3.

2.2 Riemann–Liouville Universal Map

The generating fractional differential equation for the Riemann–Liouville universal map can be written as

$$ {}_0D_t^{\alpha}x(t) + G_K(x(t-\Delta)) \sum_{n=-\infty}^{\infty} \delta\Big(t-(n+\varepsilon)\Big) = 0, \tag{28} $$

where $\varepsilon > \Delta > 0$, $\varepsilon \to 0$, $0 \leq N-1 < \alpha \leq N$, $\alpha \in \mathbb{R}$, $N \in \mathbb{Z}$, and the initial conditions

$$ ({}_0D_t^{\alpha-k}x)(0+) = c_k, \quad k = 1, \dots, N. \tag{29} $$

The left-sided Riemann–Liouville fractional derivative ${}_0D_t^{\alpha}x(t)$ is defined for $t > 0$ [40, 63, 67] as

$$ {}_0D_t^{\alpha}x(t) = D_t^n \, {}_0I_t^{n-\alpha}x(t) = \frac{1}{\Gamma(n-\alpha)} \frac{d^n}{dt^n} \int_0^t \frac{x(\tau)d\tau}{(t-\tau)^{\alpha-n+1}}, \tag{30} $$

where $n-1 \leq \alpha < n$, $D_t^n = d^n/dt^n$, and ${}_0I_t^{\alpha}$ is a fractional integral.

For a wide class of functions $G_K(x)$ Eq. (28) is equivalent to the Volterra integral equation of the second kind ($t > 0$) (see [38–40, 78])

$$ \begin{aligned} x(t) = & \sum_{k=1}^{N} \frac{c_k}{\Gamma(\alpha-k+1)} t^{\alpha-k} \\ & - \frac{1}{\Gamma(\alpha)} \int_0^t d\tau \frac{G_K(x(\tau-\Delta))}{(t-\tau)^{1-\alpha}} \sum_{k=-\infty}^{\infty} \delta\Big(\tau-(k+\varepsilon)\Big). \end{aligned} \tag{31} $$

Due to the presence of the delta function the integral on the right side of Eq. (31) can be easily calculated [20, 75, 76, 78] for $t > 0$:

$$ \begin{aligned} x(t) = & \sum_{k=1}^{N-1} \frac{c_k}{\Gamma(\alpha-k+1)} t^{\alpha-k} \\ & - \frac{1}{\Gamma(\alpha)} \sum_{k=0}^{[t-\varepsilon]} \frac{G_K(x(k+\varepsilon-\Delta))}{(t-(k+\varepsilon))^{1-\alpha}} \Theta(t-(k+\varepsilon)), \end{aligned} \tag{32} $$

where $\Theta(t)$ is the Heaviside step function. In Eq. (32) we took into account that boundedness of $x(t)$ at $t = 0$ requires $c_N = 0$ and $x(0) = 0$. After the introduction (see [80])

$$p(t) = {}_0D_t^{\alpha-N+1}x(t) \tag{33}$$

and

$$p^{(s)}(t) = D_t^s p(t), \tag{34}$$

where $s = 0, 1, \ldots, N-2$, Eq. (32) leads to

$$\begin{aligned} p^{(s)}(t) = & \sum_{k=1}^{N-s-1} \frac{c_k}{(N-s-1-k)!} t^{N-s-1-k} \\ & - \frac{1}{(N-s-2)!} \sum_{k=0}^{[t-\varepsilon]} G_K(x(k+\varepsilon-\Delta))(t-k)^{N-s-2}, \end{aligned} \tag{35}$$

where $s = 0, 1, \ldots, N-2$. Assuming $x_n = x(n)$, for $\varepsilon > \Delta > 0$ Eqs. (32) and (35) in the limit $\varepsilon \to 0$ give the equations of the Riemann–Liouville universal map

$$\begin{aligned} x_{n+1} = & \sum_{k=1}^{N-1} \frac{c_k}{\Gamma(\alpha-k+1)}(n+1)^{\alpha-k} \\ & - \frac{1}{\Gamma(\alpha)} \sum_{k=0}^{n} G_K(x_k)(n-k+1)^{\alpha-1}, \end{aligned} \tag{36}$$

$$\begin{aligned} p_{n+1}^s = & \sum_{k=1}^{N-s-1} \frac{c_k}{(N-s-1-k)!}(n+1)^{N-s-1-k} \\ & - \frac{1}{(N-s-2)!} \sum_{k=0}^{n} G_K(x_k)(n-k+1)^{N-s-2}. \end{aligned} \tag{37}$$

2.3 Caputo Universal Map

Similar to (28), the generating fractional differential equation for the Caputo universal map can be written as

$${}_0^C D_t^\alpha x(t) + G_K(x(t-\Delta)) \sum_{n=-\infty}^{\infty} \delta\Big(t-(n+\varepsilon)\Big) = 0, \tag{38}$$

where $\varepsilon > \Delta > 0$, $\varepsilon \to 0$, $0 \le N-1 < \alpha \le N$, $\alpha \in \mathbb{R}$, $N \in \mathbb{Z}$, and the initial conditions

$$(D_t^k x)(0+) = b_k, \;\; k = 0, \ldots, N-1. \tag{39}$$

The left-sided Caputo fractional derivative ${}_0^C D_t^\alpha x(t)$ is defined for $t > 0$ [40, 63, 67] as

$${}_0^C D_t^\alpha x(t) = {}_0 I_t^{n-\alpha} \, D_t^n x(t) = \frac{1}{\Gamma(n-\alpha)} \int_0^t \frac{D_\tau^n x(\tau) d\tau}{(t-\tau)^{\alpha-n+1}}, \tag{40}$$

where $n-1 < \alpha \le n$.

For a wide class of functions $G_K(x)$ Eq. (38) is equivalent to the Volterra integral equation of the second kind ($t > 0$) (see [38–40, 78])

$$x(t) = \sum_{k=0}^{N-1} \frac{b_k}{k!} t^k - \frac{1}{\Gamma(\alpha)} \int_0^t d\tau \frac{G_K(x(\tau-\Delta))}{(t-\tau)^{1-\alpha}} \sum_{k=-\infty}^{\infty} \delta\Big(\tau - (k+\varepsilon)\Big). \tag{41}$$

Integration of this equation gives for $t > 0$

$$x(t) = \sum_{k=0}^{N-1} \frac{b_k}{k!} t^k - \frac{1}{\Gamma(\alpha)} \sum_{k=0}^{[t-\varepsilon]} \frac{G_K(x(k+\varepsilon-\Delta))}{(t-(k+\varepsilon))^{1-\alpha}} \Theta(t-(k+\varepsilon)). \tag{42}$$

After the introduction $x^{(s)}(t) = D_t^s x(t)$ the Caputo universal map can be derived in the form (see [78])

$$x_{n+1}^{(s)} = \sum_{k=0}^{N-s-1} \frac{x_0^{(k+s)}}{k!} (n+1)^k - \frac{1}{\Gamma(\alpha-s)} \sum_{k=0}^{n} G_K(x_k)(n-k+1)^{\alpha-s-1}, \tag{43}$$

where $s = 0, 1, \ldots, N-1$.

3 α-Families of Maps

We'll call Eqs. (28) and (29) with various map generating functions $G_K(x)$ the Riemann–Liouville universal map generating equations and Eqs. (36) and (37) the Riemann–Liouville α-families of maps corresponding to the functions $G_K(x)$. We'll call Eqs. (38) and (39) with various map generating functions $G_K(x)$ the Caputo universal map generating equations and Eq. (43) the Caputo α-families of maps corresponding to the functions $G_K(x)$.

Fractional maps Eqs. (36), (37), and (43) are maps with memory in which the next values of map variables depend on all previous values. An increase in

α corresponds to the increase in a map dimension. It also corresponds to the increased power in the power-law dependence of weights of previous states which imply increased memory effects. For $\alpha = 1$ and $\alpha = 2$ the corresponding maps are given by Eqs. (25), (15), and (16) with $T = 1$ and $G_K(x)$ instead of $G(x)$. Equations (36), (37), and (43) with $\alpha = 3$ and variables $y = p$ and $z = \dot{p}$ produce the full-memory 3D Universal Map

$$
\begin{aligned}
x_{n+1} &= \frac{z_0}{2}(n+1)^2 + y_0(n+1) + x_0 - \frac{1}{2}\sum_{k=0}^{n} G_K(x_k)(n-k+1)^2, \\
y_{n+1} &= z_0(n+1) + y_0 - \sum_{k=0}^{n} G_K(x_k)(n-k+1), \qquad (44) \\
z_{n+1} &= z_0 - \sum_{k=0}^{n} G_K(x_k),
\end{aligned}
$$

which is equivalent to the one-step memory (Sect. 1.2) 3D universal map

$$
\begin{aligned}
x_{n+1} &= x_n - \frac{1}{2}G_K(x_n) + y_n + \frac{1}{2}z_n, \\
y_{n+1} &= -G_K(x_n) + y_n + z_n, \qquad (45) \\
z_{n+1} &= -G_K(x_n) + z_n,
\end{aligned}
$$

or

$$
\begin{aligned}
x_{n+1} &= x_n + y_{n+1} - \frac{1}{2}z_{n+1}, \\
y_{n+1} &= y_n + z_{n+1}, \qquad (46) \\
z_{n+1} &= -G_K(x_n) + z_n,
\end{aligned}
$$

which is a volume preserving map. This map has fixed points $z_0 = y_0 = G_K(x_0) = 0$ and stability of these points can be analyzed by considering the eigenvalues λ of the matrix (corresponding to the tangent map)

$$
\begin{pmatrix} 1 - 0.5\dot{G}_K(x_0) & 1 & 0.5 \\ -\dot{G}_K(x_0) & 1 & 1 \\ -\dot{G}_K(x_0) & 0 & 1 \end{pmatrix}.
$$

The only case in which the fixed points could be stable is $\dot{G}_K(x_0) = 0$, when $\lambda_1 = \lambda_2 = \lambda_3 = 1$. From Eq. (46) it follows that the only $T = 2$ points are the fixed points.

The investigation of the integer members of the α-families of maps is a subject of ongoing research. From the examples of maps with the values of α equal to one, two, and three we see that integer values of α correspond to the degenerate cases in which map equations can be written as maps with full memory. They are equivalent to α-dimensional one-step memory maps in which map variables at each step accumulate information about all previous states of the corresponding systems.

Corresponding to the fact that in the $\alpha = 2$ case the 2D universal family of maps produces the standard map if $G_K(x) = K\sin(x)$ (see Eq. (17)) and in the $\alpha = 1$ case the logistic map results from $G_K(x) = x - Kx(1-x)$ (see Eqs. (25) and (26)), we'll call:

- the Riemann–Liouville α-family of maps Eqs. (36) and (37) with $G_K(x) = K\sin(x)$ the *standard α-RL-family of maps*;
- the Caputo α-family of maps Eq. (43) with $G_K(x) = K\sin(x)$ the *standard α-Caputo-family of maps*;
- the Riemann–Liouville α-family of maps with $G_K(x) = x - Kx(1-x)$ the *logistic α-RL-family of maps*;
- the Caputo α-family of maps with $G_K(x) = x - Kx(1-x)$ the *logistic α-Caputo-family of maps*.

For $\alpha = 0$ the solution of Eq. (27) and correspondingly, the universal map is identically zero. For $\alpha < 1$ the Riemann–Liouville α-families of maps Eqs. (36) and (37) corresponding to the functions $G_K(x)$ satisfying the condition $G_K(0) = 0$, which is true for the standard and logistic α-RL-families of maps, also produces identically zero.

3.1 Integer-Dimensional Standard and Logistic Maps

In general, properties of fractional maps converge to the corresponding properties of integer maps when α approaches integer values. To better understand the properties of fractional maps we'll start with the consideration of the integer members of the corresponding families of maps.

3.1.1 One-Dimensional Logistic and Standard Maps

The one-dimensional logistic map Eq. (18) is one of the best investigated maps. This map has been used as a playground for investigation of the essential property of nonlinear systems—transition from order to chaos through a sequence of period-doubling bifurcations, which is called cascade of bifurcations, and scaling properties of the corresponding systems (see [2,15,24,42,84]). In our investigation of fractional maps we'll use the well-known stability properties of the logistic map (see [56]), which for $0 < K < 4$ are summarized in the bifurcation diagram in Fig. 2a. The $x = 0$ fixed point (sink) is stable for $K < 1$, the $(K-1)/K$ fixed point (sink) is

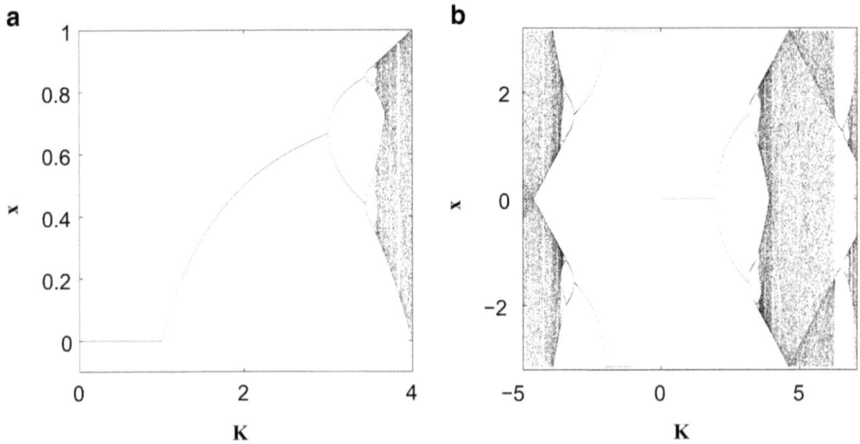

Fig. 2 (**a**) The bifurcation diagram for the logistic map $x = Kx(1-x)$. (**b**) The bifurcation diagram for the 1D standard map (circle map) Eq. (47)

stable for $1 < K < 3$, the $T = 2$ sink is stable for $3 \le K < 1 - \sqrt{6} \approx 3.449$, the $T = 4$ sink is stable when $3.449 < K < 3.544$, and the onset of chaos as a result of the period-doubling cascade of bifurcations occurs at $K \approx 3.56995$.

The one-dimensional standard map ($\alpha = 1$) considered on a circle

$$x_{n+1} = x_n - K\sin(x_n), \quad (\mathrm{mod}\ 2\pi) \tag{47}$$

is a particular form of the circle map with zero driving phase. It has attracting fixed points $2\pi n$ for $0 < K \le K_{c1}(1) = 2$ and $\pi + 2\pi n$ when $-2 \le K < 0$ (for the bifurcation diagram of the 1D standard map, see Fig. 2b). The antisymmetric $T = 2$ sink

$$x_{n+1} = -x_n \tag{48}$$

is stable for $2 < |K| < \pi$, while $x_{n+1} = x_n + \pi$ two sinks ($T = 2$) are stable when $\pi < |K| < \sqrt{\pi^2 + 2} \approx 3.445$. The stable $T = 4$ sinks appear at $|K| \approx 3.445$ and the sequence of bifurcations $T = 4 \to T = 8$ at $K \approx 3.513$, $T = 8 \to T = 16$ at $K \approx 3.526$, and so on leads to the transition to chaos at $K \approx 3.532$. Antisymmetric $T = 2$ trajectories ($K = 2.4$), $T = 4$ trajectories ($K = 3.49$), and two cases of chaotic trajectories ($K = 4.1$ and $K = 5.1$) are presented in Fig. 3. In the 1D standard map with $K > 0$ the full phase space $x \in [-\pi, \pi]$ becomes involved in chaotic motion (we'll call this case "improper attractor") when the maximum of the function $f_K(x) = x - K\sin x$ is equal to π which occurs at $K_{max1D} = 4.603339$ when $x_{max1D} = -1.351817$ (see Fig. 3c, d). Narrow bands with $|K|$ above $2\pi|n|$ (see Fig. 2b for $K > 2\pi$) are accelerator mode bands with zero acceleration within which in the unbounded space (no mod 2π) x is increasing/decreasing with the rate equal approximately to $2\pi|n|$.

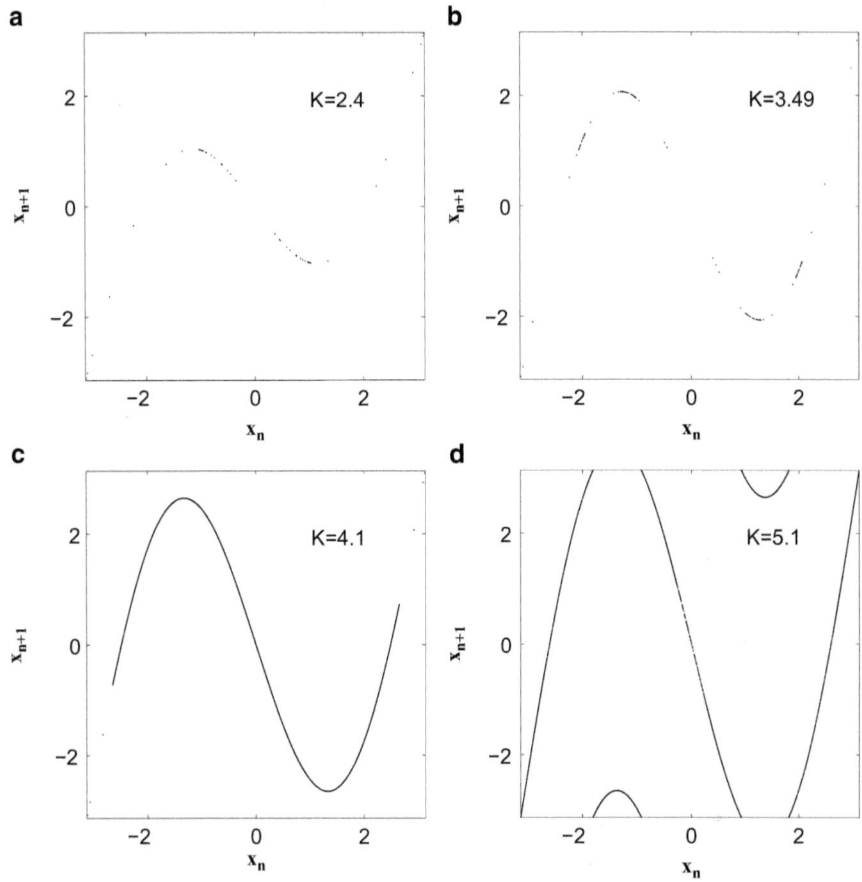

Fig. 3 Attractors in the one-dimensional standard map; x_n vs. x_{n+1} plots (seven trajectories with different initial conditions in each plot): (**a**) $K = 2.4$; antisymmetric $T = 2$ sink. (**b**) $K = 3.49$; $T = 4$ trajectories. (**c**) $K = 4.1$; proper attractor (width of the chaotic area is less than 2π). (**d**) $K = 5.1$; improper attractor (width of the chaotic area is 2π)

3.1.2 Two-Dimensional Logistic and Standard Maps

The two-dimensional logistic map

$$p_{n+1} = p_n + Kx_n(1 - x_n) - x_n,$$
$$x_{n+1} = x_n + p_{n+1} \tag{49}$$

is a quadratic area preserving map. Its phase space contains stable elliptic islands and chaotic areas (no attractors). Quadratic area preserving maps which have a stable fixed point at the origin were investigated by Hénon [33] (for a recent review

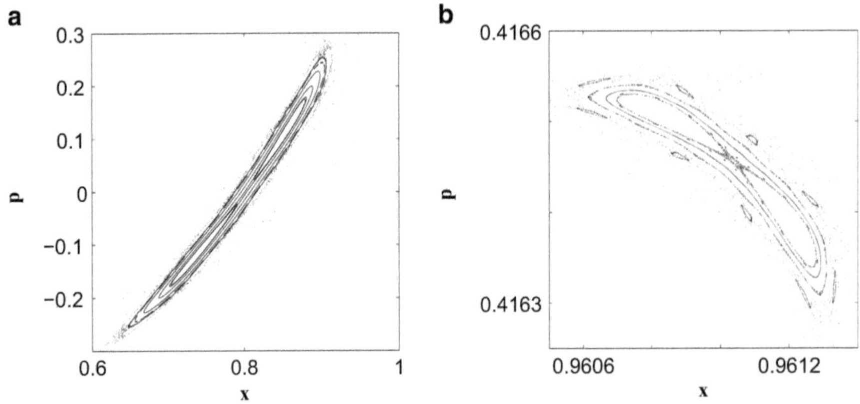

Fig. 4 Bifurcations in the 2D Logistic Map: (**a**) $T = 1 \rightarrow T = 2$ bifurcation at $K = 5$ ($K = 5.05$ on the figure). (**b**) $T = 8 \rightarrow T = 16$ bifurcation at $K \approx 5.5319$ ($K = 5.53194$ on the figure)

on 2D quadratic maps, see [94]). To investigate the logistic α-families of maps we need to know the evolution of the periodic points of the 2D logistic map with the increase of the map parameter K. For $K \in (-3, 1)$ the map Eq. (49) has the stable fixed point $(0, 0)$ which turns into the fixed point $((K-1)/K, 0)$ stable for $K \in (1, 5)$. The $T = 2$ elliptic point

$$x = \frac{K+3 \pm \sqrt{(K+3)(K-5)}}{2K},$$
$$p = \pm\frac{\sqrt{(K+3)(K-5)}}{K} \tag{50}$$

is stable for $-2\sqrt{5}+1 < K < -3$ and $5 < K < 2\sqrt{5}+1$. The period doubling cascade of bifurcations (for $K > 0$) follows the scenario of the elliptic-hyperbolic point transitions with the births of the double periodicity islands inside the original island which has been investigated in [68] and applied to investigate the standard map stochasticity at low values of the map parameter. Further bifurcations in the 2D logistic map, $T = 2 \rightarrow T = 4$ at $K \approx 5.472$, $T = 4 \rightarrow T = 8$ at $K \approx 5.527$, $T = 8 \rightarrow T = 16$ at $K \approx 5.5319$, $T = 16 \rightarrow T = 32$ at $K \approx 5.53253$, etc., and the corresponding decrease in the areas of the islands of stability (see Fig. 4) lead to chaos.

The two-dimensional standard map on a torus Eq. (17) (Chirikov standard map) is one of the best investigated 2D maps. It demonstrates a universal generic behavior of the area-preserving maps whose phase space is divided into elliptic islands of stability and areas of chaotic motion (see, e.g., [11, 45]). Elliptic islands of the standard map in the case of the standard α-families of maps with $1 < \alpha < 2$ evolve into periodic sinks (see Sect. 3.3). Properties of phase space and appearance of different types of attractors in the fractional case, as in the case of the fractional logistic map, are connected to the evolution (with the increase in parameter K) of the 2D standard map's islands originating from the stable (for $K < 4$) fixed point (0,0). At $K = 4$ the fixed point becomes unstable (elliptic-hyperbolic point transition [68]) and two elliptic islands around the stable for $4 < K < 2\pi$ period 2 antisymmetric point

$$p_{n+1} = -p_n, \quad x_{n+1} = -x_n \tag{51}$$

appear. At $K = 2\pi$ this point transforms into two $T = 2$ points

$$p_{n+1} = -p_n, \quad x_{n+1} = x_n - \pi, \tag{52}$$

which are stable when $2\pi < K < 6.59$. These points transform into $T = 4$ stable elliptic points at $K \approx 6.59$ and the period doubling cascade of bifurcations leads to the disappearance of islands of stability in the chaotic sea at $K \approx 6.6344$ [11,45]. The 2D standard map has a set of bands for K above $2\pi n$ of the accelerator mode sticky islands in which the momentum p increases proportionally to the number of iterations k and the coordinate x increases as k^2. The role of accelerator mode islands (for K above 2π) in the anomalous diffusion and the corresponding fractional kinetics is well investigated (see, for example, [90, 93]).

3.1.3 Three-Dimensional Logistic and Standard Maps

Equation (46) with $G_K(x) = x - Kx(1-x)$ (see Eq. (26)) produces the 3D logistic map

$$\begin{aligned} x_{n+1} &= x_n + y_{n+1} - \frac{1}{2} z_{n+1}, \\ y_{n+1} &= y_n + z_{n+1}, \\ z_{n+1} &= Kx_n(1-x_n) - x_n + z_n. \end{aligned} \tag{53}$$

Three-dimensional quadratic volume preserving maps were investigated in [46, 58]. Everything stated in Sect. 3 for the 3D universal map is still valid for the 3D logistic map.

The three-dimensional standard map with $G_K(x) = K\sin(x)$

$$\begin{aligned} x_{n+1} &= x_n + y_{n+1} - \frac{1}{2} z_{n+1}, \quad (\text{mod } 2\pi), \\ y_{n+1} &= y_n + z_{n+1}, \quad (\text{mod } 2\pi), \\ z_{n+1} &= -K\sin(x_n) + z_n, , \quad (\text{mod } 4\pi) \end{aligned} \tag{54}$$

has unstable fixed points $(2\pi n, 2\pi m, 4\pi k)$ and $(2\pi n + \pi, 2\pi m, 4\pi k)$, $n \in \mathbb{Z}$, $m \in \mathbb{Z}$, $k \in \mathbb{Z}$. Ballistic points $K\sin(x) = -4\pi n$, $y = 2\pi m$, $z = 4\pi k$, which appear for $|K| \geq 4\pi$, are also unstable.

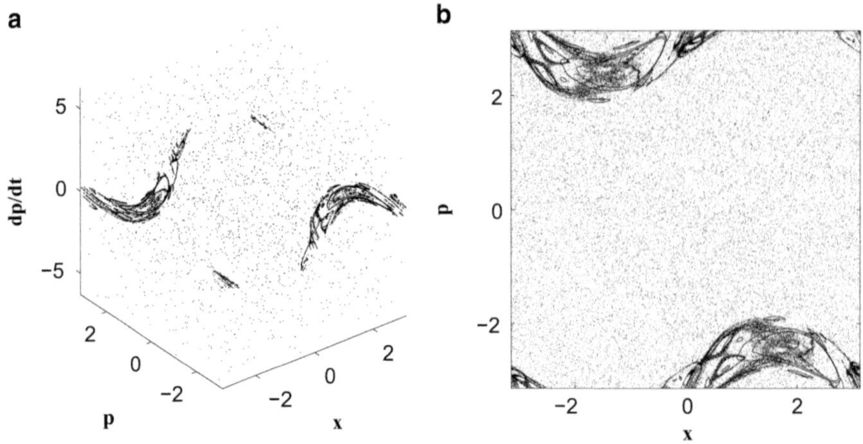

Fig. 5 Phase space of the 3D standard map (54) with $K = 3$: (**a**) Three-dimensional phase space. (**b**) A projection of the 3D phase space on the x-y plane

Stability of $T = 2$ ballistic points is defined by the eigenvalues of the matrix

$$\begin{pmatrix} 1-0.5K\cos x_1 & 1 & 0.5 \\ -K\cos x_1 & 1 & 1 \\ -K\cos x_1 & 0 & 1 \end{pmatrix} \times \begin{pmatrix} 1-0.5K\cos x_2 & 1 & 0.5 \\ -K\cos x_2 & 1 & 1 \\ -K\cos x_2 & 0 & 1 \end{pmatrix}.$$

For the period two on the torus ballistic points

$$\begin{aligned} &z_1, \;\; y_1 = \frac{z_1}{2} - \pi(2n+1), \;\; K\sin x_1 = 2z_1, \\ &z_2 = -z_1, \;\; y_2 = -\frac{z_1}{2} - \pi(2n+1), \;\; x_2 = x_1 - \pi(2n-1), \end{aligned} \tag{55}$$

where $n \in \mathbb{Z}$, the eigenvalues are

$$\left\{1, \; \frac{1}{8}\left(8 - K^2\cos^2 x_1 \pm K\cos x_1 \sqrt{K^2\cos^2 x_1 - 16}\right)\right\} \tag{56}$$

Ballistic $T = 2$ points are stable along a line defined by Eq. (55) for all values of z satisfying the condition

$$K^2 - 16 < 4z^2 < K^2. \tag{57}$$

An example of the phase space for $K = 3$ in three dimensions and its projection on the x-y plane is given in Fig. 5. For this value of K ballistic $T = 2$ points are stable when $-1.5 < z < 1.5$ and the space around the line of stability presents a series of islands (invariant curves), islands around islands, and separatrix layers.

When $K \to 0$, the volume of the regular motion shrinks. When K is small, the line of the stable $T = 2$ ballistic points exists for $-K/2 < z < K/2$. A different form of the 3D volume preserving standard map was introduced and investigated in detail in [18].

3.2 α-Families of Maps ($0 < \alpha < 1$)

As we mentioned at the end of Sect. 3, members of the logistic and standard α-families of maps corresponding to $\alpha = 0$ and RL-families' members with $0 < \alpha < 1$ are identically zeros. The only fractional logistic and standard maps with $0 < \alpha < 1$ which are not identically zeros are α-Caputo-families of maps. The α-Caputo-universal map ($0 < \alpha < 1$)

$$x_{n+1} = x_0 - \frac{1}{\Gamma(\alpha)} \sum_{k=0}^{n} G_K(x_k)(n-k+1)^{\alpha-1} \tag{58}$$

in the limit $\alpha \to 1$ is identical to the one-dimensional universal map Eq. (25) and in this limit properties of fractional maps are similar to properties of the corresponding 1D maps. Equation (58) with $G_K(x) = x - Kx(1-x)$ is the logistic α-Caputo-family of maps for $0 < \alpha < 1$

$$x_n = x_0 + \frac{1}{\Gamma(\alpha)} \sum_{k=0}^{n-1} \frac{Kx_k(1-x_k) - x_k}{(n-k)^{1-\alpha}} \tag{59}$$

and with $G_K(x) = K\sin(x)$ is the standard α-Caputo-family of maps for $0 < \alpha < 1$

$$x_n = x_0 - \frac{K}{\Gamma(\alpha)} \sum_{k=0}^{n-1} \frac{\sin x_k}{(n-k)^{1-\alpha}}, \quad (\text{mod } 2\pi) \tag{60}$$

These maps are one-dimensional maps with power-law decreasing memory [20]. The bifurcation diagrams for these maps are similar to the corresponding diagrams for the $\alpha = 1$ case (Fig. 2). A decrease in α and the corresponding decrease in weights of the earlier states (decrease in memory effects) leads to the stretchiness of the corresponding bifurcation diagrams along the parameter K-axis and this stretchiness increases as α gets smaller (Fig. 6).

Within a band of values of K, above the value which corresponds to the appearance of $T = 4$ trajectories, map trajectories are attracting cascade of bifurcations type trajectories (CBTT) (see Fig. 7). On CBTT an increase in the number of map iterations leads to the change in the map's stability properties. A trajectory which converges to a $T > 4$ periodic point or becomes a chaotic trajectory (depending on the value of K) evolves according to a certain scenario:

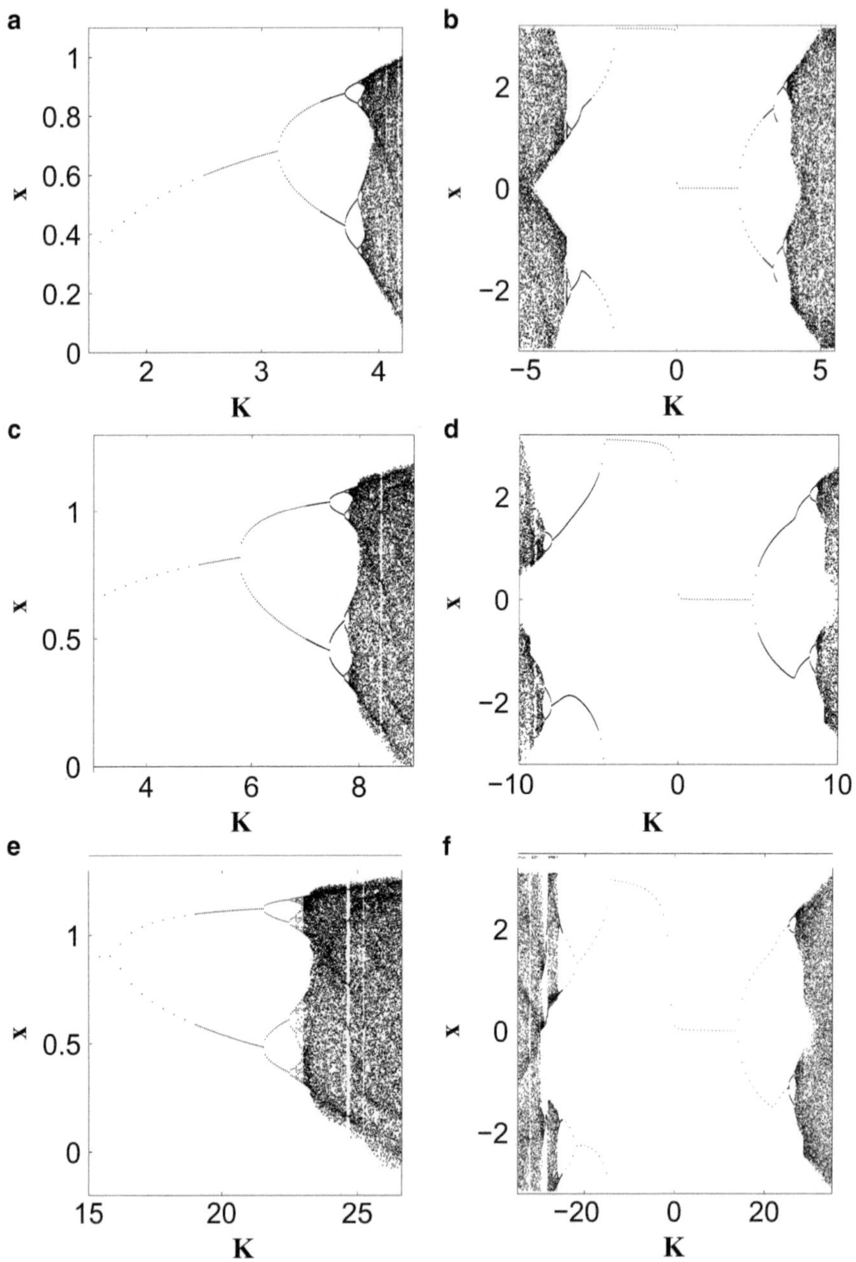

Fig. 6 Bifurcation diagrams for the logistic and standard α-Caputo-families of maps with $0 < \alpha < 1$. In (**a**)–(**f**) the bifurcation diagrams obtained after performing 10^4 iterations on a single trajectory with $x_0 = 0.1$ for various values of K. (**a**), (**c**), and (**e**)—the logistic α-Caputo-family. (**b**), (**d**), and (**f**)—the standard α-Caputo-family. In (**a**) and (**b**) $\alpha = 0.8$. In (**c**) and (**d**) $\alpha = 0.3$. In (**e**) and (**f**) $\alpha = 0.1$

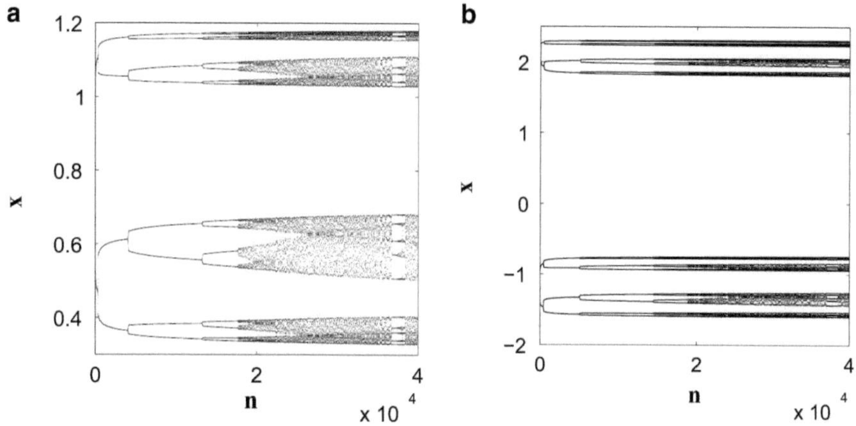

Fig. 7 Cascade of bifurcations type trajectories in the logistic and standard α-Caputo-families of maps with $\alpha = 0.1$. (**a**) The fractional logistic map with $\alpha = 0.1$ and $K = 22.65$. (**b**) The fractional standard map with $\alpha = 0.1$ and $K = 26.65$

it first converges to a $T = 4$ point; then it bifurcates, always at the same place for the given values of the parameter K and the order α, and converges to a $T = 8$ trajectory; then to a $T = 16$ trajectory; and so on. Power-law decaying memory with power $\beta \approx 0.9$ corresponding to small values of $\alpha \approx 0.1$ (see Sect. 1.1) appears in biological applications. Attracting CBTT in, for example, adaptive biological systems may represent not simply a change of a state of a biological system according to a change in a parameter, but rather a change in the evolution of the system according to the change in the parameter. Examples of CBTT in the logistic and standard α-Caputo-families of maps with $\alpha = 0.1$ are presented in Fig. 7.

It also should be noted that bifurcation diagrams of the fractional maps depend on the number of iterations used in their calculations. This is a consequence of the existence of CBTT. Trajectories which after 100 iterations converged to a fixed point in Fig. 8b after 10,000 iterations became $T = 2$ trajectories in Fig. 8a. With an increase in the number of iterations the whole bifurcation diagram shifts to the left.

3.3 α-Families of Maps ($1 < \alpha < 2$)

For $1 < \alpha < 2$ the logistic and standard α-families of maps assume the following forms:

- The RL-standard map on a cylinder

$$p_{n+1} = p_n - K \sin x_n, \tag{61}$$

$$x_{n+1} = \frac{1}{\Gamma(\alpha)} \sum_{i=0}^{n} p_{i+1} V_{\alpha}^{1}(n - i + 1), \quad (\text{mod } 2\pi), \tag{62}$$

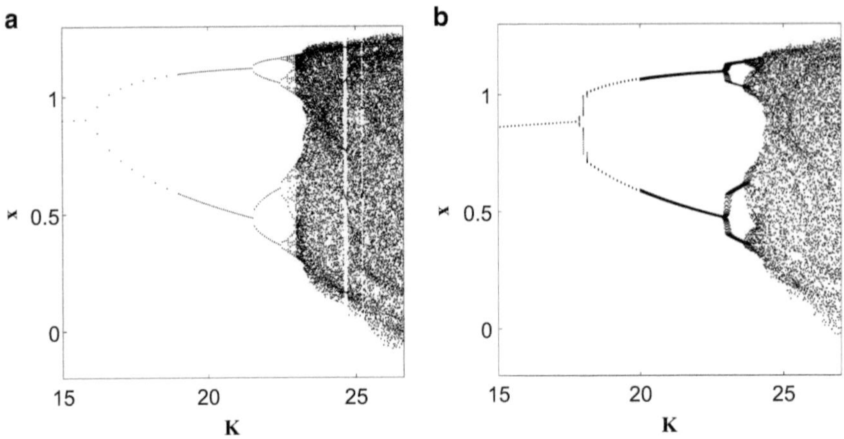

Fig. 8 Dependence of bifurcation diagrams of the fractional maps on the number of iterations on a single trajectory used in their calculation. Bifurcation diagrams for the fractional logistic map with $\alpha = 0.1$. (**a**) 10,000 iterations on each trajectory. (**b**) 100 iterations on each trajectory

where

$$V_{\alpha}^{k}(m) = m^{\alpha-k} - (m-1)^{\alpha-k}. \tag{63}$$

This map requires the initial condition $x_0 = 0$ and can't be considered on a torus.

- The Caputo-standard map on a torus

$$p_{n+1} = p_n - \frac{K}{\Gamma(\alpha-1)}\left[\sum_{i=0}^{n-1} V_{\alpha}^{2}(n-i+1)\sin x_i + \sin x_n\right], \pmod{2\pi}, \tag{64}$$

$$x_{n+1} = x_n + p_0 - \frac{K}{\Gamma(\alpha)}\sum_{i=0}^{n} V_{\alpha}^{1}(n-i+1)\sin x_i, \pmod{2\pi}. \tag{65}$$

- The RL-logistic map

$$p_{n+1} = p_n - Kx_n(1-x_n) - x_n, \tag{66}$$

$$x_{n+1} = \frac{1}{\Gamma(\alpha)}\sum_{i=0}^{n} p_{i+1}V_{\alpha}^{1}(n-i+1), \tag{67}$$

which requires the initial condition $x_0 = 0$.

- The Caputo-logistic map

$$x_{n+1}=x_0+p(n+1)^k-\frac{1}{\Gamma(\alpha)}\sum_{k=0}^{n}[x_k-Kx_k(1-x_k)](n-k+1)^{\alpha-1}, \tag{68}$$

$$p_{n+1} = p_0 - \frac{1}{\Gamma(\alpha-1)}\sum_{k=0}^{n}[x_k - Kx_k(1-x_k)](n-k+1)^{\alpha-2}. \tag{69}$$

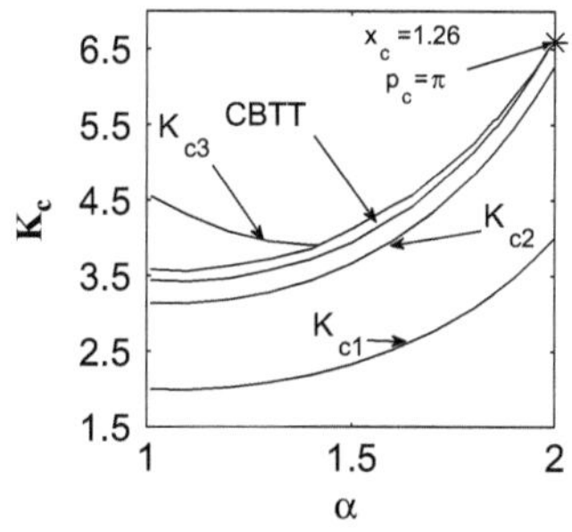

Fig. 9 Bifurcations in the standard α-families of maps with $1 < \alpha < 2$. Below $K = K_{c1}$ curve the fixed point $(0, 0)$ is stable. It becomes unstable at $K = K_{c1}$ and gives birth to the antisymmetric $T = 2$ sink which is stable for $K_{c1} < K < K_{c2}$. A pair of the $T = 2$ sinks with $x_{n+1} = x_n - \pi$, $p_{n+1} = -p_n$ is stable in the band above $K = K_{c2}$ curve. Cascade of bifurcations type trajectories (CBTT) appear and exist in the narrow band which ends at the cusp at the top right corner of the figure. (x_c, p_c) is the point at which the standard map's ($\alpha = 2$) $T = 2$ elliptic points with $x_{n+1} = x_n - \pi$, $p_{n+1} = -p_n$ become unstable and bifurcate. In the area below K_{c3} (above the CBTT band) the chaotic attractor is restricted to a band whose width is less than 2π. On the upper curves and above them the full phase space is chaotic

Here and in Eqs. (64) and (65) we assumed $x \equiv x^0$ and $p \equiv x^1$ in the Caputo universal map Eq. (43).

The fractional standard maps Eqs. (61), (62), (64), and (65) are well investigated (see [19–22]) and the logistic maps are the subject of ongoing research.

Evolution of trajectories in fractional maps depends on two parameters: the map parameter K and the fractional order α. Figure 9 reflects this dependence in the case of the standard α-families of maps with $1 < \alpha < 2$.

3.3.1 $T = 2$ Antisymmetric Sink

It is obvious that the fractional standard and logistic maps have the fixed points at the origin $(0, 0)$. But we'll start the fractional maps' phase space analysis with the consideration of the $T = 2$ antisymmetric sinks. We'll present most of the analysis for the fractional RL-standard map (Fig. 9). Results of numerical simulations suggest that the fractional Caputo-standard map has similar properties and the results for the logistic map are submitted for publication.

The 1D standard map has the $T = 2$ antisymmetric sink Eq. (48) and the 2D standard map has the $T = 2$ antisymmetric elliptic point Eq. (51). Numerical experiments (Fig. 10) show that the antisymmetric $T = 2$ sinks persist in the fractional standard maps with $1 < \alpha < 2$. In the RL-standard map these sinks attract most of the trajectories with small p_0. Assuming the existence of an antisymmetric $T = 2$ sink

$$p_n = p_l(-1)^n, \;\; x_n = x_l(-1)^n, \tag{70}$$

it is possible to calculate the coordinates of its attracting points (x_l, p_l) and $(-x_l, -p_l)$. In the limit $n \to \infty$ Eqs. (61) and (62) can be written as

$$p_l = \frac{K}{2}\sin(x_l), \tag{71}$$

$$x_l = \lim_{n\to\infty} x_{2n} = \frac{p_l}{\Gamma(\alpha)} \lim_{n\to\infty} \sum_{k=0}^{2n-1} (-1)^{k+1} V_\alpha^1(2n-i) = \frac{p_l}{\Gamma(\alpha)} V_{\alpha l}(k), \tag{72}$$

where

$$V_{\alpha l} = \sum_{k=1}^{\infty} (-1)^{k+1} V_\alpha^1(k). \tag{73}$$

Finally, the equation for the x_l takes the form

$$x_l = \frac{K}{2\Gamma(\alpha)} V_{\alpha l} \sin(x_l). \tag{74}$$

The numerical solution of Eqs. (74) and (71) for $K = 4.5$ when $1 < \alpha < 2$ is presented in Fig. 10b–d. Figure 10e, f show how well this solution agrees with the results of numerical simulations of individual trajectories. After 1,000 iterations presented in Fig. 10e, f the values of deviations $|p_n - p_l|$ and $|x_n - x_l|$ are less than 10^{-7}.

The condition of the existence of a solution for Eq. (74)

$$K > K_{c1}(\alpha) = \frac{2\Gamma(\alpha)}{V_{\alpha l}} \tag{75}$$

is the condition of the existence of the antisymmetric $T = 2$ sink. This sink exists above the curve $K = K_{c1}$ in Fig. 9. For $\alpha = 2$ Eq. (75) produces the standard map condition $K > 4$ (see Sect. 3.1.2) and for $\alpha = 1$ it gives $K > 2$ (see Sect. 3.1.1).

3.3.2 Fixed Points

Numerical simulations show that as in the 1D and 2D cases, in the case of fractional maps with $1 < \alpha < 2$ the condition of the appearance of $T = 2$ trajectories coincides with the condition of the disappearance of the stable fixed point. This

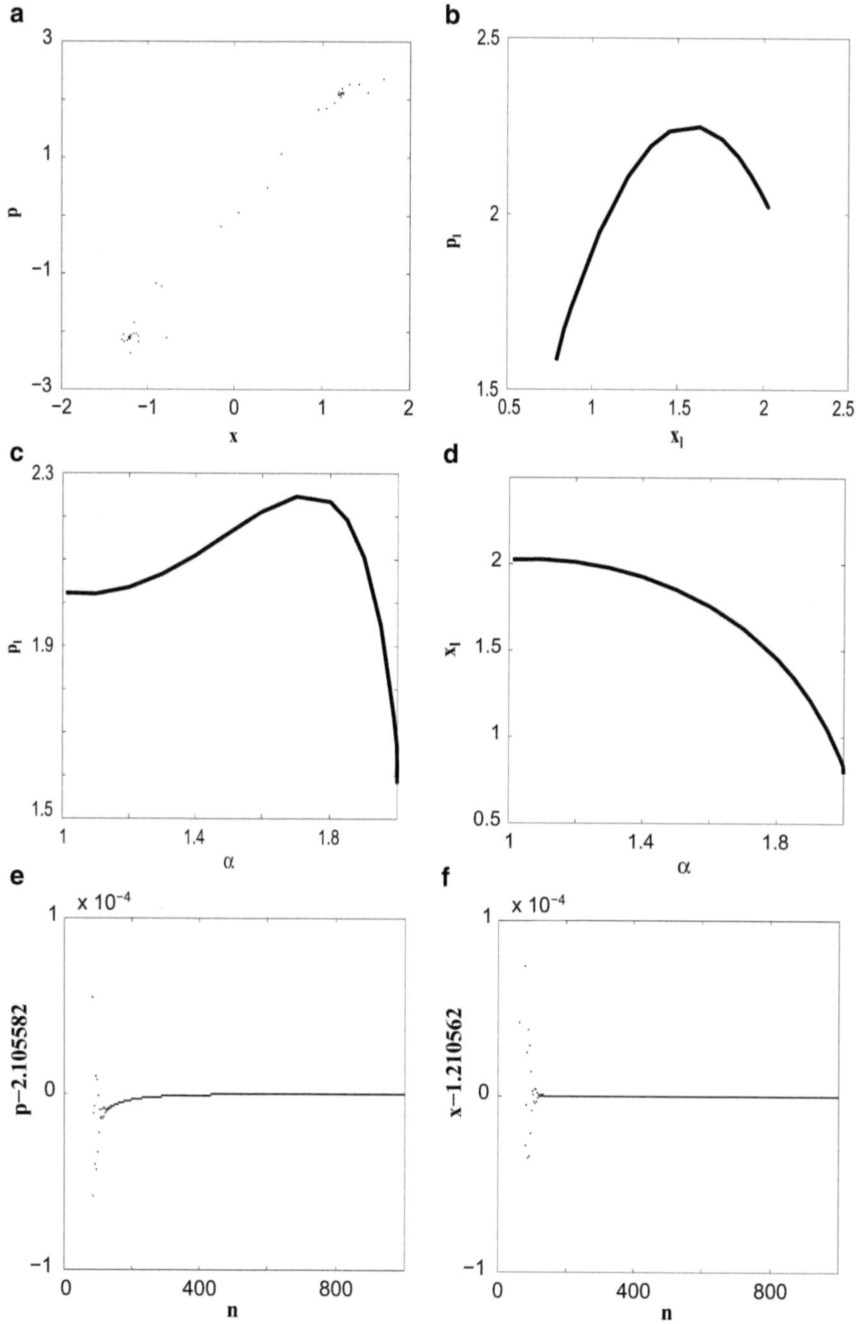

Fig. 10 The RL-standard map's period 2 sink: (**a**) An example of the $T = 2$ attractor for $K = 4.5$, $\alpha = 1.9$. One trajectory with $x_0 = 0$, $p_0 = 0.513$. (**b**) p_l of x_l for the case of $K = 4.5$. (**c**) p_l of α for the case of $K = 4.5$. (**d**) x_l of α for the case of $K = 4.5$. (**e**) $p_n - p_l$ for the trajectory in (**a**). After 1,000 iterations $|p_n - p_l| < 10^{-7}$. (**f**). $x_n - x_l$ for the trajectory in (**a**). After 1,000 iterations $|x_n - x_l| < 10^{-7}$

result for the fractional standard map was demonstrated in [22] and for the fractional logistic map was submitted for publication. Below we present two ways in which stability of the RL-standard map's $(0, 0)$ fixed point can be investigated.

In the vicinity of the fixed point $(0, 0)$ the equation for the deviation of a trajectory from the fixed point can be written as

$$\delta p_{n+1} = \delta p_n - K\delta x_n, \tag{76}$$

$$\delta x_{n+1} = \frac{1}{\Gamma(\alpha)} \sum_{i=0}^{n} \delta p_{i+1} V_\alpha(n - i + 1). \tag{77}$$

Based on the results of Sect. 3.3.1 let's look for a solution in the form

$$\delta p_n = p_0 \sum_{i=0}^{n-1} p_{n,i} \Big(\frac{2}{V_{\alpha l}}\Big)^i \Big(\frac{K}{K_{c1}(\alpha)}\Big)^i, \quad (n > 0), \tag{78}$$

$$\delta x_n = \frac{p_0}{\Gamma(\alpha)} \sum_{i=0}^{n-1} x_{n,i} \Big(\frac{2}{V_{\alpha l}}\Big)^i \Big(\frac{K}{K_{c1}(\alpha)}\Big)^i, \quad (n > 0), \tag{79}$$

where $p_{n,i}$ and $x_{n,i}$ satisfy the following iterative equations

$$x_{n+1,i} = -\sum_{m=i}^{n} (n - m + 1)^{\alpha-1} x_{m,i-1}, \quad (0 < i \le n), \tag{80}$$

$$p_{n+1,i} = -\sum_{m=i}^{n} x_{m,i-1}, \quad (0 < i < n), \tag{81}$$

for which the initial and boundary conditions are

$$p_{n+1,n} = x_{n+1,n} = (-1)^n, \quad p_{n+1,0} = 1, \quad x_{n+1,0} = (n + 1)^{\alpha-1}. \tag{82}$$

To verify the convergence of the alternating series Eqs. (78) and (79) we apply the Dirichlet's test by considering the totals

$$S_n = \sum_{i=0}^{n-1} x_{n,i} \Big(\frac{2}{V_{\alpha l}}\Big)^i, \quad I_n = \sum_{i=0}^{n-1} p_{n,i} \Big(\frac{2}{V_{\alpha l}}\Big)^i. \tag{83}$$

They obey the following iterative rules

$$S_n = n^{\alpha-1} - \frac{2}{V_{\alpha l}} \sum_{i=1}^{n-1} (n - i)^{\alpha-1} S_i, \quad I_n = 1 - \frac{2}{V_{\alpha l}} \sum_{i=1}^{n-1} S_i, \tag{84}$$

where $S_1 = 1$. Numerical simulations demonstrate that values of S_n and I_n converge to the values $(-1)^{n+1} S_\infty$ and $(-1)^{n+1} I_\infty$ presented in Fig. 11b. Figure 11c, d show an example of the typical evolution of S_n and I_n over the first 20,000 iterations.

There is still no strict mathematical proof of the convergence. From the boundedness of S_n and I_n the convergence of δp_n and δx_n requires the following condition

$$\frac{K}{K_{c1}(\alpha)} < 1, \tag{85}$$

which, as we expected, is exactly opposite to the condition of the existence of the antisymmetric $T = 2$ sink Eq. (75). Hundreds of runs of computer simulations confirmed that the transition from the stable fixed point $(0, 0)$ to the stable antisymmetric $T = 2$ sink in both the RL-standard map and the Caputo-standard map occurs on the curve $K = K_{c1}$ as depicted in Fig. 11a.

The second way to investigate stability of the $(0, 0)$ fixed point is by using generating functions [25], which in the case of convolutions allows transformations of sums of products into products of sums. After the introduction

$$\tilde{W}_\alpha(t) = \frac{K}{\Gamma(\alpha)}\sum_{i=0}^{\infty}[(i+1)^{\alpha-1} - i^{\alpha-1}]t^i, \quad \tilde{X}(t) = \sum_{i=0}^{\infty}\delta x_i t^i, \quad \tilde{P}(t) = \sum_{i=0}^{\infty}\delta p_i t^i \tag{86}$$

system Eqs. (76) and (77) can be written as

$$\tilde{X}(t) = \frac{p_0 \tilde{W}_\alpha(t)}{K}\frac{t}{1 - t\left(1 - \tilde{W}_\alpha(t)\right)}, \tag{87}$$

$$\tilde{P}(t) = p_0\frac{1 + \tilde{W}_\alpha(t)}{1 - t\left(1 - \tilde{W}_\alpha(t)\right)}. \tag{88}$$

We see that the original problem can be solved by investigating the asymptotic behavior at $t = 0$ of the derivatives of the analytic functions $\tilde{X}(t)$ and $\tilde{P}(t)$. This is still a complex unresolved problem.

When $K < K_{c1}$ and the fixed point is stable, in phase space it is surrounded by a finite basin of attraction, whose width w depends on the values of K and α. For example, for $K = 3$ and $\alpha = 1.9$ the width of the basin of attraction is $1.6 < w < 1.7$. Numeric simulations of thousands of trajectories with $p_0 < 1.6$ performed by the authors of [22], of which only 200 (with $1.5 < p_0 < 1.6$) are presented in Fig. 11e, show only converging trajectories, whereas among 50 trajectories with $1.6 < p_0 < 1.7$ in Fig. 12a there are trajectories converging to the fixed point as well as some trajectories converging to attracting slow diverging trajectories, whose properties will be discussed in the following section (Sect. 3.3.3). Figure 11e shows fast converging trajectories. In the case $K = 2$ and $\alpha = 1.4$ in addition to the fast converging trajectories and attracting slow diverging trajectories there exist attracting slow converging trajectories (Fig. 11f).

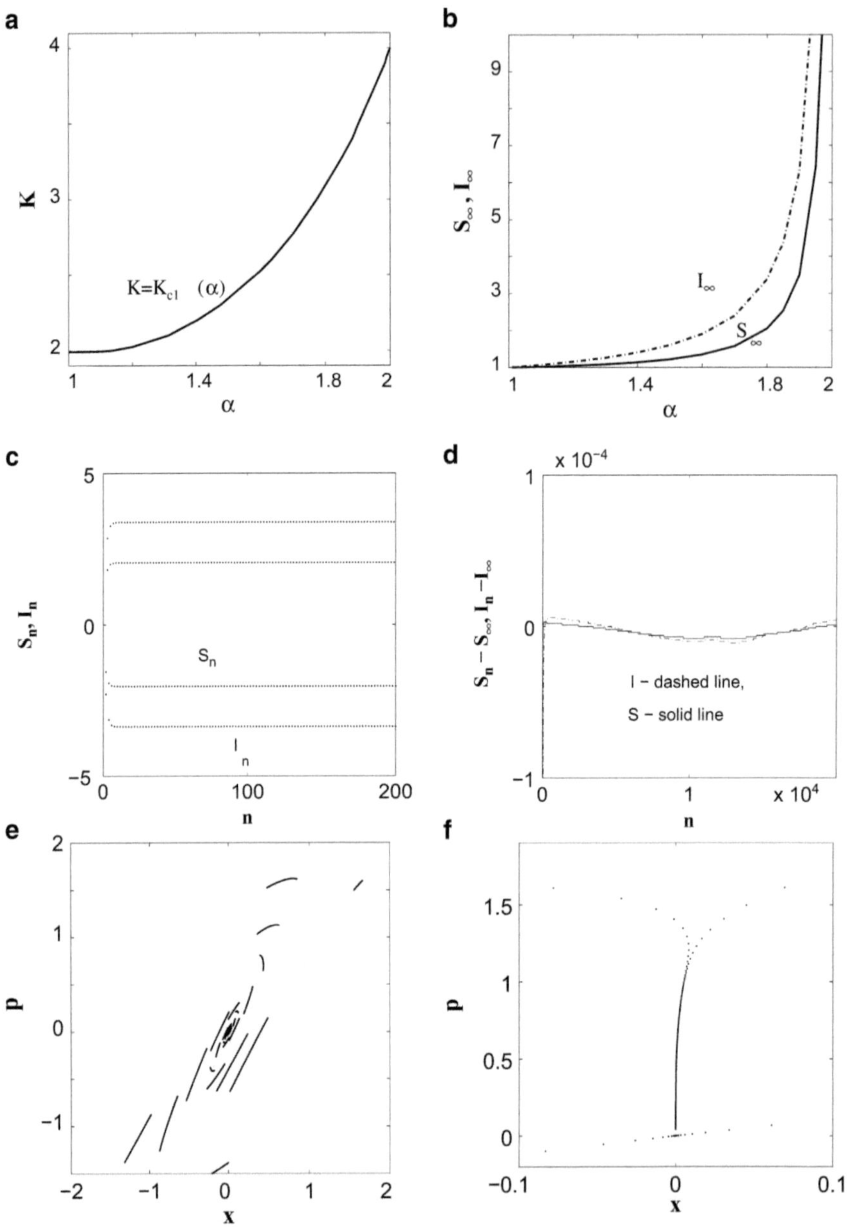

Fig. 11 Stability of the fixed point (0, 0) in the RL-standard map with $1 < \alpha < 2$: (**a**) The fixed point is stable below the curve $K = K_c(\alpha)$. (**b**) Values of S_∞ and I_∞ obtained after 20,000 iterations of Eq. (84). The values of S_∞ and I_∞ increase rapidly when $\alpha \to 2$; for example, $S_\infty \approx 276$ and $I_\infty \approx 552$ after 20,000 iterations when $\alpha = 1.999$. (**c**) An example of the typical evolution of S_∞ and I_∞ over the first 200 iterations for $1 < \alpha < 2$. This particular figure corresponds to $\alpha = 1.8$. (**d**) Deviation of the values S_n and I_n from the values $S_\infty \approx 2.04337$

(continued)

3.3.3 Attractors Below Cascade of Bifurcations Band

In the following most of the statements are conjectures made on the basis of the results of numerical simulations performed for some values of parameters K and α which then were verified for additional parameter values.

The structure of the fractional standard map's phase space preserves some features which exist in the $\alpha = 2$ case. For example, for $K < K_{c1}$ stable higher period points, which exist in the standard map, still exist in the fractional standard maps (Fig. 12), but they exist in the asymptotic sense and they transform from elliptic points into sinks and (in the case of the RL-standard map) into attracting slow ($p_n \sim n^{2-\alpha}$) diverging trajectories. In the area preserving standard map stable fixed and periodic points are surrounded by islands of regular motion which in the case of fractional maps turn into basins of attraction associated with sinks or slowly diverging attracting trajectories. In the standard map islands are surrounded by chaotic areas. For $K < K_{c1}$ and $1 < \alpha < 2$ in the fractional standard maps there are no chaotic or regular trajectories. Chaos exists in the following sense: two initially close trajectories that start in an area between basins of attractions at first diverge, but then converge to the same or different attractors.

There are differences not only between properties of the regular and fractional standard maps but also between phase space structures of the RL- and Caputo-standard maps. There is more than one way to approach an attracting periodic or fixed point of the RL-standard map. In Fig. 13 the examples of three trajectories, two for the RL-standard map and one for the Caputo-standard map, are used to demonstrate the differences in the rates of convergence. In the RL-standard map trajectories starting from attractors' basins of attractions demonstrate fast convergence with

$$\delta x_n \sim n^{-1-\alpha}, \quad \delta p_n \sim n^{-\alpha} \tag{89}$$

and trajectories with the initial conditions from chaotic areas demonstrate slow convergence:

$$\delta x_n \sim n^{-\alpha}, \quad \delta p_n \sim n^{1-\alpha}. \tag{90}$$

There is only one type of convergence in the Caputo-standard map:

$$\delta x_n \sim n^{1-\alpha}, \quad \delta p_n \sim n^{1-\alpha}. \tag{91}$$

(continued)

and $I_\infty \approx 3.37416$ for $\alpha = 1.8$ during the first 20,000 iterations (this type of behavior remains for $1 < \alpha < 2$). (**e**) Evolution of trajectories with $p_0 = 1.5 + 0.0005i$, $0 \leq i < 200$ for the case $K = 3$, $\alpha = 1.9$. The line segments correspond to the nth iteration on the set of trajectories with close initial conditions. The evolution of the trajectories with smaller p_0 is similar. (**f**) 10^5 iterations on both of the two trajectories for $K = 2$, $\alpha = 1.4$. The one at the bottom with $p_0 = 0.3$ is a fast converging trajectory. The upper trajectory with $p_0 = 5.3$ is an example of an attracting slow converging trajectory in which $p_{100,000} \approx 0.042$

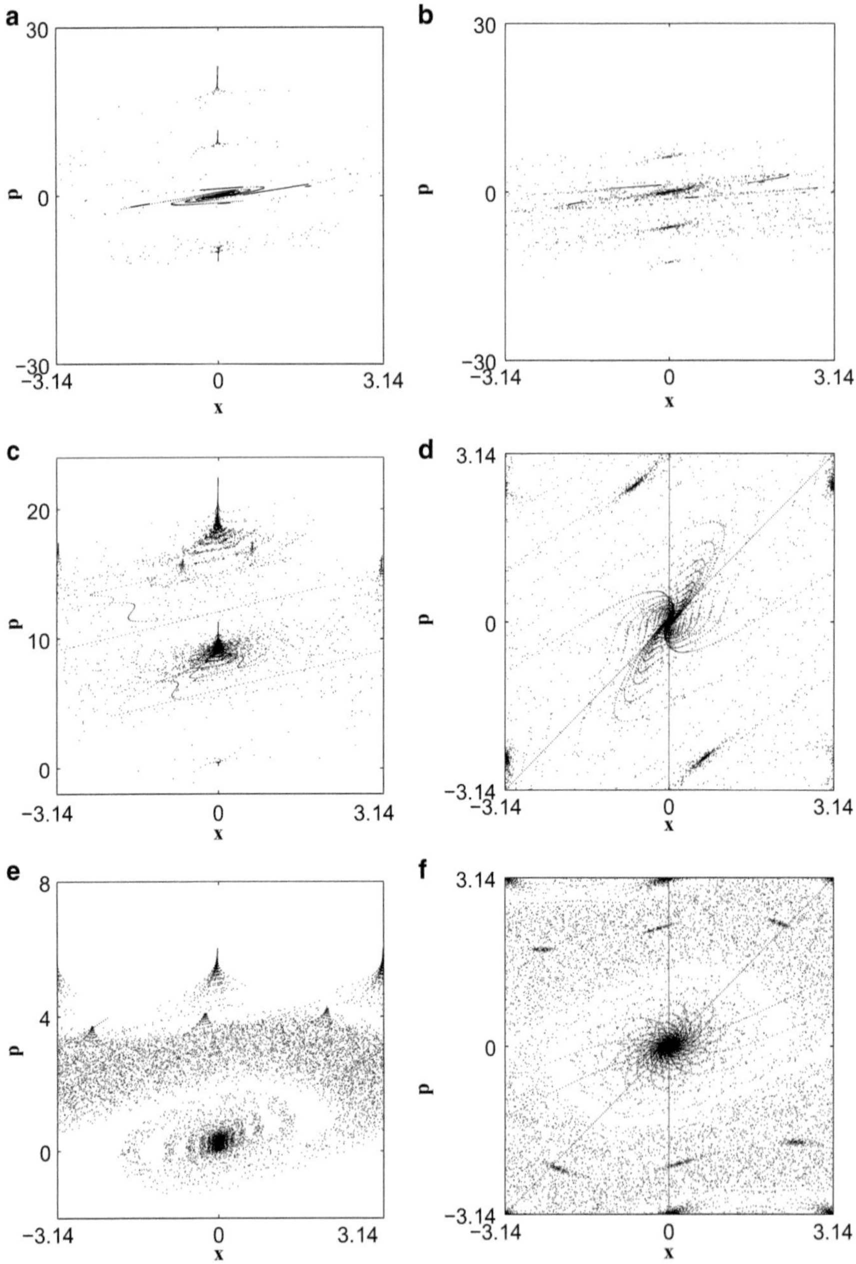

Fig. 12 The RL- and Caputo-standard maps' phase spaces for $K < Kc1$: (**a**) The RL-standard map with the same values of parameters as in Fig. 11e but $p_0 = 1.6 + 0.002i$, $0 \leq i < 50$. (**b**) The Caputo-standard map with the same values of parameters as in Fig. 11e but $p_0 = 1.7 + 0.002i$, $0 \leq i < 50$. (**c**) 400 iterations on the RL-standard map trajectories with $p_0 = 4 + 0.08i$, $0 \leq i < 125$ for the case $K = 2$, $\alpha = 1.9$. Trajectories converge to the fixed point and two types

(continued)

The same rates of convergence were observed also for antisymmetric (see Sect. 3.3.1 and Fig. 15) and $x_{n+1} = x_n - \pi$, $p_{n+1} = -p_n$ period two ($T = 2$) points (Fig. 16).

From Fig. 15a, b one can see that phase portraits on cylinders of the fractional standard maps with $K = 3$ and $\alpha = 1.9$ contain, in addition to the $(0,0)$ fixed point, attracting slow diverging trajectories (RL-case), or fixed points (Caputo-case) approximately equally spaced along the p-axis. This result agrees with the fact that the standard map with $K = 3$ has only one central island. More complex structures of the fractional standard maps' phase spaces, for $K = 2$ with $T = 4$ sinks (Fig. 15c, d) and for $K = 0.6$ with $T = 2$ and $T = 3$ sinks (Fig. 15e, f), can be explained by the presence of the islands with the same periodicity in the standard map with the same K. Numerical evaluations (see Fig. 14) lead to the suggestion that attracting slow diverging trajectories which converge to trajectories along the p-axis ($x \to x_{lim} = 0$) in the area of parameters of their stability for large n demonstrate the following asymptotic behavior

$$p_n = Cn^{2-\alpha}. \tag{92}$$

The constant C can be evaluated for $1.8 < \alpha < 2$. Consider a trajectory on a cylinder with $x_{lim} = 0$, $T = 1$, and constant step in x in the unbounded space $2\pi M$, where M is an integer. Then from Eq. (62) follows

(continued)

of attracting slow diverging trajectories: with $x_{lim} = 0$ ($T = 1$) and $T = 4$. (**d**) 100 iterations on the Caputo-standard map trajectories with $p_0 = -3.14 + 0.0314i$, $0 \le i < 200$ for the same case as in (**c**) ($K = 2$, $\alpha = 1.9$) but considered on a torus. In this case all trajectories converge to the fixed point or $T = 4$ sink. (**e**) 400 iterations on trajectories with $p_0 = 2 + 0.04i$, $0 \le i < 50$ for the RL-standard map case $K = 0.6$, $\alpha = 1.9$. Trajectories converge to the fixed point and two attracting slow diverging trajectories ($T = 2$ and $T = 3$). (**f**) 100 iterations on the Caputo-standard map trajectories with $p_0 = -3.14 + 0.0314i$, $0 \le i < 200$ for the same case as in (**e**) ($K = 0.6$, $\alpha = 1.9$) considered on a torus. In this case all trajectories converge to the fixed point, period two and period three sinks

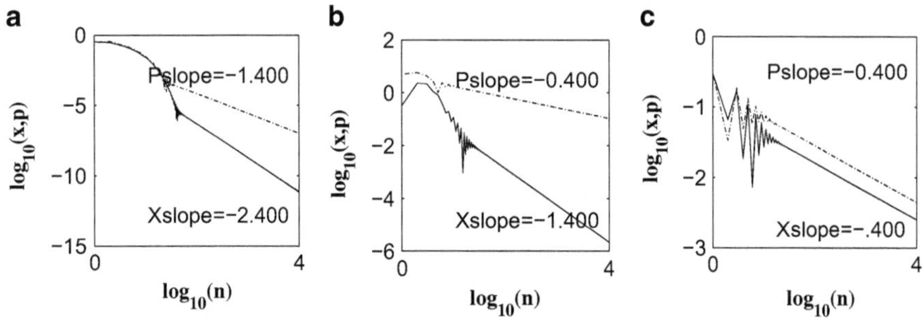

Fig. 13 Different types of convergence of trajectories to the fixed point in the RL-standard map ((**a**) and (**b**)) and the Caputo-standard map (**c**): (**a**) Time dependence of the coordinate and momentum for the fast converging trajectory with $K = 2$, $\alpha = 1.4$ and the initial conditions $x_0 = 0$ and $p_0 = 0.3$ from Fig. 11f. (**b**) The same as in (**a**) but for the attracting slow converging trajectory with the initial conditions $x_0 = 0$ and $p_0 = 5.3$. (**c**) x and p time dependence for the Caputo-standard map with $K = 2$, $\alpha = 1.4$, and the initial conditions $x_0 = 0$ and $p_0 = 0.3$

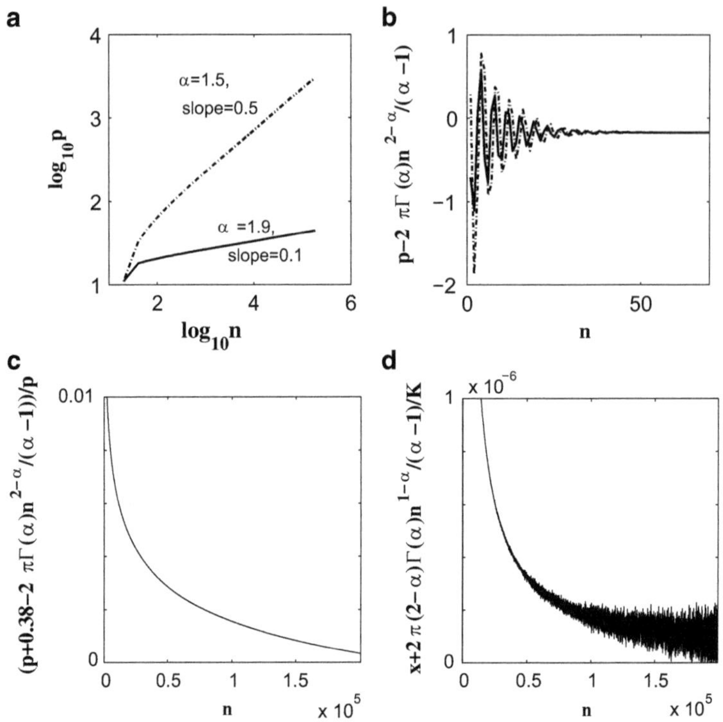

Fig. 14 Evaluation of the behavior of the attracting slow diverging trajectories: (**a**) Momenta for two trajectories with $x_n \approx 2\pi n$ in unbounded space (in this example $K = 2$). The *solid line* is related to a trajectory with $\alpha = 1.9$ and its slope is 0.1. The *dashed line* corresponds to a trajectory with $\alpha = 1.5$ and its slope is 0.5. (**b**) Deviation of momenta from the asymptotic formula for two trajectories with $x_n \approx 2\pi n$ in unbounded space, $\alpha = 1.9$, and $K = 2$. The *dashed line* has $p_0 = 7$ and the *solid* one $p_0 = 6$. (**c**) Relative deviation of the momenta for the trajectories in (**b**) from the asymptotic formula. (**d**) Deviation of the x-coordinates for the trajectories in (**b**) from the asymptotic formula

$$x_{n+1} - x_n = \frac{1}{\Gamma(\alpha)} \sum_{k=1}^{n} (p_{k+1} - p_k) V_\alpha^1(n-k+1) + \frac{p_1}{\Gamma(\alpha)} V_\alpha^1(n+1). \tag{93}$$

For large n the last term is small ($\sim n^{\alpha-2}$) and the following holds

$$\sum_{k=1}^{n} (p_{k+1} - p_k) V_\alpha^1(n-k+1) = 2\pi M \Gamma(\alpha). \tag{94}$$

It can be shown, assuming $p_n \sim n^{2-\alpha}$, that for values of $\alpha > 1.8$ the terms in the last sum with large k are small and in the series representation of $V_\alpha^1(n-k+1)$ only terms of the highest order in k/n can be kept. In this case, Eq. (94) leads to the approximations

$$p_n \approx p_0 + \frac{2\pi M\Gamma(\alpha)n^{2-\alpha}}{\alpha - 1}, \quad x_n \approx -\frac{2\pi M(2-\alpha)\Gamma(\alpha)}{K(\alpha - 1)n^{\alpha-1}}. \tag{95}$$

In the case $K = 2, \alpha = 1.9$ Fig. 14b–d show for two trajectories with $M = 1$ (initial momenta $p_0 = 6$ and $p_0 = 7$) approaching an attracting slow diverging trajectory the deviation from the asymptotic formula Eq. (95) and the relative difference with respect to Eq. (95).

As for $K < K_{c1}$, in the case $K_{c1}(\alpha) < K < K_{c2}(\alpha)$ asymptotic existence and stability of the antisymmetric sink (Sect. 3.3.1) is a result of the gradual transformation of the standard map's elliptic point with the decrease in the order of derivative from $\alpha = 2$ (see Fig. 15). Convergence of trajectories follows Eqs. (89)–(91).

The standard map's antisymmetric $T = 2$ trajectory becomes unstable when $K = 2\pi$ and at the point $(\pi/2, 0)$ in phase space a pair of $T = 2$ trajectories with $x_{n+1} = x_n - \pi$, $p_{n+1} = -p_n$ appears. Numerical simulations of the fractional standard maps (see Fig. 16) show that they demonstrate similar behavior. With the assumption that the RL-standard map Eqs. (61) and (62) have an asymptotic solution

$$p_n = (-1)^n p_l, \quad x_n = x_l - \frac{\pi}{2}[1 - (-1)^n] \tag{96}$$

it can be shown from Eq. (61) that the relationship $p_l = K/2\sin(x_l)$ (Eq. (71)) is valid in this case too.

Numerical simulations similar to those presented in Fig. 13 show that for $K > K_{c2}$ (see Fig. 9) the RL-standard map has the asymptotic behavior

$$p_n = (-1)^n p_l + An^{1-\alpha}, \tag{97}$$

where A is the same for both even and odd values of n.

After substituting (97) in (62) in the limit $n \to \infty$ one can derive

$$\sin(x_l) = \frac{\pi\Gamma(\alpha)}{K V_{\alpha l}}, \tag{98}$$

which has solutions when

$$K > K_{c2} = \frac{\pi\Gamma(\alpha)}{V_{\alpha l}} \tag{99}$$

(see Fig. 9). The value of A can also be calculated:

$$A = \frac{2x_l - \pi}{2\Gamma(2-\alpha)}. \tag{100}$$

Results of the analytic estimations Eqs. (98)–(100) are in good agreement with the direct numerical simulations of the fractional standard maps.

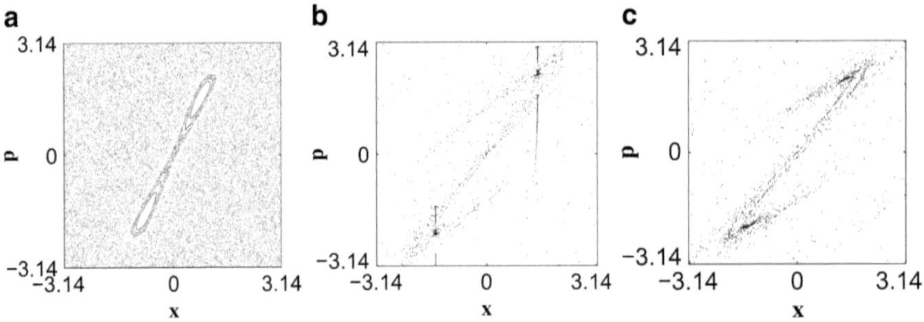

Fig. 15 Stable antisymmetric $x_{n+1} = -x_n$, $p_{n+1} = -p_n$ period $T = 2$ trajectories for $K = 4.5$: (**a**) 1,000 iterations on each of 25 trajectories for the standard map with $K = 4.5$. The only feature is a system of two islands associated with the period two elliptic point. (**b**) RL-standard map stable $T = 2$ antisymmetric sink for $\alpha = 1.8$. 500 iterations on each of 25 trajectories: $p_0 = 0.0001 + 0.08i$, $0 \leq i < 25$. Slow and fast converging trajectories. (**c**) Caputo-standard map stable $T = 2$ antisymmetric sink for $\alpha = 1.8$. 1,000 iterations on each of ten trajectories: $p_0 = -3.1415 + 0.628i$, $0 \leq i < 10$

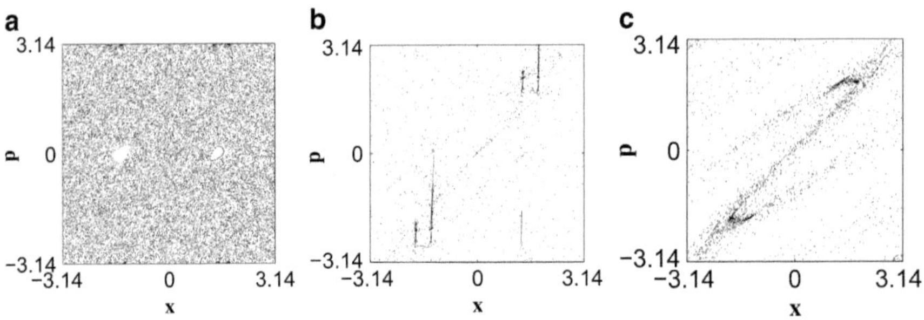

Fig. 16 Stable $x_{n+1} = x_n - \pi$, $p_{n+1} = -p_n$ period $T = 2$ trajectories for $K > K_{c2}$: (**a**) 500 iterations on each of 50 trajectories for the standard map with $K = 6.4$. The main features are two accelerator mode sticky islands around points $(-1.379, 0)$ and $(1.379, 0)$ which define the dynamics. Additional features—*dark spots* at the *top* and the *bottom* of the figure (which are clear on a zoom)—two systems of $T = 2$ tiny islands associated with two $T = 2$ elliptic points: $(1.379, \pi)$, $(1.379 - \pi, -\pi)$ and $(\pi - 1.379, \pi)$, $(-1.379, -\pi)$. (**b**) Two RL-standard map's stable $T = 2$ sinks for $K = 4.5$, $\alpha = 1.71$. 500 iterations on each of 25 trajectories: $p_0 = 0.0001 + 0.08i$, $0 \leq i < 25$. (**c**) Two Caputo-standard map's stable $T = 2$ sinks for $K = 4.5$, $\alpha = 1.71$. 1,000 iterations on each of ten trajectories: $p_0 = -3.1415 + 0.628i$, $0 \leq i < 10$

3.3.4 Cascade of Bifurcations Band

At $K \approx 6.59$ in the standard map $T = 2$ points become unstable and stable $T = 4$ elliptic points appear. Further increase in K results in the period doubling cascade of bifurcations which leads to the disappearance of the corresponding islands of stability in the chaotic sea at $K \approx 6.6344$ (see Sect. 3.1.2). The cusp in Fig. 9a points to a point $\alpha = 2$ and $6.59 < K_* < 6.63$. Inside the band leading to the cusp a new type of attractors, cascade of bifurcations type trajectories (CBTT), appears

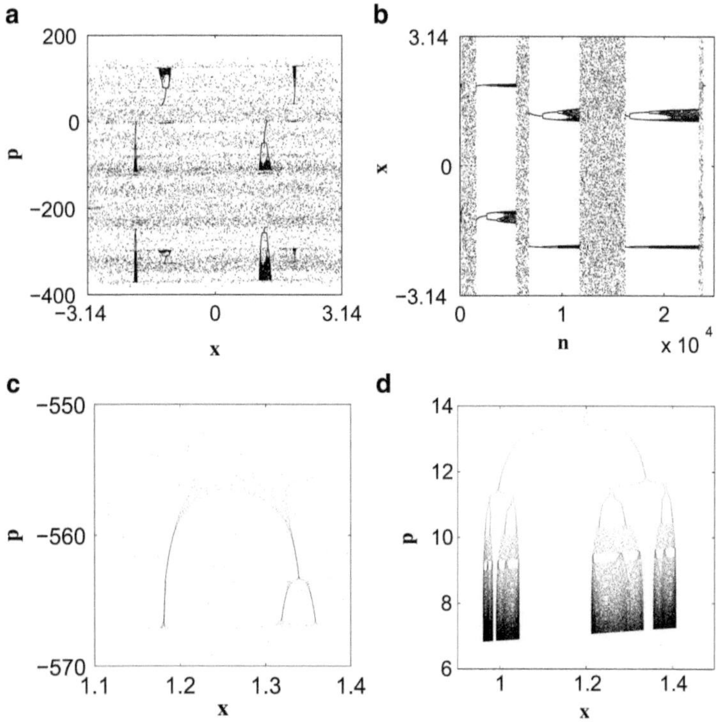

Fig. 17 Cascade of bifurcations type trajectories in the RL-standard map: (**a**) $\alpha = 1.65$, $K = 4.5$; one intermittent trajectory in phase space. (**b**) Time dependence of the coordinate x (x of n) for the case (**a**). (**c**) $\alpha = 1.98$, $K = 6.46$; zoom of a small feature for a single intermittent trajectory in phase space. (**d**) $\alpha = 1.1$, $K = 3.5$; a single trajectory enters the cascade after a few iterations and stays there during 500,000 iterations

(see Fig. 17). The lower boundary of the band approximately corresponds to the transition from the $T = 2$ sink $x_{n+1} = x_n - \pi$, $p_{n+1} = -p_n$ to the $T = 4$ sink and the upper boundary corresponds to the transition to chaos. At $\alpha = 1$ the lower and upper boundaries correspond to the $T = 2 \rightarrow T = 4$ transition and the transition to chaos in the 1D standard map (see Sect. 3.1.1). In CBTT period doubling cascade of bifurcations occurs on a single trajectory with a fixed value of the map parameter. A typical CBTT's behavior is similar to the behavior of trajectories in Hamiltonian dynamics in the presence of sticky islands: occasionally a trajectory enters a CBTT and then leaves it and enters the chaotic sea (Fig. 17a, b). With the decreases in α the relative time trajectories spend in CBTT increases. CBTT are barely distinguishable near the cusp (Fig. 17c) and trajectories spend relatively little time in CBTT. A trajectory enters a CBTT after a few iterations and stays there over the longest computational time we were running our codes—500,000 iterations when α is close to one.

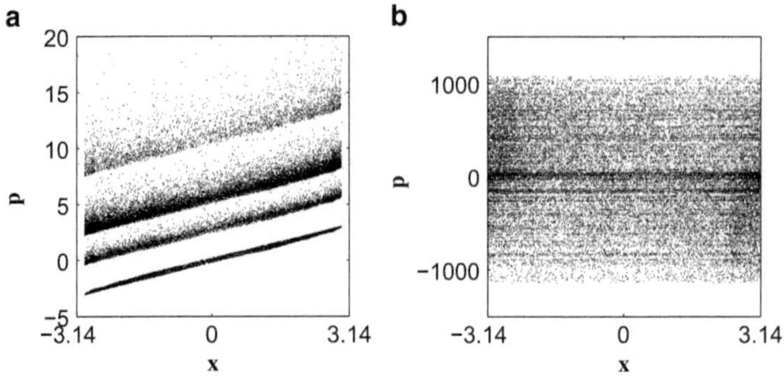

Fig. 18 "Proper" and "improper" attractors in the RL-standard map. 3,000 iterations on ten trajectories with the initial conditions $x_0 = 0$, $p_0 = 0.001 + 1.65i$, $i = 0, 1, \ldots 9$: (**a**) A "proper" chaotic attractor for $K = 4.2$, $\alpha = 1.1$. (**b**) An "improper" chaotic attractor for $K = 4.4$, $\alpha = 1.1$

The CBTT in Fig. 17 were obtained for the RL-standard map. In many cases it is difficult to find CBTT in phase space of the Caputo-standard map but they look almost the same for both fractional maps on the x vs. n plot (see Fig. 17b).

Results of numerical simulations submitted for publication show that not CBTT but inverse (in time) CBTT are present within the CBTT band (from the $T = 2 \to T = 4$ transition to the transition to chaos) of the fractional logistic maps.

3.3.5 More Fractional Attractors

In the one-dimensional standard map with $K > 0$ the "proper" chaotic attractor exists for $3.532 < K < 4.603339$ (see Sect. 3.1.1). This is the interval between the upper boundary of the CBTT band for $\alpha = 1$ and $K = K_{c3}(1)$ in Fig. 9. In the area between $K = K_{c3}(\alpha)$ curve and the upper border of the CBTT band (in Fig. 9) the fractional chaotic attractors are proper (see Fig. 18a) and above $K = K_{c3}(\alpha)$ the entire phase space is chaotic (Fig. 18b).

The standard map has a set of bands for K above $2\pi n$ of the accelerator mode sticky islands in which momentum increases proportionally to the number of iterations n and coordinate increases as n^2 (see Sect. 3.1.2). In the one-dimensional standard map the corresponding bands demonstrate cascades of bifurcations (see Fig. 2b) for $|K|$ above $2\pi|n|$. The acceleration in those bands is zero and x increases proportionally to n (see Sect. 3.1.1).

Accelerator mode attractors in the case $1 < \alpha < 2$ are not fully investigated. The standard map's accelerator mode islands evolve into the accelerator mode (ballistic) attracting sticky trajectories when α is reduced from 2 for the values of K which increase with the decrease in α (Fig. 19b). When the value of α increases from 1, the corresponding ballistic attractors evolve into the cascade of bifurcation type ballistic trajectories (see Fig. 19a) for the values of K which decrease with the increase in α.

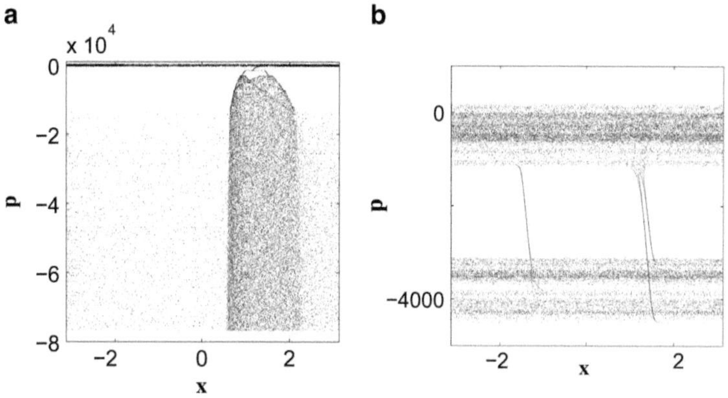

Fig. 19 RL-standard map's accelerator mode attractors. 25,000 iterations on a single trajectory with the initial conditions $x_0 = 0,\ p_0 = 0.1$: (**a**) CBTT-type accelerator mode attractor for $K = 5.7, \alpha = 1.03$. (**b**) Accelerator mode attractor for $K = 7.6, \alpha = 1.97$

This could mean that corresponding features in the one- and two-dimensional maps (at least for $K = 2\pi$) are not connected by the continuous change in α.

3.4 α-Families of Maps ($2 < \alpha < 3$)

Fractional maps for $\alpha > 2$ are not yet investigated. Here we'll present the first results [20] for the RL-standard map.

With $G_K(x) = K\sin(x)$ in Eqs. (36) and (37), the RL-standard map for $2 < \alpha \leq 3$ can be written as

$$\begin{aligned} p^1_{n+1} &= p^1_n - K\sin(x_n), \\ p_{n+1} &= p^1_n + p_n - K\sin(x_n), \quad (\text{mod } 2\pi), \qquad (101) \\ x_{n+1} &= \frac{p_0}{\Gamma(\alpha-1)}(n+1)^{\alpha-2} + \frac{1}{\Gamma(\alpha)}\sum_{k=0}^{n} p^1_{k+1} V^1_\alpha(n-k+1), \quad (\text{mod } 2\pi). \end{aligned}$$

In our simulations we did not find a stable fixed point even for small values of K (see Fig. 20c). Simulations show that for this map there are attractors in the form of the attracting multi-period lines with constant x (see Fig. 20a, b, and d). For most of the values of the map parameters the phase space is highly chaotic.

This case and the transition from the 2D standard map to the 3D standard map is not yet fully investigated.

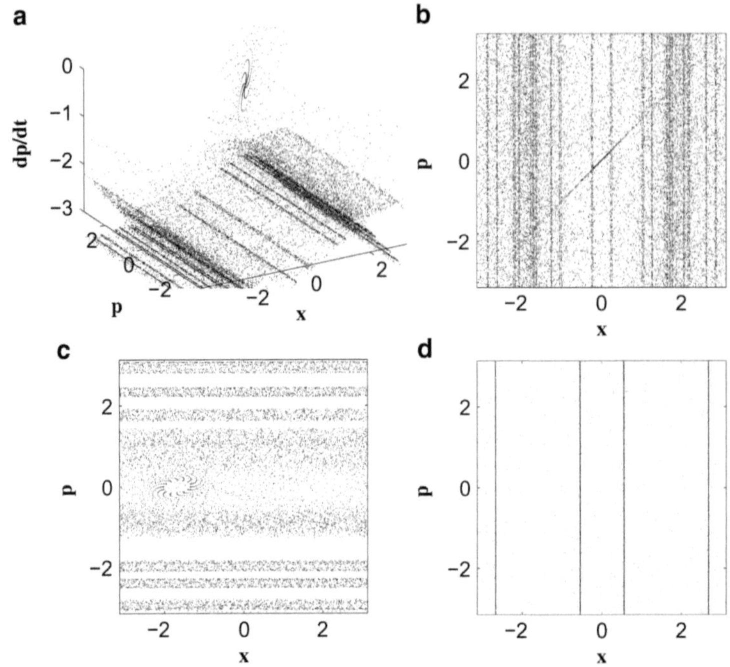

Fig. 20 RL-standard map for $2 < \alpha < 3$: (**a**) 3D phase space for $K = 1$, $\alpha = 2.01$ obtained on a single trajectory with $x_0 = p_0 = 0$ and $p_0^1 = 0.01$. (**b**) Projection of the phase space in (**a**) on the x-y plane. (**c**) Projection of the phase space for $K = 0.2$, $\alpha = 2.01$, $x_0 = p_0 = 0$ on the x-y plane obtained using 20 trajectories with different initial values of p_0^1. (**d**) The same as in (**c**), but for $K = 4$ and $\alpha = 2.9$

4 Conclusion

The systems with long-term memory that are most frequently encountered in nature are systems with power-law memory. In many applications, including biological applications, the exponent in power law, $\sim t^{-\beta}$, is $0 < \beta < 1$. This is true, in particular, for adaptive systems and for viscoelastic properties of human tissues. These systems can be described by nonlinear fractional differential equations with fractional derivatives of the order $\alpha = 1 - \beta$ with $0 < \alpha < 1$. Fractional differential equations can be modeled by discrete nonlinear maps with power-law memory. We studied maps which model fractional differential equations with $0 < \alpha < 2$ and, correspondingly, $-1 < \beta < 1$. Decrease in β and, correspondingly, increase in α means an increase in the memory effects—older states have higher weights in the definition of the present state of a system.

In Sect. 3 we showed that an increase in memory effects leads to more complicated and chaotic behavior. As can be seen in Fig. 6, systems with small α are more stable. At the values of system parameters, corresponding to the periodic behavior and transition to chaos, behavior of such systems follows a well-defined cascade of bifurcations pattern (Fig. 7). This type of evolution may mean a slow adaptation when a system changes its state long after a change in a parameter occurred.

Increase in memory effects with the transition from $0 < \alpha < 1$ to $1 < \alpha < 2$ leads to increased diversity in systems' behavior. Systems with $1 < \alpha < 2$ may demonstrate periodic sinks, attracting slow diverging trajectories, attracting accelerator mode trajectories, chaotic attractors, and cascade of bifurcations and inverse cascade of bifurcations type attracting trajectories. An intermittent cascade of bifurcations type behavior (Fig. 17a, b) may correspond to a scenario of the evolution of chronic diseases, to some mental disorders, or to the evolution of some social systems.

The way in which systems with power-law memory approach fixed and periodic points (Eqs. (89)–(91)) can be used to identify systems with memory in an analysis of experimental data.

Acknowledgements The author expresses his gratitude to V. E. Tarasov for useful discussions, to E. Hameiri and H. Weitzner for the opportunity to complete this work at the Courant Institute, and to V. Donnelly for technical help.

References

1. Anderson JR (1995) Learning and memory: An integrated approach. Wiley, New York
2. Arrowsmith DK, Place CM (1990) An introduction to dynamical systems. Cambridge University Press, Cambridge
3. Ausloos A, Dirickx M (eds) (2006) The logistic map and the route to chaos. Springer, Berlin, Heidelberg, New York
4. Bagley RL, Torvik PJ (1983a) A theoretical basis for the application of fractional calculus to viscoelasticity. J Rheol 27:201–210
5. Bagley RL, Torvik PJ (1983b) Fractional calculus: A different approach to the analysis of viscoelastically damped structures. AIAA J 21:741–748
6. Brauer F, Castillo-Chavez C (2001) Mathematical models in population biology and epidemiology. Springer, New York
7. Caponetto R, Dongola G, Fortuna L, Petras I (2010) Fractional order systems: Modeling and control applications. World Scientific, Singapore
8. Caputo M, Mainardi F (1971a) A new dissipation model based on memory mechanism. Pure Appl Geophys 91:134–147
9. Caputo M, Mainardi F (1971b) Linear models of dissipation in anelastic solids. Riv Nuovo Cimento 1:161–198
10. Cheng S, Clarke EC, Bilston LE (2008) Rheological properties of the tissues of the central nervous system: A review. Med Eng Phys 30:1318–1337
11. Chirikov BV (1979) A universal instability of many dimensional oscillator systems. Phys Rep 52:263–379
12. Coussot C, Kalyanam S, Yapp R, Insana MF (2009) Fractional derivative models for ultrasonic characterization of polymer and breast tissue viscoelasticity. IEEE Trans Ultrason Ferroelectr Freq Contr 56:715–726
13. Craiem DO, Armentano RL (2006) Arterial viscoelasticity: a fractional derivative model. Engineering in Medicine and Biology Society, EMBS '06. 28th Annual International Conference of the IEEE, New York, NY, pp 1098–1101

14. Craiem DO, Rojo FJ, Atienza JM, Guinea GV, Armentano RL (2006) Fractional calculus applied to model arterial viscoelasticity. Lat Am Appl Res 38:141–145
15. Cvitanovic P (1989) Universality in chaos. Adam Hilger, Bristol
16. Doehring TC, Freed AD, Carew EO, Vesely I (2005) Fractional order viscoelasticity of the aortic valve cusp: An alternative to quasilinear viscoelasticity. J Biomech Eng 127:700–708
17. Duck FA (1990) Physical properties of tissue: A comprehensive reference book. Academic Press, San Diego
18. Dullin HR, Meiss JD (2012) Resonances and twist in volume-preserving maps. SIAM J Appl Dyn Syst 11:319–359
19. Edelman M (2011) Fractional standard map: Riemann-Liouville vs. Caputo. Comm Nonlinear Sci Numer Simulat 16:4573–4580
20. Edelman M, (2013) Fractional maps and fractional attractors. Part I: α-families of maps. Discontinuity, Nonlinearity, and Complexity 1:305–324
21. Edelman M, Taieb LA (2013) New types of solutions of non-linear fractional differential equations. In: Almeida A, Castro, L, Speck F-O (eds) Advances in harmonic analysis and operator theory. Operator theory: Advances and applications, vol 229. Springer, Basel, pp 139–155
22. Edelman M, Tarasov VE (2009) Fractional standard map. Phys Lett A 374:279–285
23. Fairhall AL, Lewen GD, Bialek W, de Ruyter van Steveninck RR (2001) Efficiency and Ambiguity in an Adaptive Neural Code. Nature 412:787–792
24. Feigenbaum M (1978) Quantitative universality for a class of non-linear transformations. J Stat Phys 19:25–52
25. Feller W (1968) An introduction to probability theory and its applications. Wiley, New York
26. Fick E, Fick M, Hausmann G (1991) Logistic equation with memory. Phys Rev A 44: 2469–2473
27. Fulinski A, Kleczkowski AS (1987) Nonlinear maps with memory. Phys Scripta 335:119–122
28. Gallas JAC (1993a) Simulating memory effects with discrete dynamical systems. Phys A 195:417–430
29. Gallas JAC (1993b) Simulatingmemory effects with discrete dynamical systems. Phys A 198:339–339 (erratum)
30. Giona M (1991) Dynamics and relaxation properties of complex systems with memory. Nonlinearity 4:911–925
31. Grahovac NM, Zigic MM (2010) Modelling of the hamstring muscle group by use of fractional derivatives. Comput Math Appl 59:1695–1700
32. Hartwich K, Fick E (1993) Hopf bifurcations in the logistic map with oscillating memory. Phys Lett A 177:305–310
33. Hénon M (1969) Numerical study of quadratic area-preserving mappings. Q Appl Math 27:291–312
34. Holm S, Sinkus R (2010) A unifying fractional wave equation for compressional and shear waves. J Acoust Soc Am 127:542–548
35. Hoppensteadt F (1975) Mathematical theories of populations: Demographics, genetics, and epidemics. SIAM, Philadelphia
36. Hoyt K, Castaneda B, Zhang M, Nigwekar P, di SantAgnese A, Joseph JV, Strang J, Rubens DJ, Parker KJ (2008) Tissue elasticity properties as biomarkers for prostate cancer. Canc Biomarkers 4:213–225
37. Kahana MJ (2012) Foundations of human memory. Oxford University Press, New York
38. Kilbas AA, Bonilla B, Trujillo JJ (2000a) Nonlinear differential equations of fractional order in space of integrable functions. Dokl Math 62:222–226
39. Kilbas AA, Bonilla B, Trujillo JJ (2000b) Existence and uniqueness theorems for nonlinear fractional differential equations. Demonstratio Math 33:583–602
40. Kilbas AA, Srivastava HM, Trujillo JJ (2006) Theory and application of fractional differential equations. Elsevier, Amsterdam

41. Kobayashi Y, Watanabe H, Hoshi T, Kawamura K, Fujie MG (2012) Viscoelastic and nonlinear liver modeling for needle insertion simulation. Soft tissue biomechanical modeling for computer assisted surgery. Studies in mechanobiology, tissue engineering and biomaterials, vol 11, Springer, Berlin, Heidelberg, pp 41–67
42. Landford OE (1982) A computer-assisted proof of the Feigenbaum conjectures. Bull Am Math Soc 6:427–434
43. Leopold DA, Murayama Y, Logothetis NK (2003) Very slow activity fluctuations in monkey visual cortex: implications for functional brain imaging. Cerebr Cortex 13:422–433
44. Libertiaux V, Pascon F (2010) Differetial versus integral formulation of fractional hyperviscoelastic constitutive laws for brain tissue modeling. J Comput Appl Math 234:2029–2035
45. Lichtenberg AJ, Lieberman MA (1992) Regular and chaotic dynamics. Springer, Berlin
46. Lomeli HE, Meiss JD (1998) Quadratic volume-preserving maps. Nonlinearity 11:557–574
47. Lundstrom BN, Higgs MH, Spain WJ, Fairhall AL (2008) Fractional differentiation by neocortical pyramidal neurons. Nat Neurosci 11:1335–1342
48. Lundstrom BN, Fairhall AL, Maravall M (2010) Multiple time scale encoding of slowly varying whisker stimulus envelope incortical and thalamic neurons in vivo. J Neurosci 30:5071–5077
49. Mace E, Cohen I, Montaldo G, Miles R (2011) In vivo mapping of brain elasticity in small animals using shear wave imaging. IEEE Trans Med Imag 30:550–558
50. Magin RL (2010) Fractional calculus models of complex dynamics in biological tissues. J Comput Appl Math 59:1586–1593
51. Mainardi F (1994) Fractional relaxation in anelastic solids. J Alloy Comp 211:534–538
52. Mainardi F (2010) Fractional calculus and waves in linear viscoelesticity: An introduction to mathematical models. Imperial College Press, London
53. Mainardi F (2012) An historical perspective on fractional calculus in linear viscoelasticity. Fract Calc Appl Anal 15:712–717
54. Mainardi F, Gorenflo R (2007) Time fractional derivatives in relaxation processes: a tutorial survey. Fract Calc Appl Anal 10:269–308
55. Mariappan YK, Glaser KJ, Ehman RL (2010) Magnetic resonance elastography: A review. Clin Anat 23:497–511
56. May RM (1976) Simple mathematical models with very complicated dynamics. Nature 261:459–467
57. Min W, Luo G, Cherayil BJ, Kou SC, Xie XS (2005) Observation of a power-law memory kernel for fluctuations within a single protein molecule. Phys Rev Lett 94:198302
58. Moser J (1994) On quadratic symplectic mappings. Math Z 216:417–430
59. Nasholm SP, Holm S (2013) On a rractional zener elastic wave equation. Fract Calc Appl Anal 16:26–50
60. Nicolle S, Vezin P, Palierne J-F (2010) A strain-hardening bi-power law for the nonlinear behaviour of biological soft tissues. J Biomech 43:927–932
61. Nicolle S, Noguera L, Paliernea J-F (2012) Shear mechanical properties of the spleen: Experiment and analytical modelling. J Mech Behav Biomed Mater 9:130–136
62. Petras I (2011) Fractional-order nonlinear systems. Springer, Heidelberg
63. Podlubny I (1999) Fractional differential equations. Academic Press, San Diego
64. Prieur F, Holm S (2011) Nonlinear acoustic wave equations with fractional loss operators. J Acoust Soc Am 130:1125–1132
65. Prieur, F., Vilenskiy, G., and Holm, S., 2012, A more fundamental approach to the derivation of nonlinear acoustic wave equations with fractional loss operators. J Acoust Soc Am 132: 2169–2172
66. Rubin DC, Wenzel AE (1996) One hundred years of forgetting: A quantitative description of retention. Psychol Rev 103:743–760
67. Samko SG, Kilbas AA, Marichev OI (1993) Fractional integrals and derivatives theory and applications. Gordon and Breach, New York
68. Schmidt G (1980) Stochasticity and fixed-point transitions. Phys Rev A 22:2849–2854

69. Sommacal L, Melchior P, Oustaloup A, Cabelguen J-M, Ijspeert AJ (2008) Fructional multi-model of the frog gastrocnemius muscle. J Vibration Contr 14:1415–1430
70. Stanislavsky AA (2006) Long-term memory contribution as applied to the motion of discrete dynamical system. Chaos 16:043105
71. Szabo TL, Wu J (2000) A model for longitudinal and shear wave propagation in viscoelastic media. J Acoust Soc Am 107:2437–2446
72. Takeuchi Y, Iwasa Y, Sato K (eds) (2007) Mathematics for life science and medicine. Springer, Berlin, Heidelberg, New York
73. Tarasov VE (2008a) Fractional equations of Curie-von Schweidler and Gauss laws. J Phys Condens Matter 20:145212
74. Tarasov VE (2008b) Universal electromagnetic waves in dielectrics. J Phys Condens Matter 20:175223
75. Tarasov VE (2009a) Differential equations with fractional derivative and universal map with memory. J Phys Math Theor 42:465102
76. Tarasov VE (2009b) Discrete map with memory from fractional differential equation of arbitrary positive order. J Math Phys 50:122703
77. Tarasov VE (2009c) Fractional integro-differential equations for electromagnetic waves in dielectric media. Theor Math Phys 158:355–359
78. Tarasov VE (2011) Fractional dynamics: application of fractional calculus to dynamics of particles, fields, and media. Springer, HEP, New York
79. Tarasov VE, Edelman M (2010) Fractional dissipative standard map. Chaos 20:023127
80. Tarasov VE, Zaslavsky GM (2008) Fractional equations of kicked systems and discrete maps. J Phys Math Theor 41:435101
81. Taylor LS, Lerner AL, Rubens DJ, Parker KJ (2002) A Kelvin-Voight fractional derivative model for viscoelastic characterization of liver tissue. In: Scott EP (ed) ASME international mechanical engineering congress and exposition, New Orleans, LA
82. Toib A, Lyakhov V, Marom S (1998) Interaction between duration of activity and recovery from slow inactivationin mammalian brain Na+ channels. J Neurosci 18:1893–1903
83. Ulanovsky N, Las L, Farkas D, Nelken I (2004) Multiple time scales of adaptation in auditory cortex neurons. J Neurosci 24:10440–10453
84. Vul EB, Sinai YG, Khanin KM (1984) Feigenbaum universality and the thermodynamic formalism. Russ Math Surv 39:1–40
85. Wineman A (2007) Nonlinear viscoelastic membranes. Comput Math Appl 53:168–181
86. Wineman A (2009) Nonlinear viscoelastic solids: A review. Math Mech Solid 14:300–366
87. Wixted JT (1990) Analyzing the empirical course of forgetting. J Exp Psychol Learn Mem Cognit 16:927–935
88. Wixted JT, Ebbesen E (1991) On the form of forgetting. Psychol Sci 2:409–415
89. Wixted JT, Ebbesen E (1997) Genuine power curves in forgetting. Mem Cognit 25:731–739
90. Zaslavsky GM (2008) Hamiltonian chaos and fractional dynamics. Oxford University Press, Oxford
91. Zaslavsky GM, Edelman M (2000) Hierarchical structures in the phase space and fractional kinetics: I. Classical systems. Chaos 10:135–146
92. Zaslavsky GM, Edelman M (2004) Fractional kinetics: From pseudochaotic dynamics to Maxwell's Demon. Phys D 193:128–147
93. Zaslavsky GM, Edelman M, Niyazov BA (1997) Self-similarity, renormalization, and phase space nonuniformity of Hamiltonian chaotic dynamics. Chaos 7:159–181
94. Zeraoulia E, Sprott JC (2010) 2-D quadratic maps and 3-D ODE systems: A rigorous approach. World Scientific, Singapore
95. Zhang M, Nigwekar P, Castaneda B, Hoyt K, Joseph JV, di SantAgnese A, Messing EM, Strang J, Rubens DJ, Parker KJ (2008) Quantitative characterization of viscoelastic properties of human prostate correlated with histology. Ultrasound Med Biol 34:1033–1042
96. Zilany MS, Bruce IC, Nelson PC, Carney LH (2009) A phenomenological model of the synapse between the inner hair cell and auditory nerve: long-term adaptation with power-law dynamics. J Acoust Soc Am 126:2390–2412

Period-1 Motions in a Quadratic Nonlinear Oscillator

Albert C.J. Luo and Bo Yu

Abstract Analytical solutions for period-1 motions in a periodically forced, quadratic nonlinear oscillator are presented through the Fourier series solutions with finite harmonic terms, and the stability and bifurcation analyses of the corresponding period-1 motions are carried out. The parameter map for excitation amplitude and frequency is developed for different period-1 motions. For a better understanding of complex period-1 motions in such a quadratic nonlinear oscillator, trajectories and amplitude spectrums are illustrated numerically.

1 Introduction

To obtain analytical solutions of nonlinear dynamical systems is an important issue for a better understanding of nonlinear behaviors and multiplicity. So far, one cannot find an efficient way to determine complex periodic motions in nonlinear dynamical systems yet. In the nineteenth century, Poincare [1] further developed the perturbation theory for celestial bodies. In 1920, van der Pol [2] used the method of averaging to determine the periodic solutions of oscillation systems in circuits. In 1964, Hayashi [3] used perturbation methods, averaging method, and principle of harmonic balance to determine the approximate solutions of nonlinear oscillators. In 1973, Nayfeh [4] presented the multi-scale perturbation method and applied such a perturbation method for obtaining approximate solutions of periodic motions in nonlinear oscillations of structures (also see, Nayfeh and Mook [5]). In 1997, Luo and Han [6] studied the stability and bifurcations of periodic solutions of Duffing oscillators through the first order harmonic balance method. In 2008, Peng et al. [7] presented the approximate period-1 solution for the Duffing oscillator by the HB3

A.C.J. Luo (✉) • B. Yu
Southern Illinois University Edwardsville, Edwardsville, IL 62026-1805, USA
e-mail: aluo@siue.edu; byu@siue.edu

V. Afraimovich et al. (eds.), *Nonlinear Dynamics and Complexity*, Nonlinear Systems and Complexity 8, DOI 10.1007/978-3-319-02353-3_4,

method and compared with the fourth-order Runge–Kutta method. In 2011, Luo and Huang [8] discussed approximate solutions of periodic motions in nonlinear systems through the harmonic balance method. In 2012, Luo and Huang [9] developed the approximate analytical solutions of period-m motions and chaos. In Luo [10], the methodology and procedure for analytical solutions of periodic motions in general nonlinear systems were presented. Such a method will be used to develop analytical solutions of periodic motions in a quadratic nonlinear system under a periodic excitation. Such an oscillator can be used to describe ship motion under periodic ocean waves. The analytical solutions will include enough harmonic terms to give an appropriate solution of period-1 motion from which the analytical bifurcations and period-m motions can be achieved.

2 Analytical Solutions

As in Luo and Yu [11, 12], consider a periodically forced, nonlinear oscillator as

$$\ddot{x} + \delta\dot{x} + \alpha x + \beta x^2 = Q_0 \cos\Omega t \tag{1}$$

where δ is the linear damping coefficient. α and β are linear and quadratic spring coefficients, respectively. Q_0 and Ω are excitation amplitude and frequency, respectively. In Luo [10], the standard form of Eq. (1) can be written as

$$\ddot{x} + f(x, \dot{x}, t) = 0 \tag{2}$$

where

$$f(\dot{x}, x, t) = \delta\dot{x} + \alpha x + \beta x^2 - Q_0 \cos\Omega t. \tag{3}$$

The analytical solution of period-1 motion for the above equation is

$$x^*(t) = a_0(t) + \sum_{k=1}^{N} b_k(t)\cos(k\Omega t) + c_k(t)\sin(k\Omega t) \tag{4}$$

where $a_0(t)$, $b_k(t)$ and $c_k(t)$ vary slowly with time. The first and second order of derivatives of $x^*(t)$ are

$$\dot{x}(t) = \dot{a}_0(t) + \sum_{k=1}^{N} (\dot{b}_k + k\Omega c_k)\cos(k\Omega t) + (\dot{c}_k - k\Omega b_k)\sin(k\Omega t), \tag{5}$$

$$\ddot{x}(t) = \ddot{a}_0(t) + \sum_{k=1}^{N} \left[\ddot{b}_k + 2\,(k\Omega)\,\dot{c}_k - (k\Omega)^2 b_k\right] \cos\,(k\Omega t) \\ + \left[\ddot{c}_k - 2\,(k\Omega)\,\dot{b}_k - (k\Omega)^2 c_k\right] \sin\,(k\Omega t)\,. \tag{6}$$

Substitution of Eqs. (4)–(6) into Eq. (1) and averaging for the harmonic terms of $\cos(k\Omega t)$ and $\sin(k\Omega t)$ $(k = 0, 1, 2, \ldots)$ gives

$$\ddot{a}_0 + F_0(a_0, \mathbf{b}, \mathbf{c}, \dot{a}_0, \dot{\mathbf{b}}, \dot{\mathbf{c}}) = 0$$

$$\begin{aligned} &\ddot{b}_k + 2(k\Omega)\dot{c}_k - (k\Omega)^2 b_k + F_{1k}(a_0, \mathbf{b}, \mathbf{c}, \dot{a}_0, \dot{\mathbf{b}}, \dot{\mathbf{c}}) = 0 \\ &\ddot{c}_k - 2(k\Omega)\dot{b}_k - (k\Omega)^2 c_k + F_{2k}(a_0, \mathbf{b}, \mathbf{c}, \dot{a}_0, \dot{\mathbf{b}}, \dot{\mathbf{c}}) = 0 \\ &k = 1, 2, \cdots, N \end{aligned} \tag{7}$$

where

$$\begin{aligned} F_0(a_0, \mathbf{b}, \mathbf{c}, \dot{a}_0, \dot{\mathbf{b}}, \dot{\mathbf{c}}) &= \delta \dot{a}_0 + \alpha a_0 + \beta {a_0}^2 + \tfrac{\beta}{2} \sum_{l=1}^{N} ({b_l}^2 + {c_l}^2) \\ F_{1k}(a_0, \mathbf{b}, \mathbf{c}, \dot{a}_0, \dot{\mathbf{b}}, \dot{\mathbf{c}}) &= \delta(\dot{b}_k + c_k k\Omega) + \alpha b_k + 2\beta a_0 b_k + f_{1k} \\ F_{2k}(a_0, \mathbf{b}, \mathbf{c}, \dot{a}_0, \dot{\mathbf{b}}, \dot{\mathbf{c}}) &= \delta(\dot{c}_k - b_k k\Omega) + \alpha c_k + 2\beta a_0 c_k + f_{2k} \end{aligned} \tag{8}$$

and

$$\begin{aligned} f_{1k} &= \beta \sum_{l=1}^{N} \sum_{j=1}^{N} [(b_l b_j + c_l c_j)\delta_{j-l}^k + \tfrac{1}{2}(b_l b_j - c_l c_j)\delta_{j+l}^k] - Q_0 \delta_k^1, \\ f_{2k} &= \beta \sum_{l=1}^{N} \sum_{j=1}^{N} b_l c_j (\delta_{j+l}^k + \delta_{j-l}^k - \delta_{l-j}^k). \end{aligned} \tag{9}$$

Define

$$\begin{aligned} \mathbf{z} &\triangleq (a_0, \mathbf{b}^{\mathrm{T}}, \mathbf{c}^{\mathrm{T}})^{\mathrm{T}} \\ &= (a_0, b_1, \cdots, b_N, c_1, \cdots, c_N)^{\mathrm{T}} \equiv (z_0, z_1, \cdots, z_{2N})^{\mathrm{T}}, \\ \mathbf{z}_1 &= \dot{\mathbf{z}} = (\dot{a}_0, \dot{\mathbf{b}}^{\mathrm{T}}, \dot{\mathbf{c}}^{\mathrm{T}})^{\mathrm{T}} \\ &= (\dot{a}_0, \dot{b}_1, \cdots, \dot{b}_N, \dot{c}_1, \cdots, \dot{c}_N)^{\mathrm{T}} \equiv (\dot{z}_0, \dot{z}_1, \cdots, \dot{z}_{2N})^{\mathrm{T}} \end{aligned} \tag{10}$$

where

$$\mathbf{b} = (b_1, b_2, \cdots, b_N)^{\mathrm{T}} \text{ and } \mathbf{c} = (c_1, c_2, \cdots, c_N)^{\mathrm{T}}. \tag{11}$$

Equation (7) can be expressed in the form of vector field as

$$\dot{\mathbf{z}} = \mathbf{z}_1 \text{ and } \dot{\mathbf{z}}_1 = \mathbf{g}(\mathbf{z}, \mathbf{z}_1) \tag{12}$$

where

$$\mathbf{g}(\mathbf{z}, \mathbf{z}_1) = \begin{pmatrix} -F_0(\mathbf{z}, \mathbf{z}_1) \\ -\mathbf{F}_1(\mathbf{z}, \mathbf{z}_1) - 2\mathbf{k}_1\Omega\dot{\mathbf{c}} + \mathbf{k}_2\Omega^2\mathbf{b} \\ -\mathbf{F}_2(\mathbf{z}, \mathbf{z}_1) + 2\mathbf{k}_1\Omega\dot{\mathbf{b}} + \mathbf{k}_2\Omega^2\mathbf{c} \end{pmatrix} \tag{13}$$

and

$$\begin{aligned} &\mathbf{k}_1 = diag(1, 2, \cdots, N) \text{ and } \mathbf{k}_2 = diag\left(1, 2^2, \cdots, N^2\right) \\ &\mathbf{F}_1 = (F_{11}, F_{12}, \cdots, F_{1N})^{\mathrm{T}} \text{ and } \mathbf{F}_2 = (F_{21}, F_{22}, \cdots, F_{2N})^{\mathrm{T}} \\ &\text{for } N = 1, 2, \cdots, \infty. \end{aligned} \tag{14}$$

Introduce

$$\mathbf{y} \equiv (\mathbf{z}, \mathbf{z}_1) \text{ and } \mathbf{f} = (\mathbf{z}_1, \mathbf{g})^{\mathrm{T}} \tag{15}$$

Equation (12) becomes

$$\dot{\mathbf{y}} = \mathbf{f}(\mathbf{y}). \tag{16}$$

The steady-state solutions for periodic motion in Eq. (1) can be obtained by setting $\dot{\mathbf{y}} = \mathbf{0}$, i.e.,

$$\begin{aligned} &F_0\left(a_0^*, \mathbf{b}^*, \mathbf{c}^*, 0, \mathbf{0}, \mathbf{0}\right) = 0, \\ &\mathbf{F}_1\left(a_0^*, \mathbf{b}^*, \mathbf{c}^*, 0, \mathbf{0}, \mathbf{0}\right) - \Omega^2\mathbf{k}_2\mathbf{b}^* = \mathbf{0}, \\ &\mathbf{F}_2\left(a_0^*, \mathbf{b}^*, \mathbf{c}^*, 0, \mathbf{0}, \mathbf{0}\right) - \Omega^2\mathbf{k}_2\mathbf{c}^* = \mathbf{0}. \end{aligned} \tag{17}$$

The $(2N+1)$ nonlinear equations in Eq. (17) are solved by the Newton–Raphson method. In Luo [10], the linearized equation at the equilibrium point $\mathbf{y}^* = (\mathbf{z}^*, \mathbf{0})^{\mathrm{T}}$ is given by

$$\Delta\dot{\mathbf{y}} = D\mathbf{f}\left(\mathbf{y}^*\right)\Delta\mathbf{y} \tag{18}$$

where

$$D\mathbf{f}\left(\mathbf{y}^*\right) = \partial\mathbf{f}(\mathbf{y})/\partial\mathbf{y}|_{\mathbf{y}^*} = \begin{bmatrix} \mathbf{0}_{(2N+1)\times(2N+1)} & \mathbf{I}_{(2N+1)\times(2N+1)} \\ \mathbf{G}_{(2N+1)\times(2N+1)} & \mathbf{H}_{(2N+1)\times(2N+1)} \end{bmatrix} \tag{19}$$

and

$$\mathbf{G} = \frac{\partial\mathbf{g}}{\partial\mathbf{z}} = \left(\mathbf{G}^{(0)}, \mathbf{G}^{(c)}, \mathbf{G}^{(s)}\right)^{\mathrm{T}} \tag{20}$$

$$\begin{aligned}\mathbf{G}^{(0)} &= (G_0^{(0)}, G_1^{(0)}, \cdots, G_{2N}^{(0)}),\\ \mathbf{G}^{(c)} &= (\mathbf{G}_1^{(c)}, \mathbf{G}_2^{(c)}, \cdots, \mathbf{G}_N^{(c)})^{\mathrm{T}},\\ \mathbf{G}^{(s)} &= (\mathbf{G}_1^{(s)}, \mathbf{G}_2^{(s)}, \cdots, \mathbf{G}_N^{(s)})^{\mathrm{T}}\end{aligned} \tag{21}$$

for $N = 1, 2, \cdots \infty$ with

$$\begin{aligned}\mathbf{G}_k^{(c)} &= (G_{k0}^{(c)}, G_{k1}^{(c)}, \cdots, G_{k(2N)}^{(c)}),\\ \mathbf{G}_k^{(s)} &= (G_{k0}^{(s)}, G_{k1}^{(s)}, \cdots, G_{k(2N)}^{(s)})\end{aligned} \tag{22}$$

for $k = 1, 2, \cdots N$. The corresponding components are

$$\begin{aligned}G_r^{(0)} &= -\alpha\delta_0^r - \beta g_{2r}^{(0)},\\ G_{kr}^{(c)} &= (k\Omega)^2\delta_k^r - \alpha\delta_k^r - \delta k\Omega\delta_{k+N}^r - \beta g_{2kr}^{(c)},\\ G_{kr}^{(s)} &= (k\Omega)^2\delta_{k+N}^r + \delta k\Omega\delta_k^r - \alpha\delta_{k+N}^r - \beta g_{2kr}^{(s)}\end{aligned} \tag{23}$$

where

$$g_{2r}^{(0)} = 2a_0\delta_0^r + b_k\delta_k^r + c_k\delta_{k+N}^r \tag{24}$$

$$\begin{aligned}g_{2kr}^{(c)} = 2b_k\delta_r^0 + 2a_0\delta_k^r + \sum_{i=1}^{N}\sum_{j=1}^{N}[b_j(\delta_{j-i}^k + \delta_{i-j}^k + \delta_{i+j}^k)\delta_i^r\\ + c_j(\delta_{j-i}^k + \delta_{i-j}^k - \delta_{i+j}^k)\delta_{i+N}^r]\end{aligned} \tag{25}$$

$$\begin{aligned}g_{2kr}^{(s)} = 2c_k\delta_0^r + 2a_0\delta_{k+N}^r + \sum_{i=1}^{N}\sum_{j=1}^{N}c_j(\delta_{i+j}^k + \delta_{j-i}^k - \delta_{i-j}^k)\delta_i^r\\ + b_i(\delta_{i+j}^k + \delta_{j-i}^k - \delta_{i-j}^k)\delta_{j+N}^r\end{aligned} \tag{26}$$

for $r = 0, 1, \cdots 2N$.

$$\mathbf{H} = \frac{\partial \mathbf{g}}{\partial \mathbf{z}_1} = \left(\mathbf{H}^{(0)}, \mathbf{H}^{(c)}, \mathbf{H}^{(s)}\right)^{\mathrm{T}} \tag{27}$$

where

$$\begin{aligned}\mathbf{H}^{(0)} &= (H_0^{(0)}, H_1^{(0)}, \cdots, H_{2N}^{(0)}),\\ \mathbf{H}^{(c)} &= (\mathbf{H}_1^{(c)}, \mathbf{H}_2^{(c)}, \cdots, \mathbf{H}_N^{(c)})^{\mathrm{T}},\\ \mathbf{H}^{(s)} &= (\mathbf{H}_1^{(s)}, \mathbf{H}_2^{(s)}, \cdots, \mathbf{H}_N^{(s)})^{\mathrm{T}}\end{aligned} \tag{28}$$

for $N = 1, 2, \cdots \infty$, with

$$\begin{aligned} \mathbf{H}_k^{(c)} &= (H_{k0}^{(c)}, H_{k1}^{(c)}, \cdots, H_{k(2N)}^{(c)}), \\ \mathbf{H}_k^{(s)} &= (H_{k0}^{(s)}, H_{k1}^{(s)}, \cdots, H_{k(2N)}^{(s)}) \end{aligned} \tag{29}$$

for $k = 1, 2, \cdots N$. The corresponding components are

$$\begin{aligned} H_r^{(0)} &= -\delta\delta_0^r, \\ H_{kr}^{(c)} &= -2k\Omega\delta_{k+N}^r - \delta\delta_k^r, \\ H_{kr}^{(s)} &= 2k\Omega\delta_k^r - \delta\delta_{k+N}^r \end{aligned} \tag{30}$$

for $r = 0, 1, \cdots, 2N$.

The corresponding eigenvalues are determined by

$$\left| D\mathbf{f}\left(\mathbf{y}^*\right) - \lambda \mathbf{I}_{2(2N+1)\times 2(2N+1)} \right| = 0. \tag{31}$$

From Luo [10], the eigenvalues of $D\mathbf{f}(\mathbf{y}^*)$ are classified as

$$\left(n_1, n_2, n_3 \middle| n_4, n_5, n_6\right) \tag{32}$$

where n_1 is the total number of negative real eigenvalues, n_2 is the total number of positive real eigenvalues, n_3 is the total number of zero eigenvalues; n_4 is the total pair number of complex eigenvalues with negative parts, n_5 is the total pair number of complex eigenvalues with positive real parts, n_6 is the total pair number of complex eigenvalues with zero real parts. If $\mathrm{Re}(\lambda_k) < 0$ $(k = 1, 2, \cdots, 2(2N+1))$, the approximate steady-state solution $\mathbf{y}^*$ with truncation of $\cos(N\Omega t)$ and $\sin(N\Omega t)$ is stable. If $\mathrm{Re}(\lambda_k) > 0$ $(k \in \{1, 2, \cdots, 2(2N+1)\})$, the truncated approximate steady-state solution is unstable. The corresponding boundary between the stable and unstable solution is given by the saddle-node bifurcation and Hopf bifurcation.

3 Analytic Predictions

The exact steady-state solutions of periodic motions in the nonlinear oscillator can be obtained through the infinite harmonic terms. Unfortunately, it is impossible to compute the exact solution of periodic motions in such oscillator. Thus, one uses the truncated solutions to obtain the approximate solutions of the nonlinear oscillator with enough precision ($A_N \leq \varepsilon$), where the number N is the total number of harmonic terms used in the approximate solution and ε is prescribed the precision (i.e., $\varepsilon = 10^{-8}$). If more terms are used in the Fourier series solution of periodic motions, the better prediction of the periodic motions can be obtained. However, the computational workload will dramatically increase. It is very important that the suitable precision ε is selected. The eigenvalue analysis of such approximate, analytical solutions can be done through dynamics of time-varying coefficients in the Fourier series expression of periodic motion, and the stability and bifurcation analysis can be completed. The equilibrium solution of Eq. (12) can be obtained from Eq. (17) by using Newton–Raphson method, and the stability analysis will be discussed.

The curves of amplitude varying with excitation frequency Ω are illustrated. The harmonic amplitude and phase are defined by

$$A_k \equiv \sqrt{b_k^2 + c_k^2} \text{ and } \varphi_k = \arctan \frac{c_k}{b_k}. \tag{33}$$

The corresponding solution in Eq. (4) becomes

$$x^*(t) = a_0 + \sum_{k=1}^{N} A_k \cos(k\Omega t - \varphi_k). \tag{34}$$

Consider system parameters as

$$\delta = 0.05,\ \ \alpha = 10.0,\ \ \beta = 5.0, Q_0 = 4.5 \tag{35}$$

Without losing generality, as in Luo and Yu [11], the analytical approximate solutions for periodic motion based on two harmonic terms (HB2) are presented first. The constant term a_0, the first and second harmonic term amplitudes A_1 and A_2 are presented in Fig. 1a–c, respectively. In Fig. 1, the stable and unstable solutions of period-1 motion for the quadratic nonlinear oscillator with excitation are predicted analytically. The period-1 motion possesses four branches of solutions for different frequency range. The stability and bifurcation analysis are completed. For the HB2 analytical solutions, eigenvalue analysis provides the possible conditions of stability and bifurcation. The acronyms "HB", "SN," and "UHB" are used to represent the Hopf bifurcation, saddle-node bifurcation, and unstable Hopf bifurcation, respectively. Solid and dashed curves represent the stable and unstable period-1 motions, respectively. The corresponding phase angles versus excitation frequency are presented in Fig. 2a–c. The corresponding stability and bifurcation points are labeled. To consider effects of excitation amplitude, $Q_0 = 1.5, 2.5, 3.5, 4.0, 5.0$ are employed to show the response harmonic frequency–amplitude curves. The stability range for period-1 motions can be observed clearly in Fig. 3a–c. The red dashed curves and the black solid curves are unstable and stable periodic solutions, respectively. The arrow direction represents how the curves changes with excitation frequency. In Fig. 3b, for the second harmonic term, $A_2 < 10^{-3}$ for $\Omega > 6$. For $\Omega < 6$, more harmonic terms should be considered to get a good prediction of period-1 motion. Thus, the 30 harmonic terms (HB30) for period-1 motion will be considered.

As in Luo and Yu [12], the analytical prediction of period-1 motions based on 30 harmonic terms (HB30) is presented in Fig. 4a–f. In Fig. 4a, the constant term of period-1 motion is presented. For $\Omega < 2.0$, there are many branches of

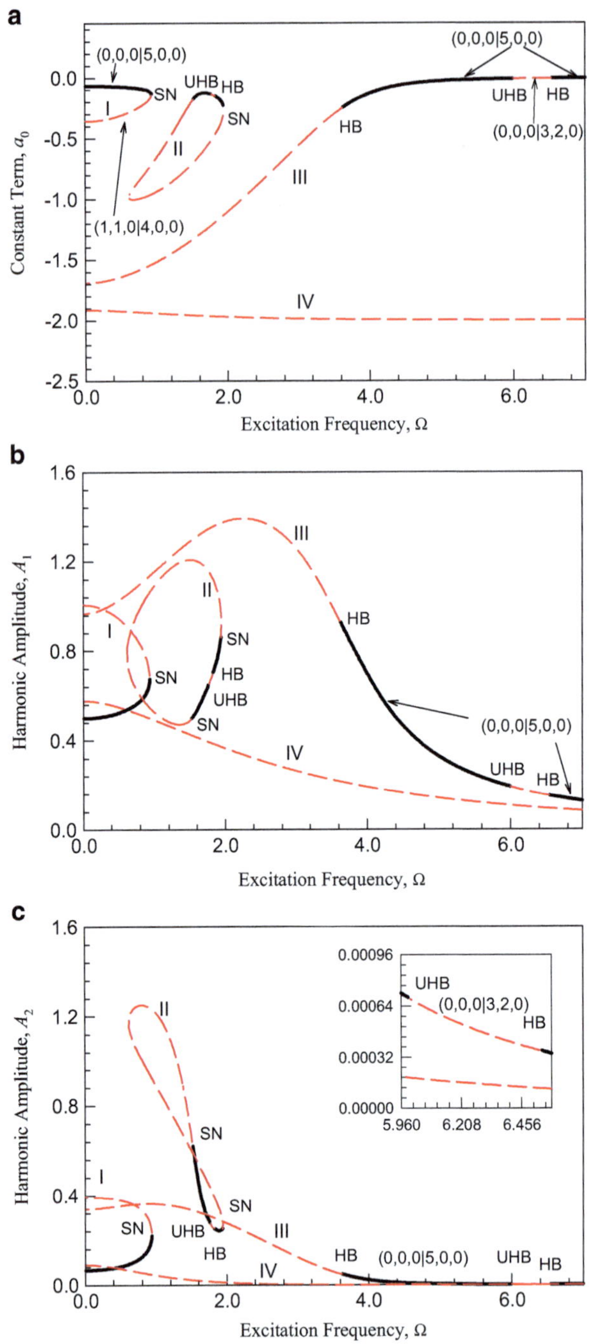

Fig. 1 Analytical prediction of period-1 motions from two harmonic terms (HB2): (**a**) a_0, (**b**) A_1, (**c**) A_2. Parameters ($\delta = 0.05$, $\alpha = 10.0$, $\beta = 5.0$, $Q_0 = 4.5$)

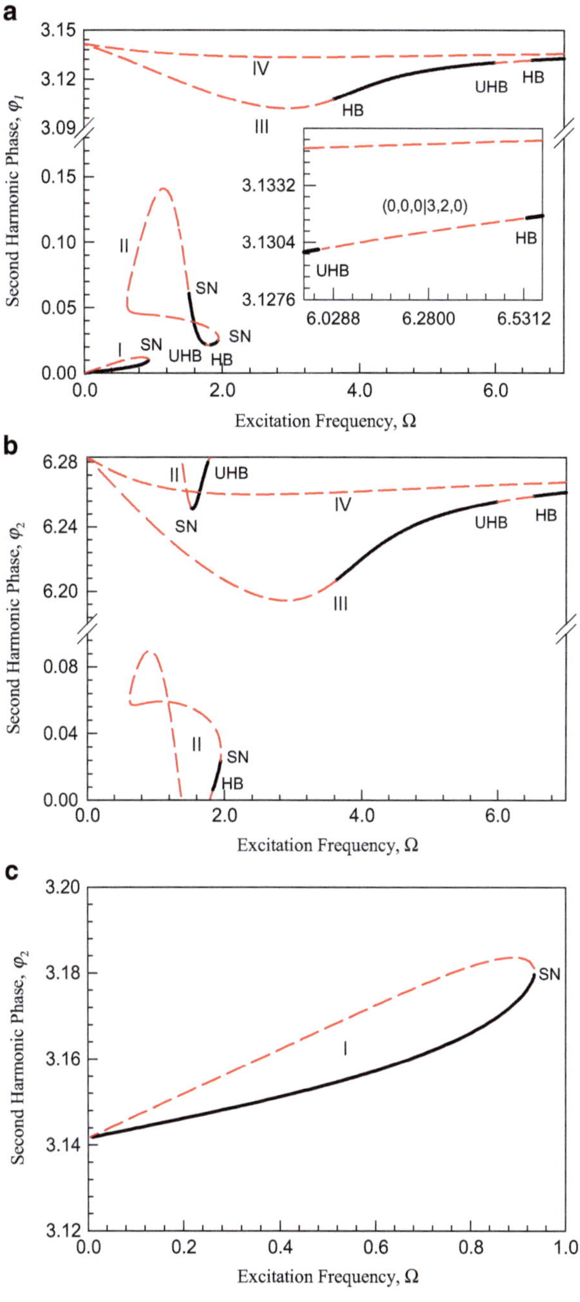

Fig. 2 Analytical prediction of period-1 motions from two harmonic terms (HB2): (**a**) φ_1, (**b**) φ_2, (**c**) a zoomed view of φ_2. Parameters ($\delta = 0.05$, $\alpha = 10.0$, $\beta = 5.0$, $Q_0 = 4.5$)

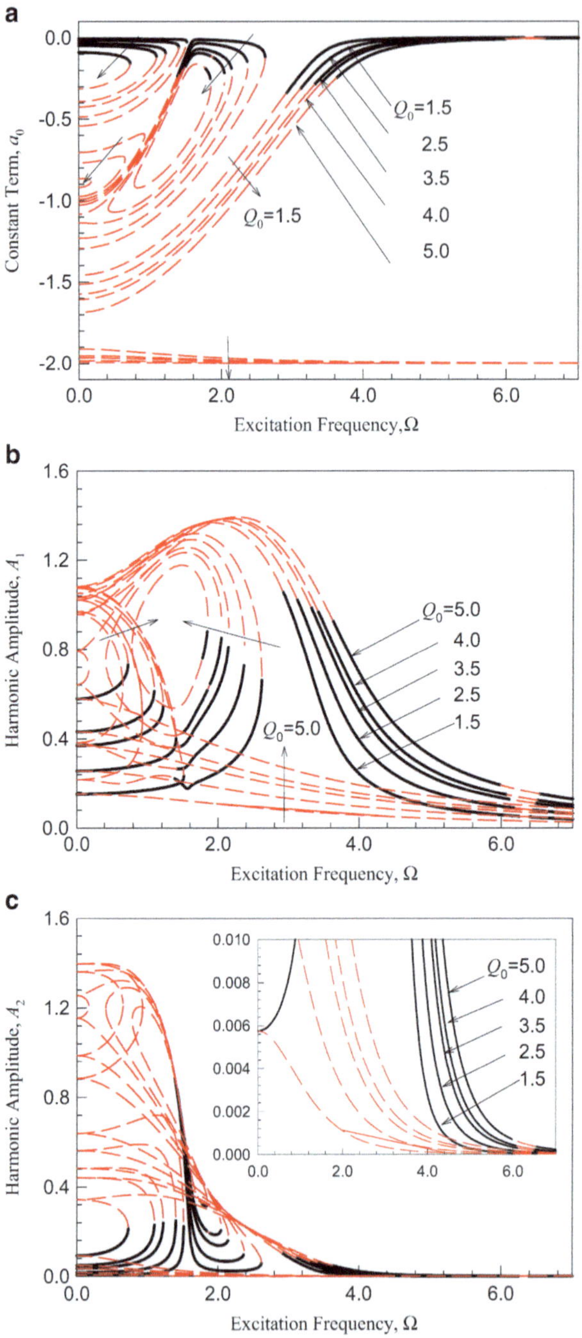

Fig. 3 Excitation amplitudes effects on periodic motions from two harmonic terms (HB2) for $Q_0 = 1.5, 2.5, 3.5, 4.0, 5.0$: (**a**) a_0, (**b**) A_1 and (**c**) A_2. Parameters ($\delta = 0.05$, $\alpha = 10.0$, $\beta = 5.0$, $Q_0 = 4.5$)

period-1 motions and bifurcations, and they have similar structures. The unstable Hopf bifurcation is observed, which is also called the subcritical Hopf bifurcation. The stable Hopf bifurcation is also called the supercritical Hopf bifurcation. For $\Omega > 1.0$, the curves of constant terms varying excitation frequency is very simple. In Fig. 4b, the harmonic amplitude A_1 versus excitation frequency Ω is presented, and the frequency-amplitude curves for $\Omega > 1.0$ are very clearly presented. However, for $\Omega < 2.0$, many branches of period-1 solutions are crowded, but the similar structures for each branch of period-1 motion are observed. In Fig. 4c, the frequency-amplitude curve for harmonic amplitude A_2 is presented. For $\Omega > 5.0$, $A_2 < 3 \times 10^{-3}$ from the zoomed window and for $\Omega \in (2.0,5.0)$, $A_2 \in (10^{-4},0.3)$ is observed. For $\Omega \in (0,2.0)$, $A_2 \sim 10^0$ with many branches of period-1 motion and they become more crowded. The aforementioned three plots are based on linear scale. Once the amplitude quantity level changes with the power laws, it is very difficult to present the changes of the amplitude with excitation frequency. For $\Omega \in (0,2.0)$, the harmonic amplitudes in the frequency-amplitude curves of period-1 motions are more crowded. Thus, the common logarithm scale is used to plot the harmonic amplitude. The overview of amplitude quantity levels can clearly be observed. In Fig. 4d, the harmonic amplitude A_3 is plotted through the common logarithm scale because quantity level changes too big for $\Omega \in (0.0,7.0)$. The braches of period-1 motions in $\Omega \in (0,2.0)$ are obviously presented and the zoomed window for $\Omega \in (2.0,7.0)$ shows that the harmonic amplitude quantity level changes very clear with $A_3 \in (10^{-7},10^{-2})$. Similarly, the harmonic amplitude A_4 varying with excitation frequency is presented in Fig. 4e, and for $\Omega \in (2.0,7.0)$ the harmonic amplitude A_4 lies in the range of $A_4 \in (10^{-10},10^{-3})$. For $\Omega < 2$, the harmonic amplitude A_4 is $A_4 \in (10^{-4},10^0)$. Due to the limitation of the number of pages, the harmonic amplitude A_{30} for $\Omega \in (0.0,1.0)$ is presented in Fig. 4f with $A_{20} \in (10^{-15},10^0)$ and for $\Omega > 1.0$, the harmonic amplitude $A_{30} < 10^{-15}$ can be ignored. For excitation frequency close to zero, more harmonic terms should be included to the analytical expression in the Fourier series solution of period-1 motion. To obtain the entire picture of period-1 motion, the parameter map (Ω, Q_0) is presented in Fig. 5a, b. For $\Omega < 2.5$, the parameter map is zoomed, which shows the similar bifurcation and stability patterns. The solid and dashed curves are for stable and unstable Hopf bifurcations, respectively. The dash–dotted curve is for the saddle-node bifurcation. The notation S^mU^n $(m, n = 0, 1, 2, \cdots)$ represents m stable solutions and m unstable solutions for period-1 motions in the corresponding region.

4 Illustrations

To illustrate the approximate analytical solutions of periodic motion in such a quadratic nonlinear oscillator, numerical simulations are carried out by the symplectic scheme. The initial conditions for numerical simulation are computed from the approximate analytical solutions. The numerical results are depicted by

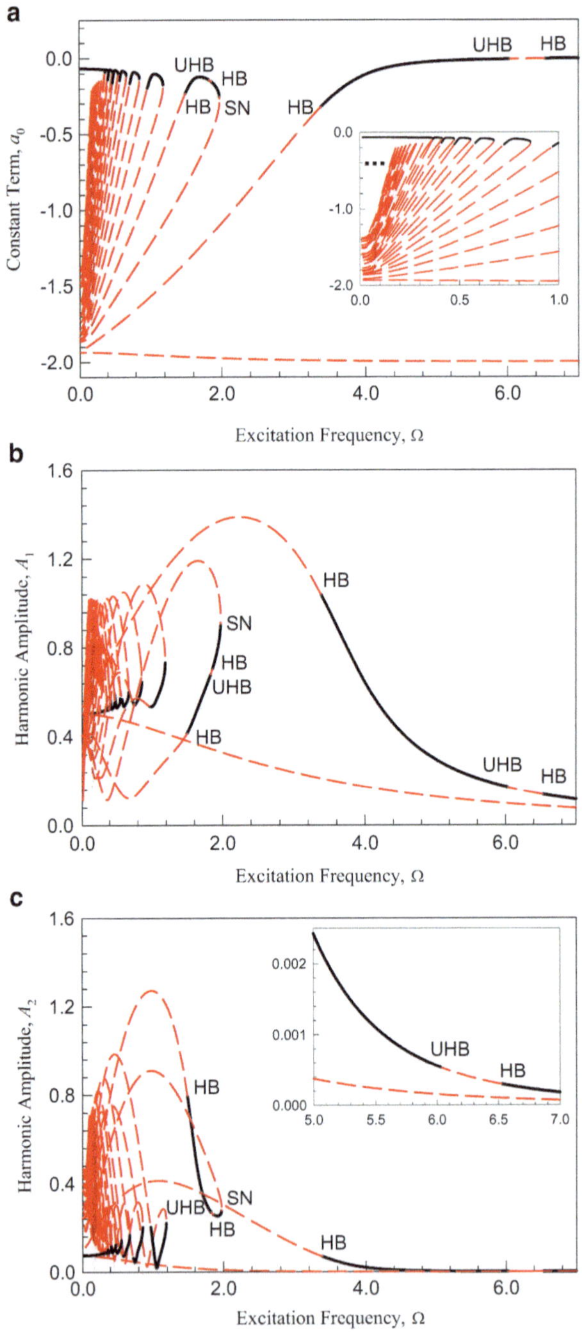

Fig. 4 Analytical prediction of period-1 motions based on 30 harmonic terms (HB30): (**a**) a_0, (**b**)–(**f**) A_k ($k = 1, 2, \cdots, 4, 30$). Parameters ($\delta = 0.05$, $\alpha = 10.0$, $\beta = 5.0$, $Q_0 = 4.5$)

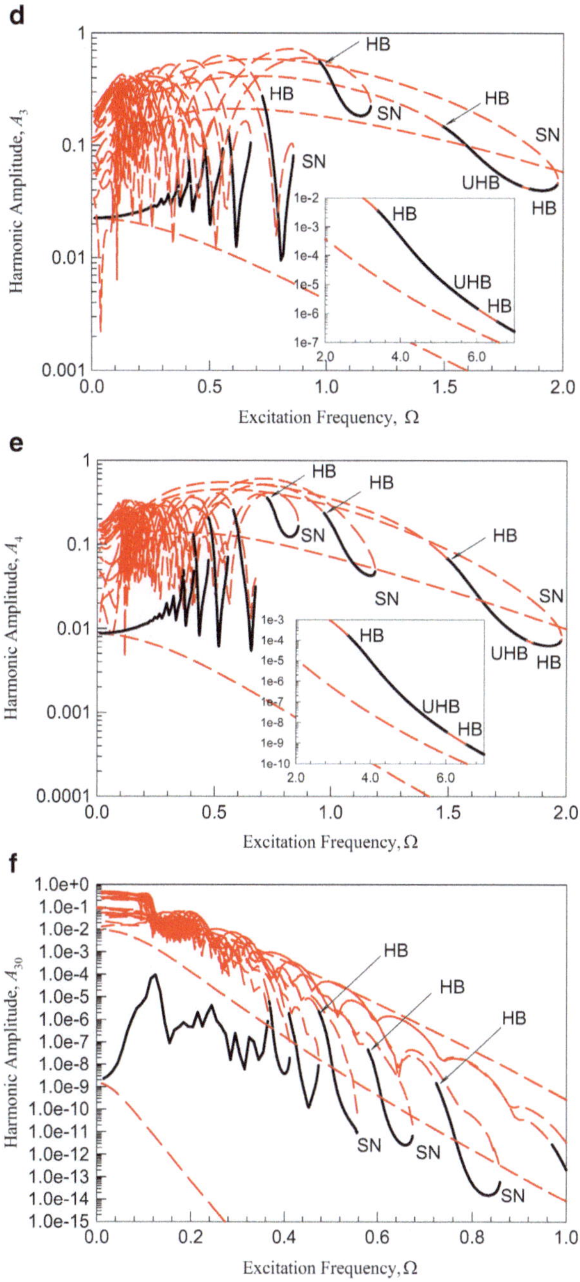

Fig. 4 (continued)

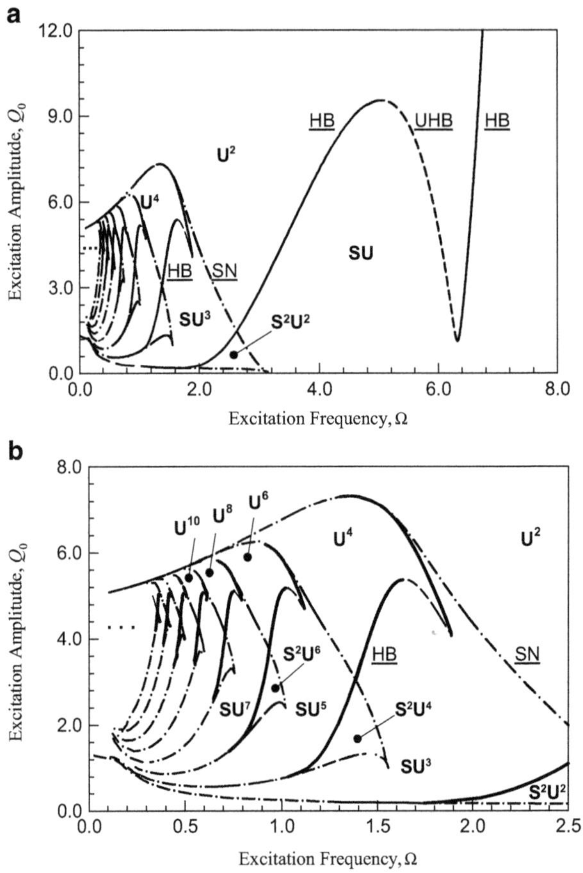

Fig. 5 Analytical parameter map for period-1 motions based on 30 harmonic terms (HB30): (**a**) overview, (**b**) zoomed view. Parameters ($\delta = 0.05$, $\alpha = 10.0$, $\beta = 5.0$)

solid curves, but the analytical solutions are given by red circular symbols. The big filled circular symbol is the initial conditions.

As in Luo and Yu [12], the displacement, velocity, trajectory, and amplitude spectrum of stable period-1 motion are presented in Fig. 6 for $\Omega = 5.8$ with initial conditions ($x_0 \approx -0.197945$, $\dot{x}_0 \approx 0.013202$). This analytical solution is based on four harmonic terms (HB4) in the Fourier series solution of period-1 motion. In Fig. 6a, b, for over 40 periods, the analytical and numerical solutions match very well. In Fig. 6c, analytical and numerical trajectories match very well. In Fig. 6d, the amplitude spectrum versus the harmonic order is presented. $a_0 \approx -9.032271\text{e-}3$, $A_1 \approx 0.189645$, $A_2 \approx 7.213358\text{e-}4$, $A_3 \approx 2.335642\text{e-}6$, $A_4 \approx 6.654057\text{e-}9$. The second harmonic term ($A_2 \sim 10^{-3}$) and higher order harmonic terms are very small and ignorable. Such solution can be easily obtained even if the perturbation method or traditional harmonic balance is used.

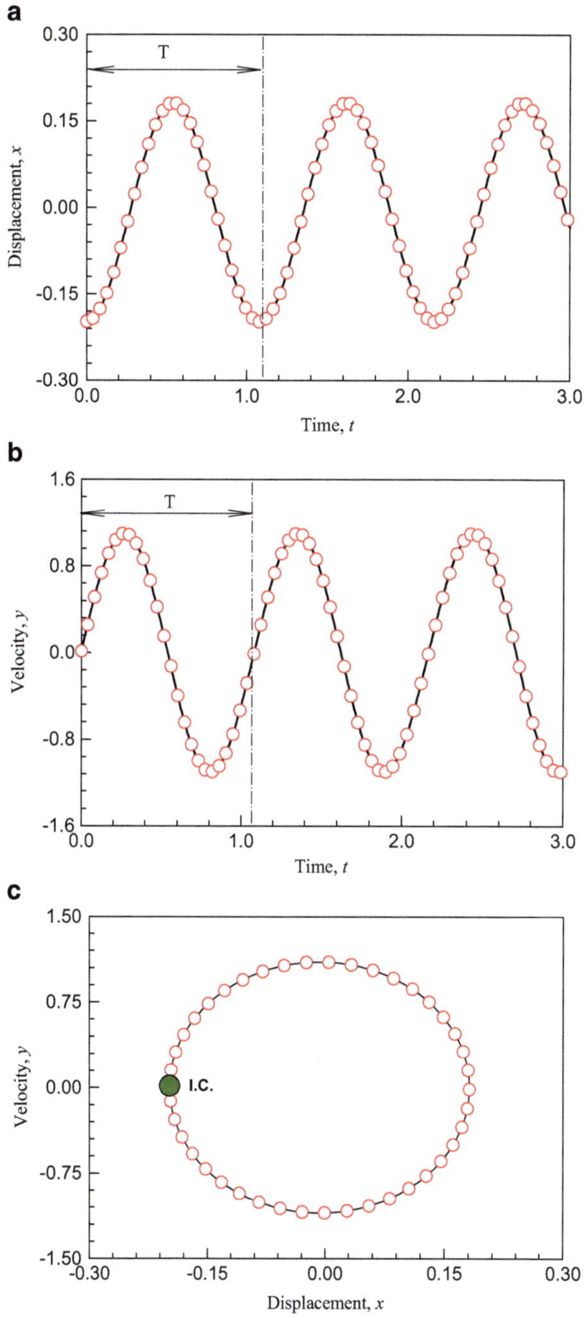

Fig. 6 Analytical and numerical solutions of stable period-1 motion based on the 4 harmonic terms (HB4): (**a**) displacement, (**b**) velocity, (**c**) phase plane, and (**d**) amplitude spectrum. Initial condition ($x_0 \approx -0.197945$, $\dot{x}_0 \approx 0.013202$). Parameters ($\delta = 0.05$, $\alpha = 10.0$, $\beta = 5.0$, $Q_0 = 4.5$, $\Omega = 6.5$)

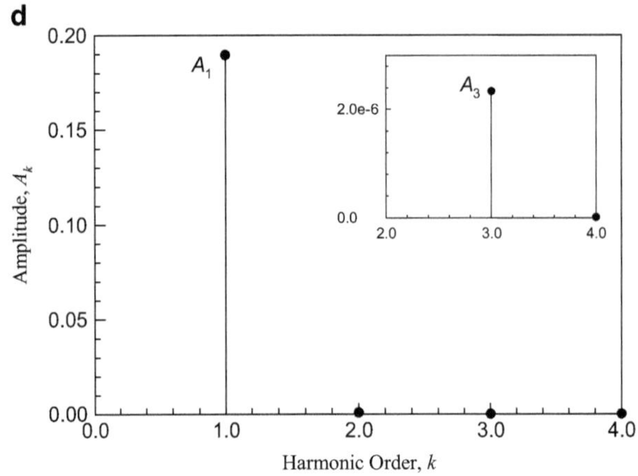

Fig. 6 (continued)

Table 1 Input data for numerical illustrations ($\delta = 0.05$, $\alpha = 10.0$, $\beta = 5.0$, $Q_0 = 4.5$)

Figure number	Ω	Initial condition $x_0, \dot{x}_0$	Type	Harmonics terms
Figure 7a, b	3.6	(−1.801123, 0.092939)	P-1	HB7 (stable)
Figure 7c, d	1.101	(0.155338, −2.175708e−3)	P-1	HB14 (stable)
Figure 7e, f	0.98	(−0.245784, 0.098412)	P-1	HB21 (stable)
Figure 8a, b	0.735	(0.815054, −0.196194)	P-1	HB21 (stable)
Figure 8c, d	0.6	(0.137234, 0.181059)	P-1	HB21 (stable)
Figure 8e, f	0.477	(0.781392, −0.2295502)	P-1	HB32 (stable)

Since there are many branches of period-1 motions, only the trajectory and amplitude spectrums are plotted to look into regularity and complexity of period-1 motions. From the bifurcation trees, period-2 and period-4 motions will be presented through trajectories and amplitude spectrums. The input data for numerical simulations are presented in Table 1. In Fig. 7a–f, the trajectories and harmonic amplitude spectrums of period-1 motions are presented for $\Omega = 3.6, 1.101,\ 0.98$. In Fig. 7a, b, the seven harmonic terms (HB7) are used in the analytical solution of period-1 motion for $\Omega = 3.6$. The analytical and numerical simulations match very well with a cycle. For this periodic motion, the second order harmonic term becomes more important except the first harmonic term. The harmonic amplitude distributions are $a_0 = -0.2280$, $A_1 \sim 10^0$, $A_2 \sim 5 \times 10^{-2}$, $A_3 \sim 2 \times 10^{-3}$, $A_4 \sim 7 \times 10^{-5}$, $A_5 \sim 2 \times 10^{-6}$, $A_6 \sim 8 \times 10^{-8}$ and $A_7 \sim 2.5 \times 10^{-9}$. The second and third harmonic terms will have relatively important contributions on the analytical solution of such a period-1 motion. In Fig. 7c, d, the 14 harmonic terms (HB14) are employed in the analytical solution of period-1 motion for $\Omega = 1.101$. The analytical and numerical simulations match very well with two cycles, which is on the second solution branch of

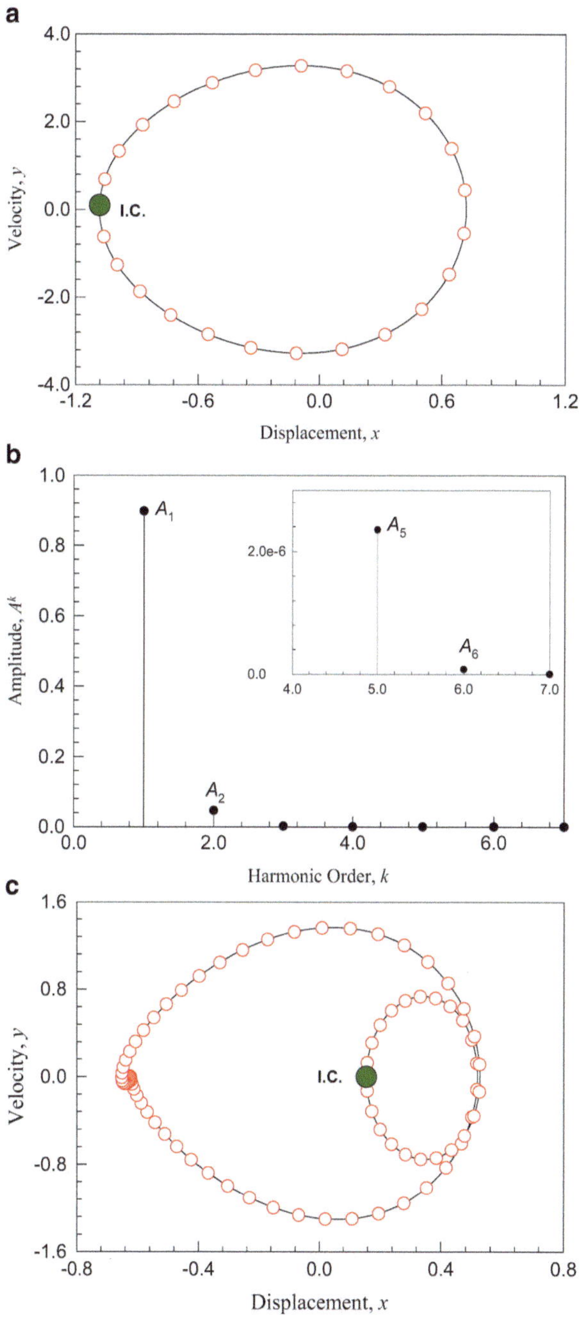

Fig. 7 Phase plane and amplitude spectrums of stable period-1 motions, (**a**, **b**): $\Omega = 3.6$ with ($x_0 \approx -1.081123$, $\dot{x}_0 \approx 0.092939$, HB7). (**c**, **d**) $\Omega = 1.101$ with ($x_0 \approx 0.155338$, $\dot{x}_0 \approx -2.175708\text{e}-3$, HB14), (**e**, **f**): $\Omega = 0.98$ with ($x_0 \approx -0.245784$, $\dot{x}_0 \approx 0.098412$, HB21). Parameters ($\delta = 0.05$, $\alpha = 10.0$, $\beta = 5.0$, $Q_0 = 4.5$)

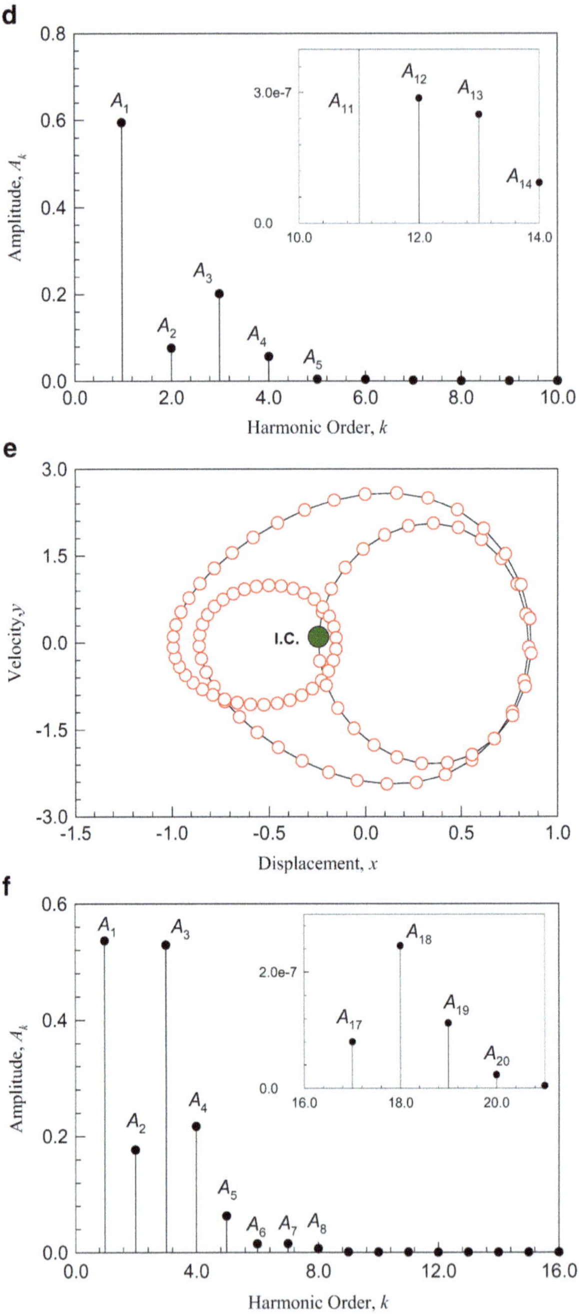

Fig. 7 (continued)

period-1 motion. The harmonic amplitude distributions are $a_0 \approx -0.106346$, $A_1 \sim 6 \times 10^{-1}$, $A_2 \sim 7.6 \times 10^{-2}$, $A_3 \sim 2 \times 10^{-1}$, $A_4 \quad 6 \times 10^{-2}$, $A_5 \sim 4 \times 10^{-3}$, $A_6 \sim 3 \times 10^{-3}$, $A_7 \sim 1.4 \times 10^{-3}$, $A_8 \sim 2.2 \times 10^{-4}$, $A_9 \sim 2.4 \times 10^{-5}$, $A_{10} \sim 2.2 \times 10^{-5}$, $A_{11} \sim 5.3 \times 10^{-6}$, $A_{12} \sim 2.9 \times 10^{-7}$, $A_{13} \sim 2.5 \times 10^{-7}$, $A_{14} \sim 9.3 \times 10^{-8}$. In Fig. 7e, f, the 21 harmonic terms (HB21) are used in the analytical solution of period-1 motion for $\Omega = 0.98$. The analytical and numerical simulations match very well with three cycles, which is still on the second solution branch of period-1 motions. The harmonic amplitude distributions are $a_0 \approx -0.178142$, $A_{1,3} \sim 5 \times 10^{-1}$, $A_2 \sim 1.8 \times 10^{-1}$, $A_4 \sim 2.2 \times 10^{-1}$, $A_5 \sim 6 \times 10^{-2}$, $A_{6,7} \sim 1.5 \times 10^{-2}$, $A_8 \sim 6.3 \times 10^{-3}$, $A_9 \sim 8.3 \times 10^{-4}$, $A_{10} \sim 4.5 \times 10^{-4}$, $A_{11} \sim 3.5 \times 10^{-4}$, $A_{12} \sim 1.1 \times 10^{-4}$, $A_{13} \sim 8 \times 10^{-6}$, $A_{14} \sim 1.2 \times 10^{-5}$, $A_{15} \sim 6.6 \times 10^{-6}$, $A_{17} \sim 1.7 \times 10^{-8}$, $A_{18} \sim 2.5 \times 10^{-7}$, $A_{19} \sim 1.1 \times 10^{-7}$, $A_{20} \sim 2.3 \times 10^{-8}$, $A_{21} \sim 3.8 \times 10^{-9}$. From the amplitude distribution, it is observed that many higher order harmonic terms contribute significantly on the period-1 motion.

To further look into the complexity of period-1 motion, in Fig. 8a–f, the trajectories and harmonic amplitude spectrums of period-1 motions are presented for $\Omega = 0.735, 0.6, 0.477$. The three period-1 motions are on three different branches of period-1 motions. The analytical and numerical solutions match very well for the three period-1 motions. The trajectory of period-1 motion with 21 harmonic terms (HB21) in the Fourier series solution for $\Omega = 0.735$ has four cycles in Fig. 8a. The main amplitude distributions for $\Omega = 0.735$ in Fig. 8b are $a_0 \approx -0.141771$, $A_1 \sim 5.5 \times 10^{-1}$, $A_2 \sim 4.9 \times 10^{-2}$, $A_3 \sim 2.3 \times 10^{-1}$, $A_4 \sim 3.3 \times 10^{-1}$, $A_5 \sim 2.3 \times 10^{-1}$, $A_6 \sim 6.7 \times 10^{-2}$, $A_7 \sim 1.0 \times 10^{-2}$. For very higher order harmonic terms, $A_{17} \sim 5.1 \times 10^{-7}$, $A_{18} \sim 2.6 \times 10^{-6}$, $A_{19} \sim 3.9 \times 10^{-6}$, $A_{20} \sim 1.9 \times 10^{-6}$, $A_{21} \sim 4.1 \times 10^{-7}$. In Fig. 8c, the trajectory of period-1 motion with 21 harmonic terms (HB21) in the Fourier series motion with 21 harmonic terms solution for $\Omega = 0.6$ experiences five cycles. In Fig. 8d, the main amplitude distributions for $\Omega = 0.6$ are $a_0 \approx -0.095182$, $A_1 \sim 5.4 \times 10^{-1}$, $A_2 \sim 7.5 \times 10^{-2}$, $A_3 \sim 4.9 \times 10^{-2}$, $A_4 \sim 1.4 \times 10^{-1}$, $A_{5,6} \sim 1.5 \times 10^{-1}$, $A_7 \sim 4.2 \times 10^{-2}$. For higher order harmonic terms, $A_{16} \sim 1.6 \times 10^{-5}$, $A_{17} \sim 2.9 \times 10^{-5}$, $A_{18} \sim 2.4 \times 10^{-5}$, $A_{19} \sim 8.5 \times 10^{-6}$, $A_{20} \sim 2.1 \times 10^{-6}$, $A_{21} \sim 9.6 \times 10^{-7}$. In Fig. 8e, the trajectory of period-1 motion with 32 harmonic terms (HB32) in the Fourier solution for $\Omega = 0.477$ possesses six cycles. In Fig. 8f, the main amplitude distributions for $\Omega = 0.477$ are $a_0 \approx -0.075150$, $A_1 \sim 5.8 \times 10^{-1}$, $A_2 \sim 9.8 \times 10^{-2}$, $A_3 \sim 1.0 \times 10^{-1}$, $A_4 \sim 1.9 \times 10^{-1}$, $A_5 \sim 2.4 \times 10^{-1}$, $A_6 \sim 7.3 \times 10^{-2}$, $A_7 \sim 2.8 \times 10^{-1}$, $A_8 \sim 1.6 \times 10^{-1}$, $A_9 \sim 1.7 \times 10^{-2}$. For the zoomed window with higher order harmonic amplitudes, $A_{26} \sim 2 \times 10^{-6}$, $A_{27} \sim 3.8 \times 10^{-6}$, $A_{28} \sim 9.1 \times 10^{-5}$, $A_{29} \sim 2.5 \times 10^{-6}$, $A_{30} \sim 1.8 \times 10^{-6}$, $A_{31} \sim 9.1 \times 10^{-7}$, $A_{32} \sim 4.8 \times 10^{-7}$. The complexity of period-1 motions is strongly dependent on the harmonic amplitude contributions. With reduction of excitation frequency, more harmonic terms should be included in the Fourier series solution of period-1 motion.

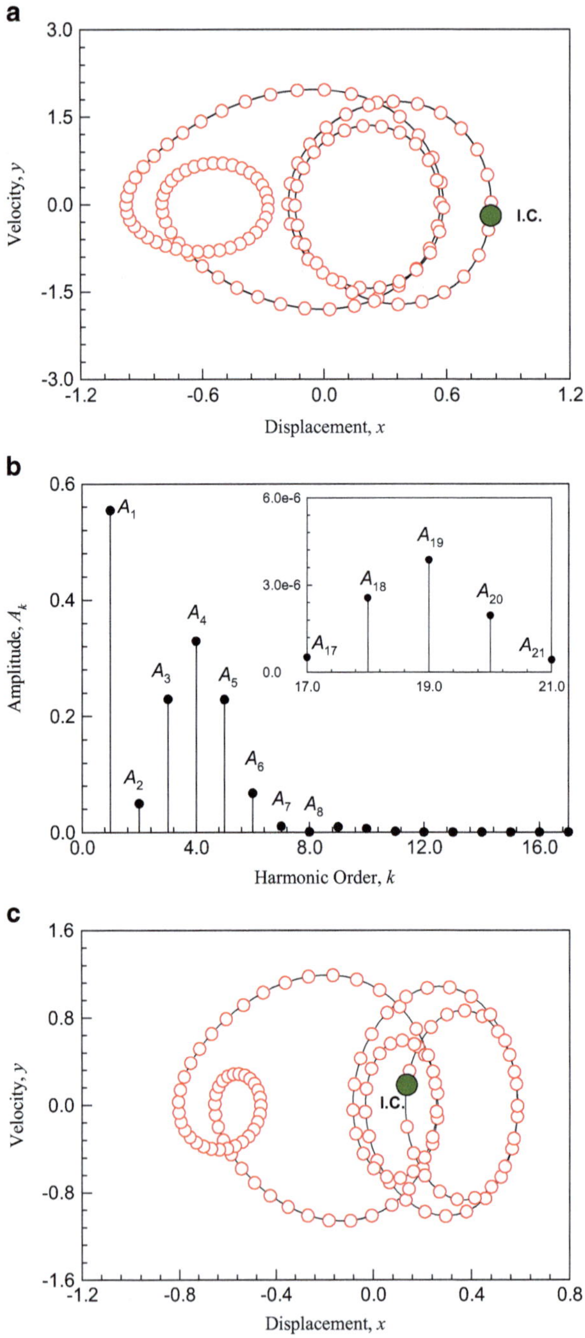

Fig. 8 Phase plane and amplitude spectrums of stable period-1 motions for lower frequency: (**a**, **b**) $\Omega = 0.735$ with ($x_0 \approx 0.815054$, $\dot{x}_0 \approx -0.196194$, HB21). (**c**, **d**): $\Omega = 0.6$ with ($x_0 \approx 0.137234$, $\dot{x}_0 \approx 0.181059$, HB21), (**e**, **f**): $\Omega = 0.477$ with ($x_0 \approx 0.781392$, $\dot{x}_0 \approx -0.229502$, HB32). Parameters ($\delta = 0.05$, $\alpha = 10.0$, $\beta = 5.0$, $Q_0 = 4.5$)

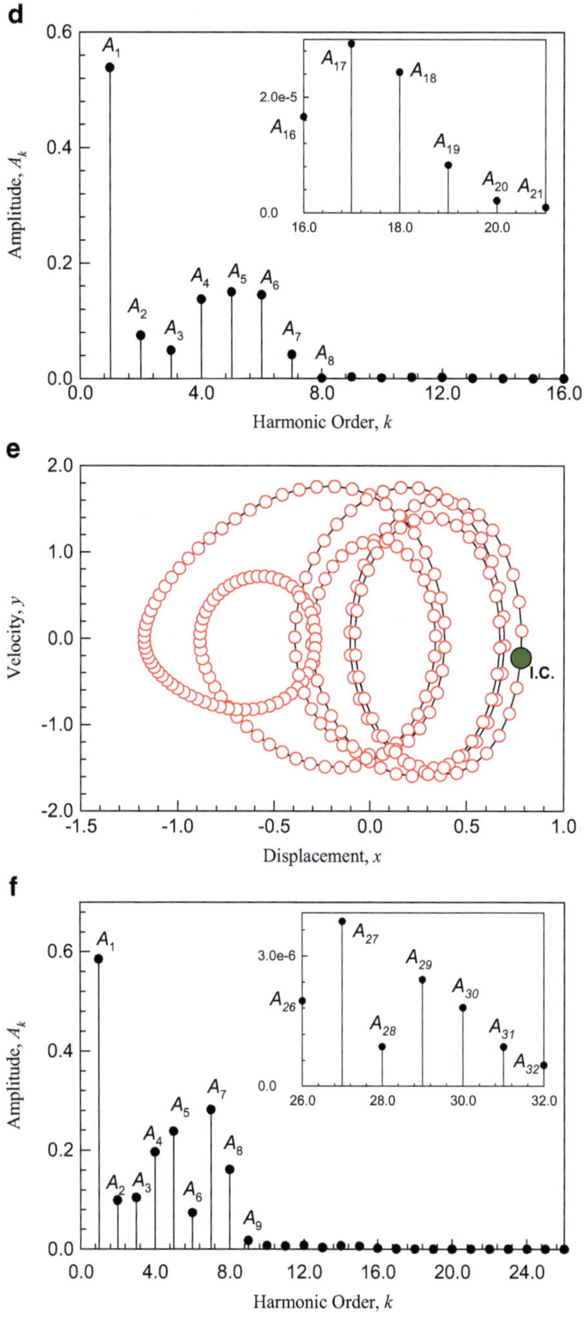

Fig. 8 (continued)

References

1. Poincare H (1899) Methodes Nouvelles de la Mecanique Celeste, vol 3. Gauthier-Villars, Paris
2. van der Pol B (1920) A theory of the amplitude of free and forced triode vibrations. Radio Rev 1:701–710, 754–762
3. Hayashi G (1964) Nonlinear oscillations in physical systems. McGraw-Hill Book Company, New York
4. Nayfeh AH (1973) Perturbation methods. Wiley, New York
5. Nayfeh AH, Mook DT (1979) Nonlinear oscillation. Wiley, New York
6. Luo ACJ, Han RPS (1997) A quantitative stability and bifurcation analyses of a generalized Duffing oscillator with strong nonlinearity. J Franklin Inst 334B:447–459
7. Peng ZK, Lang ZQ, Billings SA, Tomlinson GR (2008) Comparisons between harmonic balance and nonlinear output frequency response function in nonlinear system analysis. J Sound Vib 311:56–73
8. Luo ACJ, Huang JZ (2012) Approximate solutions of periodic motions in nonlinear systems via a generalized harmonic balance. J Vib Control 18:1661–1674
9. Luo ACJ, Huang JZ (2012) Analytical dynamics of period-*m* flows and chaos in nonlinear systems. Int J Bifurcat Chaos 22 (Article No. 1250093) (29 p)
10. Luo ACJ (2012) Continuous dynamical systems. HEPL&H Scientific, Beijing, Glen Carbon
11. Luo ACJ, Yu B (2013) Analytical solutions for stable and unstable period-1 motions in a periodically forced oscillator with quadratic nonlinearity. ASME J Vib Acoust 135 (Article No.: 032501)
12. Luo ACJ, Yu B (2013) Complex period-1 motions in a periodically forced, quadratic nonlinear oscillator. J Vib Control. doi:10.1177/1077546313490525 (in press)

Map-Based Approach to Problems of Spiking Neural Network Dynamics

Oleg V. Maslennikov and Vladimir I. Nekorkin

1 Introduction

Mathematical modeling of phenomena in living systems by using discrete-time systems has a long history. In particular, in the 1940s N. Wiener and A. Rosenblueth developed a cellular automaton system for modeling the propagation of excitation pulses in the cardiac tissue. Cellular automata are regular lattices of elements (cells), each having a finite number of specific states. These states are updated synchronously at discrete time moments, according to some fixed rule. Recently, a new class of discrete-time systems has aroused considerable interest for studying cooperative processes in large-scale neural networks: systems of the coupled nonlinear maps. The state of a map varies at discrete time moments as a cellular automaton, but unlike the latter, takes continuous values. Map-based models hold certain advantages over continuous-time models, i.e. differential equation systems. For example, for reproducing oscillatory properties in continuous-time systems one needs at least two dimensions, and at least three dimensions for chaotic behavior. In discrete time both types of dynamics can be described even in a one-dimensional map. Such a benefit is especially important when modeling complex activity regimes of individual neurons as well as large-scale neural circuits composed of various structural units interacting with each other. For example, to simulate in continuous time the regime of chaotic spike-bursting oscillations, one of the key neural behaviors, one needs to have at least a three-dimensional system of nonlinear differential equations. On the other hand, there are discrete-time two-dimensional systems [1, 2] adequately reproducing this oscillatory activity as well as many other dynamical regimes. For example, the model of Chialvo [3] allows to simulate, among other things, the so-called normal and supernormal excitability. The model

---

O.V. Maslennikov • V.I. Nekorkin (✉)
Institute of Applied Physics of the Russian Academy of Sciences, Nizhny Novgorod, Russia
e-mail: vnekorkin@neuron.appl.sci-nnov.ru

V. Afraimovich et al. (eds.), *Nonlinear Dynamics and Complexity*, Nonlinear Systems and Complexity 8, DOI 10.1007/978-3-319-02353-3_5,

of Rulkov [4, 5] has several modifications, one of which is adjusted to simulate different spiking and bursting oscillatory regimes, the other can generate the so-called sub-threshold oscillations, i.e., small-amplitude oscillations below a threshold of excitability. Here we propose the authors' model [6–9] of neural activity in the form of a two-dimensional map and describe some activity modes which it can reproduce. We show that the model is fairly universal and generates many regimes of neuronal electrical activity. Next, we present the results of modeling the dynamics of a complex neural structure, the olivo-cerebellar system (OCS) of vertebrates, using our basic discrete-time model of neural activity.

2 Discrete-Time Model of Neural Activity

Consider the following two-dimensional map [6]:

$$x_{n+1} = x_n + F(x_n) - \beta H(x_n - d) - y_n, \tag{1a}$$

$$y_{n+1} = y_n + \varepsilon (x_n - J). \tag{1b}$$

Here the variable x describes qualitatively the membrane potential dynamics of a nerve cell, y is responsible for the total effect of all ionic currents (the so-called recovery variable). The positive parameter ε determines a time scale of the recovery variable, the parameters β, d, J control the shape of a signal generated. Note that the model is based on a discrete version of the FitzHugh–Nagumo model, well known in neurodynamics, with a cubic nonlinearity $F(x)$ and the Heaviside step function $H(x)$ additionally added:

$$F(x) = x(x-a)(1-x), \quad H(x) = \begin{cases} 1, & x \geq 0, \\ 0, & x < 0. \end{cases} \tag{2}$$

The nullclines [the curve $y = F(x) - \beta H(x-d)$ and the line $x = J$] (1) relates to the system of (1a) and (1b). (x, y) are shown qualitatively in Fig. 1.

2.1 Regular Regimes of Neural Activity

One of the basic properties of a neuron in a resting state is its ability to generate an action potential, when being stimulated by a sufficiently strong pulse of the injected current (the property of excitability). The stable equilibrium (fixed point) O in the system (1) corresponds to a resting state of a neuron. There are two thresholds on the phase plane, actually determined by the unstable invariant curves $W_1{}^u$ and $W_2{}^u$ (more precisely, by thin layers composed of slow trajectories localized in the vicinity of these invariant curves), where $W_1^u = (x, y) : y = F(x) + ..., J_{min} < x < d$, $W_2^u = (x, y) : y = F(x) - \beta + ..., d < x < J_{max}$. The system does not fire a spike in

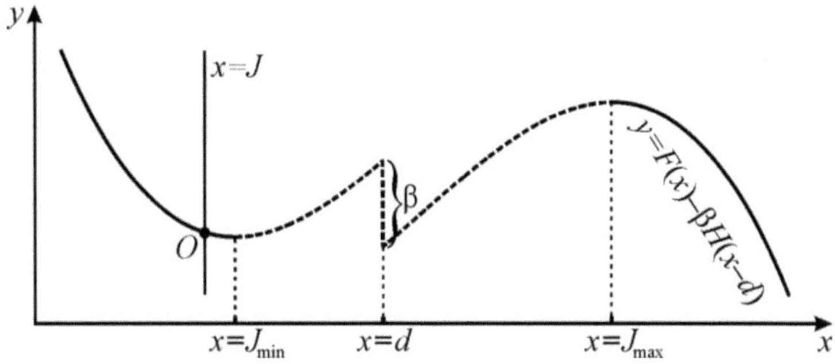

Fig. 1 The nullclines of the map (1)

response to weak stimuli which are insufficient for overcoming the first threshold ($W_1{}^u$), only a small-amplitude (sub-threshold) signal is generated [Fig. 2a, b (i)]. Stronger stimuli sufficient for overcoming the second threshold ($W_2{}^u$) result in a large-amplitude trajectory, entering the basin of attraction of the stable invariant curve $W_2{}^s$, where $W_2^s = (x, y) : y = F(x) - \beta + ..., x > J_{max}$, and terminating in the stable fixed point O. Such behavior on the phase plane corresponds to the generation of an action potential (or spike) [Fig. 2a, b (ii)].

Another important regime observed in the system is sub-threshold oscillations. A stable closed invariant curve C_{th} on the phase plane (Fig. 2c), emerged as a result of the Neimark–Sacker bifurcation when the fixed point loses its stability, corresponds to them. These quasi-sinusoidal oscillations (Fig. 2d) below the threshold of excitation are of great importance, in particular, in functioning of the OCS [10], which will be discussed further.

The next regular neuronal mode, reproduced in the system, is periodic spike oscillations. It is shown [6] that one of the conditions for their appearance is the relative smallness of the parameter β. As this takes place, a new type of trajectories, which move between the layers of slow motions, localized in the vicinity of the two stable curves $W_1{}^s$ $[W_1^s = (x, y) : y = F(x) + ..., x < J_{min}]$ and $W_2{}^s$, without changing their direction around the discontinuity line $x = d$, becomes possible. As a result, a stable closed invariant curve C_{sp} (Fig. 2e) appears on the phase plane corresponding to the periodic spike oscillations (Fig. 2f).

It is shown in [10] that another type of spiking activity is observed in the model (1) in addition to the preceding one. In this case, a discontinuous attractor A_{sp} emerges on the phase plane (Fig. 2g) which determines such sharp spike oscillations (Fig. 2h).

2.2 Chaotic Regimes of Neural Activity

It is found in [6] that there can be different chaotic attractors in the system (1). One of these attractors, A_{th}, is shown in Fig. 3a, and the corresponding mode of neural activity in Fig. 3b. In this mode, action potentials are generated alternating with sub-threshold oscillations. The dynamical mechanism of these oscillations lies in the fact that the invariant curve $W_2{}^u$ separates trajectories of (1) into two

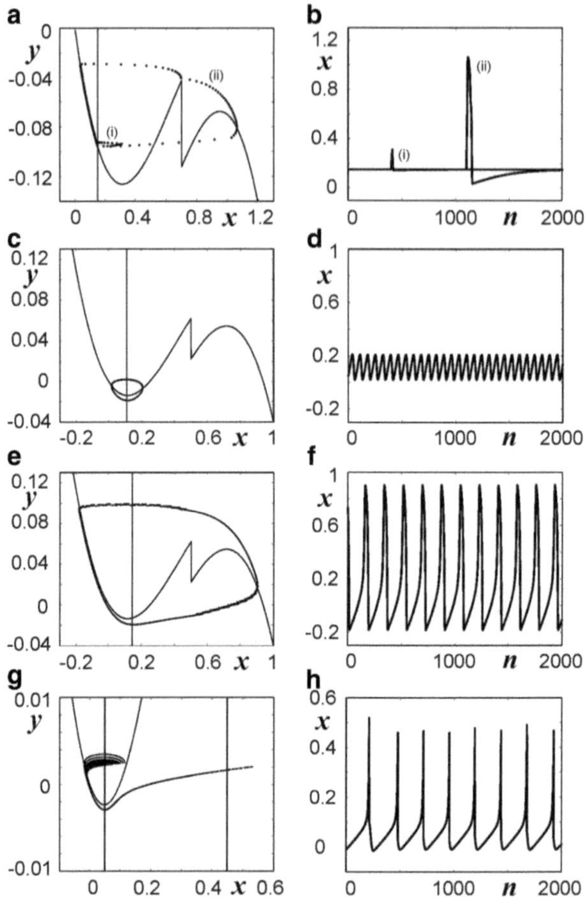

Fig. 2 Phase portraits (the *left column*) and corresponding waveforms (the *right column*) of the regular regimes of neural activity in the model (1): (**a, b**) excitable regime; (**c, d**) subthreshold oscillations; (**e, f**) periodic spike oscillations; (**g, h**) another type of spiking.

streams in the vicinity of the discontinuity line $x = d$. The first stream consists of trajectories in the vicinity of $x = d$. The second stream is formed by trajectories, overcoming the second threshold and reaching the neighborhood of $W_2{}^s$. As a result of this separation, trajectories circulate changing chaotically their direction from one stream to another, forming the attractor A_{th}.

One of the main important dynamical regimes, observed in various neuronal structures, is chaotic spike-bursting oscillations. This regime with different characteristics is reproduced in a wide range of parameters. The relaxation chaotic attractor A_{sb} on the phase plane (Fig. 3c) determines this mode; the corresponding waveform is shown in Fig. 3d. Due to the importance of chaotic spike-bursting oscillations we focus on the dynamical mechanism for their appearance.

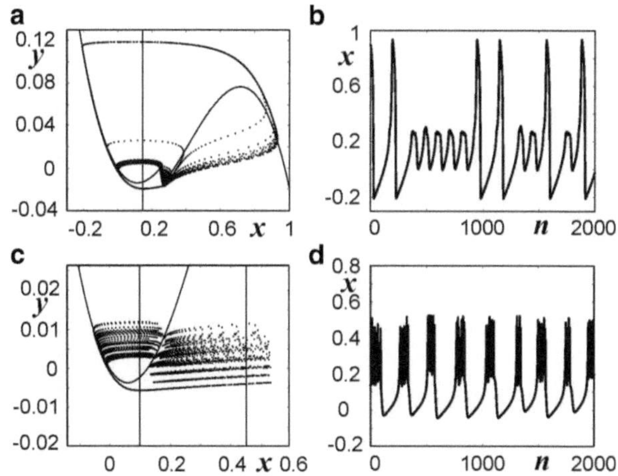

Fig. 3 Phase portraits (the *left column*) and corresponding waveforms (the *right column*) of the chaotic regimes of neural activity in the model (1): (**a, b**) alternating spiking and subthreshold activity; (**c, d**) spike-burting oscillations.

2.3 *Spike-Bursting Oscillations*

Chaotic spike-bursting oscillations can be divided into two phases: fast and slow (see Fig. 3c, d). Consider the formation of these two-scale oscillations. First, let $\varepsilon = 0$, and therefore, the variable $y = y^0 = \text{const}$ and plays the role of a parameter in (1a). Depending on the value of y^0, the map of fast motions

$$x_{n+1} = x_n + F(x_n) - \beta H(x_n - d) - y^0 \tag{3}$$

can exhibit both regular and chaotic dynamics (Fig. 4). In case of the former, the only attractor of the map (3) is the stable fixed point x_s. The quasi-static decrease of the parameter y^0 causes that the equilibrium x_s merges with the unstable fixed point x_u at some bifurcation value and disappears (the tangent or saddle-node bifurcation occurs).

In the case of chaotic dynamics, the Eq. (3) behaves like a Lorentz-type map and has an invariant interval containing a chaotic attractor. The quasi-static increase of the parameter y^0 causes that this attractor undergoes internal bifurcations, and finally, at some value $y^0 = {y_{cr}}^0$, a boundary crisis [11] occurs. One of the attractor's boundaries merges with an unstable fixed point x_u, and the attractor collapses.

Now let $\varepsilon \neq 0$. In this case, the family of the stable fixed points of (3) forms the stable invariant curve ${W_1}^s$ on the phase plane (x, y) in the region $x < J$. The family of one-dimensional chaotic attractors forms a transient chaotic set in the region $x < J$. Suppose that initially a trajectory starts in the neighborhood of ${W_1}^s$. The variable y decreases slowly along this curve, and the variable x is supported in the quasi-equilibrium state according to the equations

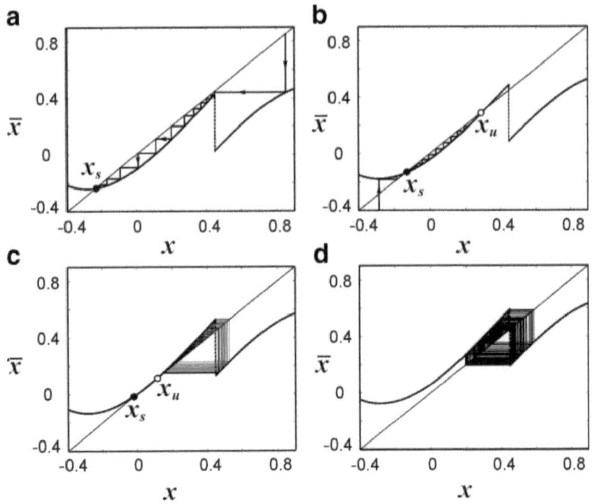

Fig. 4 Dynamics of the map (1) for different values y^0 corresponding to the existence of (**a**) a unique stable fixed point; (**b**) two fixed points, stable and unstable; (**c**) two fixed points and a chaotic attractor; (**d**) a unique chaotic attractor.

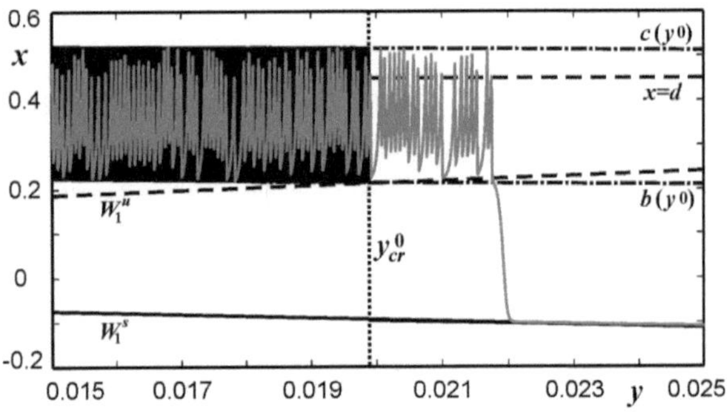

Fig. 5 Trajectories on the plane (x, y) in the stationary case (*black*) and a typical trajectory in the dynamical case when the variable y increases slowly in time (*gray*)

$$\overline{y} = y + \varepsilon \left(x - J\right), \quad y = F(x). \tag{4}$$

This phase forms a slow regular area of the two-dimensional chaotic attractor, corresponding to the passive phase of spike-bursting oscillations.

In the vicinity of somevalue of y, when the stable ${W_1}^s$ and unstable ${W_1}^u$ curves merge, the trajectory leaves the layer of slow motions and enters the basin of attraction of the transient chaotic set. Since the trajectory is now in the region $x > J$, then the variable y is slowly increasing. Iterating the map (1a) numerically, we found that chaotic oscillations persist after the value $y^0 = {y_{cr}}^0$, corresponding to the crisis of the chaotic attractor in the static case $\varepsilon = 0$ (Fig. 5). The effect of delay of the

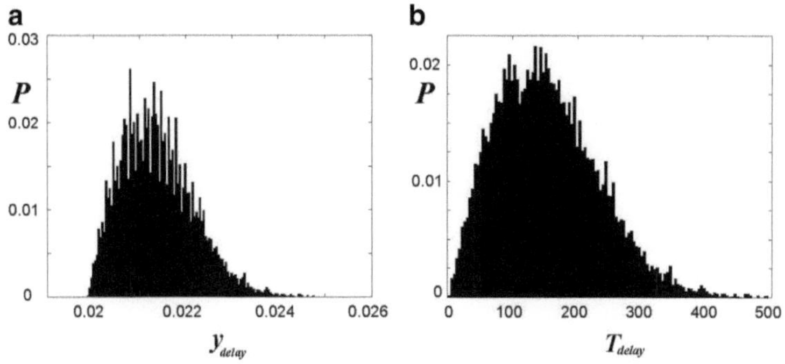

Fig. 6 Histograms of (**a**) the coordinates of escaping from the chaotic region and (**b**) time delays. The parameter $\varepsilon = 10^{-5}$, the critical value ${y_{cr}}^0 \approx 0.02$

disappearance of chaotic oscillations takes place.[1] This phase of motions forms the fast chaotic area of the attractor, which corresponds to the active phase or the phase of depolarization of spike-bursting oscillations.

We dwell on the effect of the delay of chaotic motion disappearance.

Consider an ensemble of trajectories with different initial conditions at $y < {y_{cr}}^0$, taken within the invariant interval, the boundaries of which are defined by the values $b(y^0)$ and $c(y^0)$ (see Fig. 5). Let us study how the trajectories leave the region of chaotic motions when y slowly increases. When $y > {y_{cr}}^0$ the criterion of escaping from the chaotic area is the fact that a trajectory is below the curve ${W_1}^u$ (Fig. 5). Since the trajectories of chaotic attractors are characterized by high sensitivity to initial conditions, then they escape from the region of chaotic motions at different values y_{delay} of the slow variable y. It is clear that the time delay T_{delay} for ceasing of chaotic motions depends on the initial conditions.

Figure 6 shows the histograms of the coordinates of escaping and time delays for $N = 2.5 \times 10^5$ initial conditions. The time delay is measured starting from the intersection of the value $y = {y_{cr}}^0$ by trajectories. The histograms show that these quantities are distributed in a random manner, but these are the most likely values (different from ${y_{cr}}^0$ for y_{delay} and different from zero for T_{delay}). Also, the time delay can be characterized by the curve showing how the number of trajectories in the region of chaotic motions decreases in time.

Figure 7 shows the logarithm of this number versus time, having a quadratic shape. Note that for most of the transient chaotic sets described in the literature [11], the logarithm of the number of phase points in the chaotic region (or a similar characteristic—the logarithm of the survival probability) decreases in time linearly. In this case the most probable value of the time delay in the chaotic region is zero. In other words, for these sets there is no such effect as the delay of chaotic motion

[1]Note that in the theory of dynamic bifurcations [12], i.e., bifurcations occurring during a slow drift of the control parameters, the effect of stability loss delay is now studied only for the case of regular attractors: equilibrium states and limit cycles.

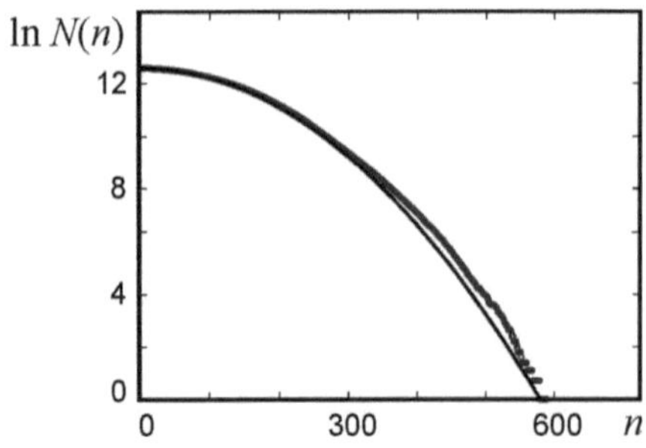

Fig. 7 The logarithm of the number of trajectories in the chaotic region as a function of time, when y varies slowly with the constant rate $\varepsilon = 10^{-5}$. The initial moment $n = 0$ is taken at $y = y_{cr}{}^0$. The plot is well approximated by the curve $\ln N(n) = \ln N_0 - \kappa n^2$

disappearance. On the contrary, in (1) this effect determines the duration of the active phases of spike-bursting oscillations (see Fig. 3d) fluctuating around some mean, depending on the average time delay.

3 Dynamics of the OCS

The formation of oscillatory structures (patterns), namely, neural groups that synchronously generate action potentials against the background of a relative rest of the neurons outside the group, is one of the most important forms of collective electrical activity of neural networks. Such forms of activity have been detected experimentally in different neural structures, including the structures in visual [13] and somatosensory [14] cortex, in the temporal lobe [15], in the inferior olives [16], etc. However, despite the significant progress and success in modern experimental neurophysiological studies of neural systems, the results achieved with the help of these studies still remain limited. For now, the research results do not give a complete understanding of the operation of large-scale neural networks and circuits requiring simultaneous record of the activity of a very large number of neurons which often belong to different structural units.

In particular, the mechanisms of the appearance and disappearance of the electrical activity patterns in large neural systems composed of several interacting structural units, i.e., neural subsystems, each being formed by cells of a certain type, remain unclear in many cases. Modeling can be one of the possible ways of identifying such mechanisms. Here we show some results of modeling the OCS, which is a special part of brain in vertebrates. A large number of experimental data point out the important role of this system in the implementation of the motor function of the body, i.e., control of motor systems, adaptation of motion under varied external conditions, and motor learning [17, 18]. Neurons of the OCS are combined into highly ordered groups, each playing its own functional role. The exclusive importance of this system stimulates its study by researchers of different directions [18–27] and modeling of its properties for the creation of artificial neuromorphic structures simulating the capabilities of the biological prototype.

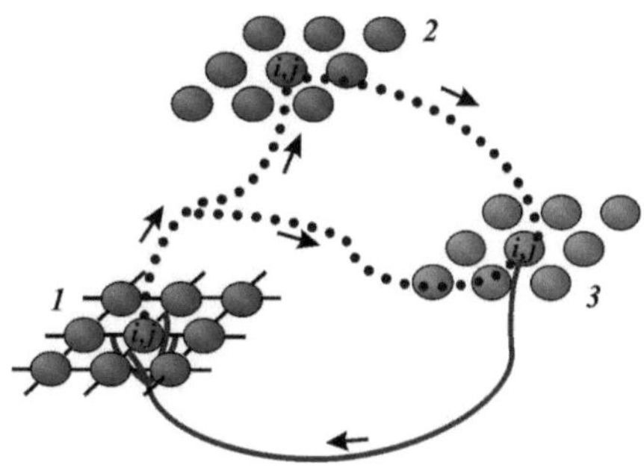

Fig. 8 Structure of the discrete model of the olivo-cerebellar system. *1* is the layer of inferior olive (IO) neurons, which is the main layer, *2* is the layer of Purkinje cells (PC), and *3* is the layer of cerebellar nuclei (CN) neurons

In our research we rely on the concept which R. Llinas, a neurophysiologist from the Medical School of the University of New York, proposed (see, in particular [24]) for functioning of the OCS. A decisive role in this concept is assigned to the processes of the formation and destruction of the electrical activity clusters in this neural network. Note that the previous works [28–31] utilized continuous models of structural units of the OCS, which were based, with varying degrees of completeness, on the Hodgkin–Huxley formalism. Here we discuss the structure and dynamics of the OCS discrete-time model. It is important to mention that in our model, all kinds of the OCS neurons, despite their different functional properties, have been described using the same system of nonlinear maps and only choosing appropriately the parameters of this model.

Our model consists of three layers (Fig. 8). The first one is the layer of inferior olive (IO) neurons, which we will call the main layer since its dynamics is output in our model. The second layer consists of Purkinje cells (PC). The third layer comprises the cerebellar nuclei (CN) neurons. The neighboring elements of the main layer are coupled with each other via electrical synapses having the property of plasticity. Each inferior olive neuron with indexes i, j has an axon, the so-called climbing fiber branching in two (see Fig. 8). Both axonal branches (collaterals) terminate with excitatory synapses, one on a Purkinje neuron with the number (i, j) and another on the nuclei neuron (i, j). Each Purkinje cell (i, j) has an axon ending with an inhibitory synapse on the corresponding nuclei neuron with indexes i, j. Thus, each nuclei neuron receives signals of different types via its synaptic inputs, namely, excitatory from the inferior olive neurons and inhibitory from the Purkinje cell. In turn, each nuclei neuron (i, j) has an axon with inhibitory synaptic ends which suppress the electrical couplings of the corresponding inferior olive neuron (i, j) with the neighboring elements of the main layer, i.e. with the neurons $(i-1, j)$, $(i+1, j)$, $(i, j-1)$, and $(i, j+1)$.

The complete system of maps describing the dynamics of the models looks as follows:

$$\overline{x}_i^{IO} = x_i^{IO} + F\left(x_i^{IO}\right) - 0.9H\left(x_i^{IO} - 0.85\right) - y_i^{IO} + I_i^{IO}, \quad i = 1..N^2, \quad (5)$$

$$\overline{y}_i^{IO} = y_i^{IO} + 0.005\left(x_i^{IO} - 0.049\right), \quad i = 1..N^2, \quad (6)$$

$$I_i^{IO} = \sum_{j=1}^{m} g_{i,j}^{IO} \left(x_j^{IO} - x_i^{IO} \right), \quad i = 1..N^2, m = 4, \tag{7}$$

$$\overline{g}_{i,j}^{IO} = \gamma g_{i,j}^{IO} + \delta \left[1 - H \left(v_i + v_j - v_{thresh} \right) \right], \quad i, j = 1..N^2, \tag{8}$$

$$v_i = x_i^F, \text{the last element of the } CN\text{-axon}, \tag{9}$$

$$\overline{x}_i^F = x_i^F + F \left(x_i^F \right) - y_i^F + I_i^F, i = 1..7N^2, \tag{10}$$

$$\overline{y}_i^F = y_i^F + 0.011 \left(x_i^{IO} - 0.04 \right), i = 1..7N^2, \tag{11}$$

$$I_i^F = 0.15 \sum_{j=1}^{m} \left(x_j^F - x_i^F \right), i = 1..7N^2, m = 1, 2, \text{or } 3, \tag{12}$$

$$\overline{x}_i^{PC} = x_i^{PC} + F \left(x_i^{PC} \right) - 0.5H \left(x - 0.6 \right) - y_i^{PC} + I_i^{PC}, i = 1..N^2, \tag{13}$$

$$\overline{y}_i^{PC} = y_i^{PC} + 0.001 \left(x_i^{PC} - 0.045 \right), i = 1..N^2, \tag{14}$$

$$I_i^{PC} = -0.3H \left(x_i^{PC,pre} - 0.48 \right) \left(x_i^{PC} + 0.6 \right), i = 1..N^2, \tag{15}$$

$$x_i^{PC,pre} = x_i^F, \text{the last element of the } IO - \text{axon}, \tag{16}$$

$$\overline{x}_i^F = x_i^F + F \left(x_i^F \right) - y_i^F + I_i^F, i = 1..15N^2, \tag{17}$$

$$\overline{y}_i^F = y_i^F + 0.011 \left(x_i^{IO} - 0.04 \right), i = 1..15N^2, \tag{18}$$

$$I_i^F = 0.15 \sum_{j=1}^{m} \left(x_j^F - x_i^F \right), i = 1..15N^2, m = 1 \text{ or } 2, \tag{19}$$

$$\overline{x}_i^{CN} = x_i^{CN} + F \left(x_i^{CN} \right) - 0.6H \left(x_i^{CN} - 0.6 \right) - y_i^{CN} + I_i^{exc} + I_i^{inh}, i = 1..N^2, \tag{20}$$

$$I_i^{exc} = -0.2H\left(x_1 - 0.3\right)\left(x_i^{CN} - 0.6\right), i = 1..N^2, \tag{21}$$

$$I_i^{inh} = -0.2H\left(x_2 - 0.7\right)\left(x_i^{CN} + 0.2\right), i = 1..N^2, \tag{22}$$

$$x_1 = x_i^F, \text{the last element of the } IO - \text{axon}, \tag{23}$$

$$x_2 = x_i^F, \text{the last element of the } PC - \text{axon}, \tag{24}$$

$$\overline{x}_i^F = x_i^F + F\left(x_i^F\right) - y_i^F, i = 1..N^2, \tag{25}$$

$$\overline{y}_i^F = y_i^F + 0.011\left(x_i^{IO} - 0.04\right), i = 1..N^2, \tag{26}$$

where the maps (5–9) correspond to the dynamics of the layer of inferior olive neurons, (10–12) to the axons of inferior olive neurons, (13–16) to the layer of Purkinje cells, (17–19) to the axons of Purkinje cells, (20–24) to the layer of nuclei neurons, (25 and 26) to their axons. We dwell on the individual parts of the system.

3.1 Inferior Olive Neurons

The inferior olive neurons have a number of exclusive dynamical properties [24–27]. Firstly, in a resting state the membrane potential of these cells is in the mode of sub-threshold oscillations. These are quasi-sinusoidal oscillations with a small amplitude and frequency about 10 Hz. The action potential is generated at the peak of these oscillations once the membrane potential reaches some threshold value. Secondly, the inferior olive neurons interact with each other via electrical synapses, i.e., gap junctions, and the strength of this coupling is dynamically varied through the agency of glomerulus which is a special mechanism. The electrical synapse is controlled by the activity of the cerebellar nuclei neurons.

One of the key properties of inferior olive neurons reproduced in the model (1) is their ability to switch from the sub-threshold oscillation mode to the action potential generation mode on the oscillation peaks during the action of varying synaptic current I_{syn} (see Fig. 9).

3.2 Purkinje and Deep Cerebellar Nuclei Neurons

In accordance with our scheme of the OCS (Fig. 8), the Purkinje and deep cerebellar nuclei neurons are the elements of the second and third layers, respectively. When

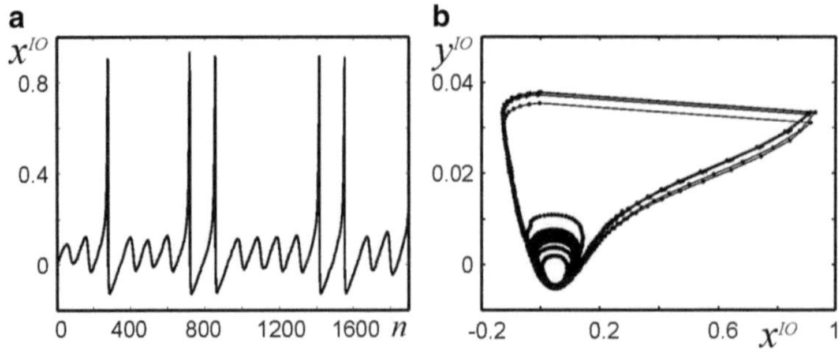

Fig. 9 (**a**) The waveform and (**b**) the corresponding phase portrait of the system (1) during the action of the variable synaptic current for $J^{IO} = 0.049$, $\varepsilon = 0.005$

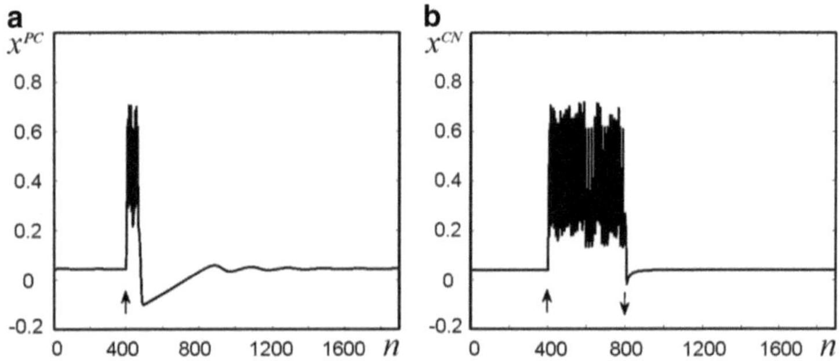

Fig. 10 Dynamical properties of (**a**) the Purkinje model neurons and (**b**) cerebellar nuclei neurons

modeling these cells, we took into account their key properties which give the main contribution to the operation of the OCS. In the model of a Purkinje cell (13 and 14), we took into account the generation of the so-called complex spike excited by a climbing fiber (see Fig. 10). In other words, an element of the second layer is a system in the excitable mode, which generates a burst of spikes during the excitation.

The model of a cerebellar nuclei neuron reflects the following property of the latter. This neuron turns into the active spiking state during excitation via the synaptic terminal of the corresponding collateral of the axon of the inferior olive neuron and comes back to the resting state during inhibition of the synaptic terminal of the axon of the Purkinje cell (see Fig. 10b). In other words, the element of the third layer is a bistable system having two states. For a qualitative modeling of these properties it suffices to have a one-dimensional map (20) for the variable x^{CN} characterizing the membrane potential of the cell.

3.3 Axons

Axons, or nerve fibers, are represented in the model in the form of chains of elements coupled electrically with the nearest neighbors. Each element is in excitable mode, generating an action potential during excitation. Successive excitation of the neighboring elements of the chain models the propagation of a pulse along the nerve fiber. Note that such a model is not only applicable for describing the dynamics of the linear fibers, but also, as in our model, is able to reproduce the propagation of a pulse through bifurcating axons (see, e.g., [32]). The system of maps describing such an element is given above numbered (10 and 11), (17 and 18), (25 and 26). Synaptic current is equal to the sum over the nearest neighboring elements of the chain, the number of which can amount to one (at the end of the fiber), two (in the middle of the fiber), and three (at the branching point). This term describes the electrical coupling, whose strength is characterized by the parameter c. It is noteworthy that the representation of a nerve fiber as a chain of discrete elements is not just convenient for analysis by mathematical abstraction. Many types of nerve fibers, in particular, myelinated axons, including those in the OCS, in fact consist of individual blocks separated by the so-called nodes of Ranvier.

3.4 Dynamical Coupling Between the Inferior Olive Neurons

As was mentioned above, the inferior olive neurons interact via gap junctions, or electrical synapses. The strength of this coupling is controlled by GABAergic synapses (GABA—gamma-aminobutyric acid) of the axons of the cerebellar nuclei cells, i.e., the electrical synapses are plastic. This control is inhibitory: in the absence of control pulses, the coupling strength is equal to some positive value, which results in synchronization of sub-threshold oscillations of the interacting inferior olive neurons. On the contrary, the coupling is suppressed and the electrical coupling between neurons breaks if the signal is fed. The equations for the dynamical coupling between two inferior olive neurons, which is controlled by the signals coming from the cerebellar nuclei neurons, is given above numbered (7–9).

3.5 Qualitative Description of the Model Dynamics

Let the membrane potential of one of the inferior olive neurons reaches a threshold value at some instant of time. This results in the action potential excitation in the neural axon. Propagating along both branches of the climbing fiber, this spike evokes activity in the Purkinje and cerebellar nuclei neurons, i.e. in the corresponding elements of the second and third layer. A short series of fast spikes, i.e. a complex spike, is generated in the Purkinje cell and a long sequence of action potentials is excited in the cerebellar nuclei neuron. A complex spike in the Purkinje cell causes the inhibitory action potential to propagate through the axon to the nuclei neuron. A long spike series in the latter is interrupted due to the inhibition by the action potential coming from the Purkinje neuron. A burst of spikes transmitted through the axon of the nuclei neuron to synaptic couplings of the corresponding inferior

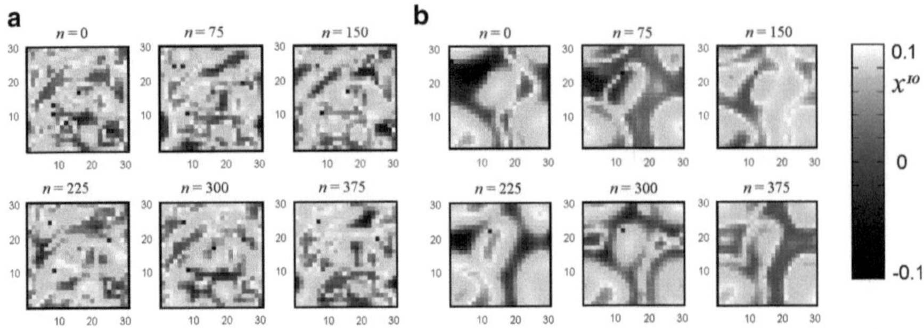

Fig. 11 Dynamics of the activity clusters in the inferior olive layer with 30×30 elements with electrical interaction between the neighboring elements and the feedback loop. The maximal coupling (**a**) $g_{\max} = 0.005$ and (**b**) 0.02

olive neuron leads to the suppression of the latter. Thus, synaptic couplings of the inferior olive neuron with the neighboring elements of the main layer break for a time equal to the duration of the burst generated in the nuclei neuron.

3.6 Spontaneous Activity Structures in the OCS

The cluster formation in the inferior olive layer of the model studied is controlled primarily by the maximal strength of coupling $g_{\max}$ between them. A physiological prototype of this parameter is the concentration of the GABAergic antagonist, whose release is controlled by inhibitory synapses of the cerebellar nuclei neurons. A higher degree of suppression of the coupling corresponds to the smaller value $g_{\max}$, and vice versa. Another important parameter is duration of the break of synaptic couplings between the inferior olive neurons. In our model, we change this characteristic time by changing the length of the Purkinje cell axons terminating at the nuclei neurons. The time of propagation of action potentials along these axons, as was mentioned above, depends on their length. The faster the action potential reaches the nuclei neuron, the sooner the neural activity is suppressed and the shorter the depolarization phase of the neural oscillation is. Therefore, the electrical coupling of the corresponding inferior olive neuron with the neighboring neurons of the main layer breaks during a shorter time.

Figure 11 shows snapshots of the activity clusters in an ensemble of inferior olive neurons for the case where coupling between the elements of the layer is dynamically controlled through a feedback loop as it is described above. The clusters are shown for the case with the Purkinje cell axons composed of 15 elements and the maximal strength of coupling between the inferior olive neurons (a) $g_{\max} = 0.005$ and (b) $g_{\max} = 0.02$. It is seen even from a visual comparison that for a small value of $g_{\max}$ the size of clusters is also small (Fig. 11a), and the characteristic time for their spatial position to change is equal approximately to the oscillation period. In case with a larger value of the maximal strength of coupling between the elements (Fig. 11b), the cluster size increases on average, and their spatial position remains constant during several oscillation periods. These qualitative results are confirmed by quantitative characteristics, as will be shown below.

3.7 Measurement of the Degree of Spatial Order

For a quantitative description of the collective activity of the inferior olive neurons, we apply the theory of Markov random fields [16]. This method, initially developed to characterize the spatial order of two-dimensional matrices, is used for analysis of the experimental data obtained by optical visualization of sections of the brain and multi-electrode measurement of neuron activity. In our case, this method is used for analysis of the numerical data obtained within the model considered.

We introduce the Markov parameter which for a two-dimensional square lattice of elements is determined as follows. Denote by $x_{i,j}$ the values of the membrane potentials of the neurons located at the lattice sites with the numbers (i, j). Then for estimation of the degree of clustering in this layer we calculate the parameter

$$\eta = \frac{\sum\limits_{i,j\in\Omega} x_{i,j}y_{i,j} - \left(\sum\limits_{i,j\in\Omega} x_{i,j}\right)\left(\sum\limits_{i,j\in\Omega} y_{i,j}\right)/m}{\sum\limits_{i,j\in\Omega} y_{i,j}^2 - \left(\sum\limits_{i,j\in\Omega} y_{i,j}\right)^2/m}, \tag{27}$$

where $y_{i,j} = x_{i-1,j} + x_{i+1,j} + x_{i,j-1} + x_{i,j+1}$. Here, Ω denotes one of the two sub-lattices that form the initial lattice, whose elements are distributed relative to each other as white and black squares of the chessboard, and m is the number of elements in Ω. The final value η results after taking the mean arithmetic value of this parameter for both sub-lattices.

The quantitative meaning of η is as follows. The more the absolute value of η is different from zero, the higher the level of spatial order of the layer is. This means that if the values of the potentials of the elements are randomly distributed, then η tends to zero. The case is opposite when the potentials are the same for all elements, i.e., all the elements of the lattice form one cluster. Then η takes its maximal value. The more the parameter η is different from zero, the larger is the characteristic size of the clusters in the lattice.

Figure 12 shows temporal dependences of the Markov parameter η and the corresponding histograms of probability of its distribution for two values of the coupling parameter $g_{\max}$. It follows from the latter that for small values of $g_{\max}$ the quantity η varies over relatively wide limits, but the absolute values of η are close

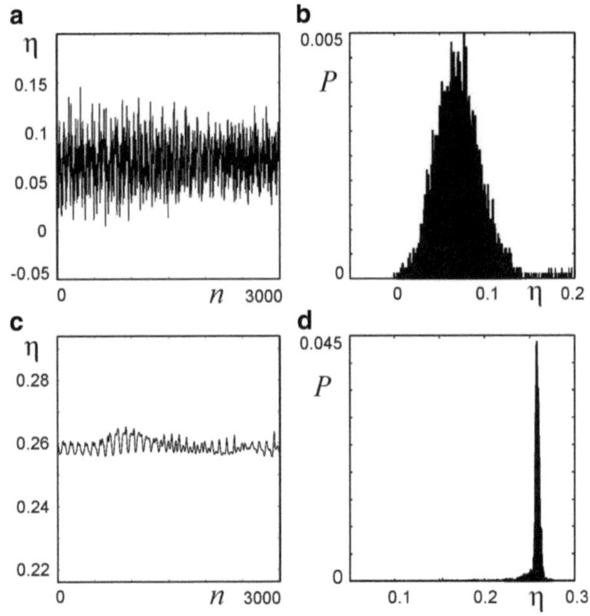

Fig. 12 (**a**, **c**) The Markov parameter η for the elements of the inferior olive layer as a function of time. (**b**, **d**) Histograms of the distribution probability of the parameter η. The maximal strength of coupling $g_{\max}$ is 10^{-3} (**a**, **b**) and 10^{-2} (**c**, **d**)

to zero (Fig. 12a, b). As the coupling parameter $g_{\max}$ increases, the mean value of η also increases and the variance of η decreases (Fig. 12c, d).

This result is in good agreement with the experimental measurements of the Markov parameter for different levels of blocking the electrical coupling between the inferior olive neurons. For a weak coupling between neurons, the layer is divided into a set of clusters of a smaller size, and the picture of such a division changes drastically with time. As the coupling is enhanced, the clusters increase in size and the spatial configuration of the synchronous activity of the layer does not disappear with time.

3.8 Stimulated Activity Patterns

For a motor control, the OCS not only generates but also preserves in time the spatiotemporal structures (patterns), or activity patterns, and changes them depending on the input signals. According to these signals, sub-groups of inferior olive neurons with synchronous or phase-shifted oscillations are formed, and this spatial structure is preserved for several periods of oscillation. When a new pattern comes, oscillations of the elements are rearranged in accordance with it. Such behavior of the system is the basis for control of the motor function of the body [24].

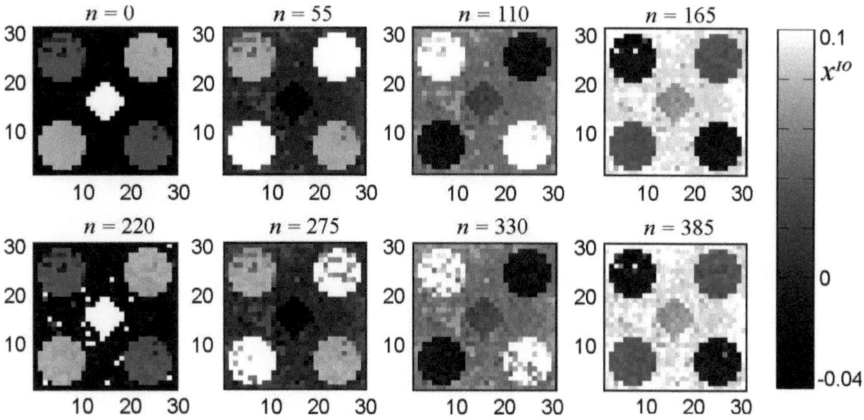

Fig. 13 Dynamics of a stimulated activity pattern in a layer of inferior olive neurons with 30 × 30 elements. The feedback is activated, the maximal strength of electrical coupling is $g_{\max} = 0.02$, and the axon length of the Purkinje cells is equal to 18 elements

Figure 13 shows the dynamics of the activity clusters in a layer of inferior olive neurons. The initial values of the variables x^{IO} were randomly chosen. Then an external stimulus in the form of a spatiotemporal activity pattern is fed to the layer. As an example, this pattern is taken in the form of five clusters shaped as circles. The stimuli acting on the elements of all clusters have the same amplitude and duration. The difference is that stimulation acts on the different clusters in different instants of time, thereby creating a phase shift of oscillatory activity between the clusters of a pattern. Since the spikes are generated at the peaks of sub-threshold oscillations, the phase shift of sub-threshold oscillations corresponds one-to-one to the phase shift of the action potentials transmitted into the climbing fiber and then to the Purkinje and cerebellar nuclei neurons.

4 Conclusion

In conclusion, models of neural activity in the form of maps are an effective tool for studying the dynamics of individual neurons as well as complex neural circuits.

We have shown that using the same two-dimensional map (1) and only choosing appropriately its parameters, one can get a wide range of different modes of neural activity from the simple excitable regime to chaotic spike-bursting oscillations. Thanks to this universality, we were able to develop a model of a very complex neural structure, the OCS, using the map (1) as a basic element. In our discrete-time model of the OCS three layers of neurons interact with each other through axonal connections, and the propagation of nerve impulses along the axons is modeled by the chains of excitable elements. Dynamical properties of individual elements of the system are examined, and the parameters corresponding to the dynamics, that is adequate to the biological prototypes, are found. A model of dynamic coupling between the inferior olive neurons is proposed and the pattern formation processes in the layer they form are explored. The relation between the size of synchronous activity patterns and the maximal coefficient of electrical coupling is established. It is shown that stimulated activity patterns of a certain configuration can be formed in the system and can be reset by an external stimulus. The latter property underlies the neuronal systems of the motor function control.

Acknowledgments The study was supported by the Ministry of Education and Science of Russian Federation (projects 14.132.21.1354, 8497, 8205) and the Russian Foundation for Basic Research (grants 12-02-31252, 12-02-00526, 12-02-31825, 12-04-31963).

References

1. Courbage M, Nekorkin VI (2010) Map based models in neurodynamics. Int J Bifurcat Chaos 20:1631–1651
2. Ibarz B, Casado JM, Sanjuán MAF (2011) Map-based models in neuronal dynamics. Phys Rep 501:1–74
3. Chialvo DR (1995) Generic excitable dynamics on a two-dimensional map. Chaos Solitons Fractals 5:461–479
4. Rulkov NF (2002) Modeling of spiking-bursting neural behavior using two-dimensional map. Phys Rev E 65:041922
5. Shilnikov AL, Rulkov NF (2004) Subthreshold oscillations in a map-based neuron model. Phys Lett A 328:177–184
6. Nekorkin VI, Vdovin LV (2007) Diskretnaya model neyronnoy aktivnosti. Izvest vys ucheb zaved Prikladnaya nelinejnaya dinamika 15(5):36–60
7. Courbage M, Nekorkin VI, Vdovin LV (2007) Chaotic oscillations in a map-based model of neural activity. Chaos 17:043109
8. Courbage M, Maslennikov OV, Nekorkin VI (2012) Synchronization in time-discrete model of two electrically coupled spike-bursting neurons. Chaos Solitons Fractals 45(5):645–659
9. Nekorkin VI, Maslennikov OV (2011) Spike-burst synchronization in an ensemble of electrically coupled discrete model neurons. Radiophys Quantum Electron 54(1):56–73
10. Maslennikov OV, Nekorkin VI (2012) Discrete model of the olivo-cerebellar system: structure and dynamics. Radiophys Quantum Electron 55(3):198–214
11. Lai Y-C, Tél T (2011) Transient chaos: complex dynamics on finite time scales, Applied mathematical sciences. Springer, Berlin
12. Benoît E (ed) (1991) Dynamic bifurcations. Lecture notes in mathematics. Springer, Berlin, p 1493
13. König P, Engel AK, Singer W (1995) Relation between oscillatory activity and long-range synchronization in cat visual cortex. Proc Natl Acad Sci U S A 92(1):290
14. Yang J-W, Hanganu-Opatz IL, Sun J-J, Luhmann HJ (2009) Three patterns of oscillatory activity differentially synchronize developing neocortical networks in vivo. J Neurosci 29(28):9011
15. Manning JR, Polyn SM, Baltuch GH et al (2011) Oscillatory patterns in temporal lobe reveal context reinstatement during memory search. PNAS 108(31):12893
16. Leznik E, Makarenko V, Llinas R (2002) Electrotonically mediated oscillatory patterns in neuronal ensembles: an in vitro voltage-dependent dye-imaging study in the inferior olive. J Neurosci 22(7):2804
17. Kandel ER, Schwartz JH, Jessel TM (2000) Principles of neural science. McGraw-Hill, New York

18. Nicholls JG, Martin RA, Fuchs PA, Moore JW, Stuart AE (2011) From neuron to brain, 5th edn. Sinauer Associates, Inc Sunderland, Massachusetts, USA.
19. Marr D (1969) A theory of cerebellar cortex. J Physiol 202:437
20. Albus JS (1971) A theory of cerebellar function. Math Sci 10:25
21. Llinas R, Baker R, Sotelo C (1974) Electrotonic coupling between neurons in cat inferior olive. J Neurophysiol 37:560–571
22. Llinas R, Yarom Y (1981) Electrophysiology of mammalian inferior olivary neurones in vitro. Different types of voltage-dependent ionic conductances. J Physiol 315(549)
23. Sotelo C, Llinas R, Baker R (1974) Structural study of inferior olivary nucleus of the cat: morphological correlates of electrotonic coupling. J Neurophysiol 37:541
24. Llinas R, Lang EJ, Welsh JP (1997) The cerebellum, LTD, and memory: alternative views. Learn Mem 3:445
25. Jacobson GA, Lev I, Yarom Y, Cohen D (2009) Invariant phase structure of olivo-cerebellar oscillations may underlie cerebellar pattern generation. Proc Natl Acad Sci U S A 106(9):3579
26. Jacobson GA, Rokni D, Yarom Y (2008) A model of the olivo-cerebellar system as a temporal pattern generator. Trends Neurosci 31(12):617
27. Devor A, Yarom Y (2002) Generation and propagation of subthreshold waves in a network of inferior olivary neurons. J Neurophysiol 87:3059
28. Velarde MG, Nekorkin VI, Makarov VA et al (2004) Clustering behavior in a three-layer system mimicking olivo-cerebellar dynamics. Neural Netw 17:191
29. Kazantsev VB, Nekorkin VI, Makarenko VI, Llinas R (2003) Olivo-cerebellar cluster-based universal control system. Proc Natl Acad Sci U S A 100:13064
30. Kazantsev VB, Nekorkin VI, Makarenko VI, Llinas R (2004) Self-referential phase reset based on inferior olive oscillator dynamics. Proc Natl Acad Sci U S A 101(52):18183
31. Katori Y, Lang EJ, Onizuka M et al (2010) Quantitative modeling of spatio-temporal dynamics of inferior olive neurons with simple conductance-based model. Int J Bifurcat Chaos 20(3):583
32. Yanagita T (2007) Input–output relation of FitzHugh–Nagumo elements arranged in a trifurcated structure. Phys Rev E 76:056215

Adaptive Landscape with Singularity in Evolutionary Processes

Shuyun Jiao, Song Xu, and Ping Ao

Abstract Adaptive landscape, proposed by Sewall Wright, has been used to find optimized solutions of a system. The optimized solution of an evolutionary system is when evolution maximizes or minimizes the value of some function of the trait under consideration, thus providing an absolute measure of fixation for a biological process in a probabilistic sense. We survey the role of adaptive landscape and give some general results concerning the question of infinite potential escaping. The results presented include complex dynamical behaviors manifested by adaptive landscape with singularity in all parameters regimes. In addition, both metaphoric and quantitative description of many complex biological phenomena is provided by adaptive landscape, such as the rare event of transition between different stable states.

1 Introduction

The concept of adaptive landscape was first proposed by Sewall Wright [1], in attempting to visualize his shifting balance theory in evolutionary dynamics. Since then, the metaphoric and visualizing part of this concept has been widely used in population genetics and evolutionary biology [2–4]. Wright's original landscape

S. Jiao
Shanghai Center for Systems Biomedicine, Department of Mathematics,
Shanghai Jiao Tong University, Shanghai, China

Xinyang Normal University, Henan, China
e-mail: jiaoshuyun@gmail.com

S. Xu
Department of Biomathematiccs, University of California, Los Angeles, USA
e-mail: sxu11@ucla.edu

P. Ao (✉)
Shanghai Center for Systems Biomedicine, Shanghai Jiao Tong University, Shanghai, China
e-mail: aoping@sjtu.edu.cn; aoping@u.washington.edu

V. Afraimovich et al. (eds.), *Nonlinear Dynamics and Complexity*,
Nonlinear Systems and Complexity 8, DOI 10.1007/978-3-319-02353-3_6,

can be interpreted as the surface of selective fitness [5]. Its definition has two main derivatives [6]: project the individual fitness over space of allele type or genotype or phenotype, or project the mean fitness of a population over the frequencies of allele type or genotype or phenotype. The primary goal of adaptive landscape in modern science is understanding, i.e., creating orderly pictures of the world compatible with our observations and can provide new predictions.

Evolutionary processes are highly nonlinear. We need to understand these processes. Adaptive landscape, when used in visualizing evolutionary processes, is remarkably suitable grasping the nature of nonlinearity conveniently.

The appeal of this approach is its analogy with a physical landscape, whose gradient predicts a rolling marbles spatial trajectory [7]. But it is complex that required to fully integrate the genetics with Darwinian natural selection. Geometric representations of microevolutionary trajectories over adaptive landscape of one or another are motivated by the desire to provide some heuristic intuition into the process of microevolution. If we can extract the population genetics by investigating the contours of such a topographic surface, we can visualize how an evolving population will behave without always resorting to the algebraic heavy lifting. So the notion of a landscape implies the existence of an exact potential function projected over some space, its predictive utility comes from the implication that at each instant, the system will shift its configuration by following the steepest gradient on the potential function [7]. The predictive capability is the basis for the hypothetico-deductive method [8].

Though the landscape has such good predictive property, it has suffered certain conceptual and theoretical problems [9]. Biologically, there is an ongoing argument about the heuristic value of Wright's landscape diagrams [10]. Some think there might not be anything like peaks separated by valleys at all [10], the problem of how a population crosses an adaptive valley in its way from one adaptive peak to another may be nonexistent [11]. Since then such a landscape has been known as the fitness landscape in some parts of literatures. However, there are a considerable amount of confusions on the definitions of fitness [12,13]. Mathematically, Pigliucci and Kaplan argue that in the end it may be impossible to articulate the metaphor in a way that is both conceptually coherent and practically fruitful [14].

To get further at the predictive side we need a more realistic handle on adaptive landscape. Biologists [9, 15] noticed a framework was needed to construct adaptive landscape. Recently considerable progress has been made on the evolutionary dynamics of transiting the fitness valley for finite populations in the presence of diverse evolutionary forces. de Vladar and Barton [15] used information entropy to analyze the process of population evolution. It is pointed out that the method works only for high mutation rates and breaks down for low mutation rates [15]. One of us proposed adaptive landscape can be quantified as potential function [13]. He identified gradual parameter changes that preserve the stationary distribution as being reversible in the thermodynamic sense, this line of thinking does not lead to any constraint on the increase in mean fitness that would correspond to the constraint identified by Carnot in classical thermodynamics [16]. These theoretical progress makes the utility of adaptive landscape to complex biological processes and understanding these processes.

In this chapter we address how the singular adaptive landscape indicates the evolutionary dynamics and predicts how long the population stays at an adaptive state. There are two reasons for embarking on such an effort. The first one is practical. We want to manifest the reach of conceptually coherent and practically fruitful adaptive landscape. In general, adaptive landscape appears working well, notwithstanding its basically singularity. The next reason comes into play when we want to know the escape problems from the infinite potential.

2 Technical Preliminaries

The tools for handling a dynamical system are diverse. Here we only base on Fokker–Planck equation and some relevant knowledge about diffusion approximation, which is a widely used theory; however, there are some shortcomings about diffusion theory such as addresses by [17].

2.1 Diffusion Approximation

In general, we first consider the deterministic effects of certain biological factors for an evolutionary process. We expect to get some useful information for an evolutionary process with infinite population size. But randomness always exists, and population size is also finite, these result in the information lacked. We have to consider the effect of stochastic factors. We briefly outline the diffusion approximation according to [18] in the following.

We only address one-dimensional process. That is, we only consider populations with haploid individuals and one locus with two allele A and a and focus on the evolution of allele A. At generation t the frequency of allele A is i/N, after driving by deterministic and stochastic factors, at generation $t+1$ the allele frequency becomes j/N. Here the probability that allele frequency becomes j/N is

$$P_j(t+1) = \sum_{i=0}^{N} W_{ij} P_i(t) \tag{1}$$

where W_{ij} is the transition probability from i/N to j/N. We replace the allele frequency i/N by real number x, $0 \le x \le 1$. And $P(x, t+1)$ denotes the probability of allele frequency x after $t+1$ generations. Given that A starts out at gene frequency x_0 and additional Markov property, then

$$P(x, t+\delta t) = \sum_{\delta x} W(x, t+\delta t | x-\delta x, t) P(x-\delta x, t). \tag{2}$$

We assume that $P(x,t)$ admits a density function $\rho(x,t)$, that is,

$$\frac{\partial P(x,t)}{\partial x} = \rho(x,t). \tag{3}$$

In order to derive forward Fokker–Planck equation, we assume the following properties hold.

1)

$$\lim_{\delta t \to 0} \frac{1}{\delta t} p(z, t + \delta t | x, t) = 0 \tag{4}$$

uniformly in x, z, and t for $|x - z| \geq \varepsilon$.

2)

$$\lim_{\delta t \to 0} \frac{1}{\delta t} \int dz (z - x) p(z, t + \delta t | x, t) = M(x,t) + o(\varepsilon) \tag{5}$$

3)

$$\lim_{\delta t \to 0} \frac{1}{\delta t} \int dz (z - x)^2 p(z, t + \delta t | x, t) = V(x,t) + o(\varepsilon) \tag{6}$$

Equations (5) and (6) are uniform in x, z, and t.

4) All higher-order coefficients are $o(\varepsilon)$.

Consider the time evolution of the expectation of a function $f(z)$ which is twice continuously differentiable. Thus,

$$\partial_t \int dx f(x) p(x,t|y,t') \tag{7}$$

$$= \lim_{\delta t \to 0} \frac{1}{\delta t} \left\{ \int dx f(x) \left[p(x, t + \delta t | y, t') - p(x,t|y,t') \right] \right\} \tag{8}$$

$$= \lim_{\delta t \to 0} \frac{1}{\delta t} \left\{ \int dx \int dz f(x) p(x, t + \delta t | z, t) p(z,t|y,t') - \int dz f(z) p(z,t|y,t') \right\} \tag{9}$$

Expand $f(x)$ at position z:

$$f(x) = f(z) + f'(z)(x - z) + \frac{1}{2} f''(z)(x - z)^2 + o((x - z)^2) \tag{10}$$

Now substitute Eq. (10) with Eq. (9):

$$
\begin{aligned}
&\lim_{\delta t\to 0}\frac{1}{\delta t}\left\{\int dxdz\left[f(z)+f'(z)(x-z)+\frac{1}{2}f''(z)(x-z)^2\right]\right.\\
&\qquad\left.\times p(x,t+\delta t|z,t)p(z,t|y,t')-\int dzf(z)p(z,t|y,t')\right\}\\
=&\lim_{\delta t\to 0}\frac{1}{\delta t}\left\{\int dxdz\left[f(z)+f'(z)(x-z)+\frac{1}{2}f''(z)(x-z)^2\right]\right.\\
&\qquad\left.\times p(x,t+\delta t|z,t)p(z,t|y,t')-\int dxdzf(z)p(x,t+\delta t|z,t)p(z,t|y,t')\right\}\\
=&\int dz\left\{\lim_{\delta t\to 0}\frac{1}{\delta t}\int dx\left[f'(z)(x-z)+\frac{1}{2}f''(z)(x-z)^2\right]\right.\\
&\qquad\left.\times p(x,t+\delta t|z,t)\right\}p(z,t|y,t')\\
=&\int dz\left[f'(z)M(z,t)+\frac{1}{2}f''(z)V(z,t)\right]p(z,t|y,t')\\
=&\int dzf(z)\left\{-\partial_z\left[M(z,t)p(z,t|y,t')\right]+\frac{1}{2}\partial_{zz}\left[V(z,t)p(z,t|y,t')\right]\right\}\\
\overset{z\to x}{=}&\int dxf(x)\left\{-\frac{\partial}{\partial x}\left[M(x,t)p(x,t|y,t')\right]+\frac{1}{2}\partial_{xx}\left[V(x,t)p(x,t|y,t')\right]\right\}.
\end{aligned}
\tag{11}
$$

Then, we obtain the Kolmogorov forward equation (Fokker–Planck equation)

$$
\partial_t p(x,t|y,t')=-\frac{\partial}{\partial x}\left[M(x,t)p(x,t|y,t')\right]+\frac{1}{2}\frac{\partial^2}{\partial x^2}\left[V(x,t)p(x,t|y,t')\right]. \tag{12}
$$

Define $M(x)$ as the probability that x increases by systematic force that includes mutation and selection. And define $V(x)$ as the probability that x changes because of random drift, either decreasing by amount δx with the probability $V(x)/2$ or increasing by the amount δx with the probability $V(x)/2$.

$$
M(x)=\lim_{\delta t\to 0}\frac{1}{\delta t}\int_0^1 \delta x W(x,t+\delta t|x-\delta x,t)d(\delta x), \tag{13}
$$

$$
V(x)=\lim_{\delta t\to 0}\frac{1}{\delta t}\int_0^1 (\delta x)^2 W(x,t+\delta t|x-\delta x,t)d(\delta x). \tag{14}
$$

2.2 Relation for the Solutions of Between Forward Fokker–Planck Equation and Backward Fokker–Planck Equation

In most physical applications of Kolmogorov diffusion equations the coefficient $V(x)$ is essentially positive. However, in certain applications we encounter equations such that $V(x)$ vanishes at one (or possibly both) of the boundaries or one of the coefficients has no finite limit. Equations with coefficients such as described are called singular diffusion equations.

The classification of boundaries depends on the Lebesgue integrability of the function

$$g_1(x) = \exp\left\{-\int_{x'}^{x} \frac{M(z)}{V(z)} dz\right\} \tag{15}$$

where $x' \in (r_1, r_2)$ is fixed, and related functions on a prescribed open interval contained in (r_1, r_2). The above function was introduced by Feller [19]. Let $I_i (i = 1, 2)$ denote the interval (x', r_i). The function $g_1(x)$ is Lebesgue integrable on I_i (written $g_1(x) \in L(I_i)$) if

$$\int_{I_i} g_1(x)dx < \infty \tag{16}$$

i.e., the integral of $g_1(x)$ over the interval I_i is bounded. Before giving the classification criteria, we introduce the following functions:

$$g_2(x) = \frac{1}{V(x)g_1(x)} \qquad h(x) = g_1(x) \int_{x'}^{x} g_2(z)dz \tag{17}$$

Then Feller classifies the boundaries as the following

1. The boundary r_i is *regular* if $g_1(x) \in L(I_i)$ and $g_2(x) \in L(I_i)$.
2. The boundary r_i is an *exit* boundary if $g_2(x) \notin L(I_i)$ and $h(x) \in L(I_i)$.
3. The boundary r_i is an *entrance* boundary if $g_2(x) \in L(I_i)$ and $g_2(x) \int_{x'}^{x} g_1(z) dz \in L(I_i)$.
4. In all other cases the boundary is called *natural*.

By utilizing the theory of semigroups, Feller has obtained the following results which relate the existence and uniqueness problem of diffusion equations to that of classifying the boundaries:

1. When none of the boundaries is regular, there exists exactly one fundamental solution (or Green's function) common to the forward and backward diffusion equations, even though the initial value problem as such may have many solutions.

2. When one boundary is regular, or when both boundaries are regular, there exist infinitely many common fundamental solutions.
3. When both boundaries are natural, the initial value problem for both forward and backward diffusion equations is uniquely determined, and the solutions are generated by a common fundamental solution.
4. When r_1 is a natural boundary and r_2 is an exit boundary, the initial value problem for the backward diffusion equation has infinitely many solutions, but that for the forward diffusion equations is uniquely determined.
5. When r_1 is a natural boundary but r_2 is a regular boundary, there exist infinitely many solutions for the initial value problem for both the forward and backward diffusion equations.
6. When neither boundary is natural, there are two sources for nonuniqueness, and in these cases two lateral conditions must be imposed.

3 Construction of Adaptive Landscape

Diffusion theory is an elegant approximation for analyzing population evolution. To construct adaptive landscape, we start from one-dimensional forward Fokker–Planck equation:

$$\frac{\partial}{\partial t}\rho(x,t) = -\frac{\partial}{\partial x}[M(x)\rho(x,t)] + \frac{1}{2}\frac{\partial^2}{\partial x^2}[V(x)\rho(x,t)]. \tag{18}$$

Among this $M(x)$ is the symbol for the average change in allele frequency [20, 21] that occurs due to systematic force. The function $V(x)$ is the average square change in allele frequency.

The diffusion process can also be expressed by the following symmetric equation

$$\partial_t\rho(x,t) = \partial_x[\epsilon D(x)\partial_x - f(x)]\rho(x,t) \tag{19}$$

with

$$\begin{aligned} f(x) &= M(x) - \epsilon D'(x), \\ 2\epsilon D(x) &= V(x). \end{aligned} \tag{20}$$

With a prime denoting differentiation of a function with respect to its argument such as $D'(x) \equiv \partial_x D(x)$ where $M(x)$ and $V(x)$ is from Eq. (18). Adaptive landscape is directly given when we consider natural boundary as Feller's classification. It is

$$\Phi(x) = \int^x dx' \frac{f(x')}{D(x')}. \tag{21}$$

The symmetric Eq. (19) has two advantages. On the one hand, the adaptive landscape is directly read out when the detailed balance is satisfied. On the other hand, the constructive method is dynamical, independent of existence and normalization of stationary distribution. We call $f(x)$ directional transition rate, integrating the effects of $M(x)$ and the derivative of $V(x)$. Directional transition rate can give equilibrium states when it appears in linear form.

The adaptive landscape not only avoids the mismatch of the fixed points of the force and those of extremals of steady state distribution, but also can be related to the dynamical behaviors.

The stationary distribution for the diffusion approximation satisfying natural boundary condition is given by

$$\rho(x, t = \infty) = \frac{1}{Z} \exp\left(\frac{\Phi(x)}{\epsilon}\right). \tag{22}$$

$$Z = \int_{-\infty}^{+\infty} dx \exp\left(\frac{\Phi(x)}{\epsilon}\right). \tag{23}$$

It has the form of Boltzmman–Gibbs distribution [12], so the scalar function $\Phi(x)$ naturally acquires the meaning of potential energy [22]. The value of Z determines the normalization of $\rho(x, t = \infty)$ from the perspective of probability, and the finite value of Z manifests the normalization of $\rho(x, t = \infty)$. The stationary distribution is not true in the face of infinite Z. It demonstrates absorbing phenomenon occurs at the boundary. Together with the flux at the boundary, the true stationary distribution could be got. The constant ϵ holds the same position as temperature of Boltzmman–Gibbs distribution in statistical mechanics. But it does not hold the nature of temperature in Boltzmman–Gibbs distribution.

4 Two Applications

As the concept and their technical implementation described, below we shall show the singular adaptive landscape describes internal equilibria for constant population size and variable biological factors such as original Wright–Fisher process and Muller ratchet.

4.1 *Wright–Fisher Process*

The simplest stochastic model of an evolving population dates from the 1930s and was introduced independently by Ronald Fisher [23] and Sewall Wright [24]. In the model, constant organisms from one generation to the next generation are considered, and each instance of a gene in one generation is an exact copy of one randomly chosen with replacement from previous generation [25]. Let us consider first of all diploid organisms which carry two copies of each gene and reproduce sexually. A new generation could be formed by each of the organisms producing two infinite sets of gametes before dying. Among this process mutation, selection and random drift could occur. Then random fixed size samples of these gametes survive to become offspring organisms. A random mating diploid population evolving in this way is often referred to as an ideal population. In reality, individuals do not mate at random. Some factors such as individuals' geographical locations and age may also lead to deviations from an ideal population. Nevertheless, in some cases it is found that the predictions of the ideal model are relevant [25].

Suppose we have an ideal population of individuals satisfying (i) Generations are taken to be discrete, so that the population evolves by a discrete-step Markov process. (ii) The population size is taken to be fixed, so that alleles compete only against other alleles and not against an external environment. (iii) Random mating is assumed. Concretely we consider a population of diploid sexual individuals with population size being N and factors such as one locus with two alleles A_1, A_2. The start point in a generation is taken to be the adult stage, after all mutation and selection has occurred and selection immediately prior to reproduction. The regulation of the population number is supposed to occur through nonoverlapping generations and randomly picking N individuals from the population, then there are always $2N$ alleles in the allele pool in any generation. The treatment is consistent with [26].

We focus on the evolution of allele A_1. Assume p_i to represent the proportion or relative frequency of allele A_1 in generation t, that is $p_i = i/2N, i = 0, \dots, 2N$. Mathematically under general diffusion approximation, frequency p_i are treated as continuous quantity x, and this also leads to the distribution of the frequency of considered allele A_1 being probability density.

4.2 *Dynamics of Original Wright–Fisher Process*

Adaptive landscape can be quantified as potential function. Potential function can give both quantitative and qualitative description of behaviors near steady states or metastable states. Great efforts have been spent to find such a potential function [13]. We here give specific potential function corresponding to diverse biological factors. It is clear how our potential function is related to the dynamical trajectories.

4.2.1 Pure Drift Balance

In the past, people also come into notice the effect of random drift, for example, [27] studied the effect through assuming that population fixed for a single genotype. Krakauer and Plotkin [28] studied the effect of random drift through analyzing small perturbations from the deterministic equilibrium. de Vladar and Barton [15] studied the same process, but the method of maximizing entropy cannot work because of non-normalized stationary distribution. Waxman [29] studied the random drift process by assuming Dirac function at the boundary. We here do not focus on the boundary condition. To avoid the difficult boundary problem, here we show our method is independent of the normalization of stationary distribution and can directly get the population behaviors by adaptive landscape. We could address it from the perspective of adaptive landscape. For random sampling variance being

$$V(x) = \frac{x(1-x)}{2N}, \tag{24}$$

at the same time the directional force reads

$$f(x) = \frac{2x-1}{4N}. \tag{25}$$

In addition, we are interested in the dynamical property of adaptive landscape, so we treat Φ and Φ/ϵ no difference in this respect, that is, for convenience we can take $\epsilon = 1$ of $\epsilon D(x)$. So adaptive landscape from Eq. (21) is expressed as

$$\Phi(x) = -\ln x(1-x). \tag{26}$$

For biological understanding, no mutation equals to no driving force, without driving force the stationary distribution should be plain. That is, each state has the same chance to be ultimate state as neutral evolution, but in fact because of inertia of driving force, there exists the effect of force in the system, this makes the system move on. This is consistent with the expression of directional force $f(x)$ in Eq. (20). The inertia of driving force can just balance out the effect of random drift and let the population move on. Here from the expression of adaptive landscape, two singular points with allele frequency $x = 0$ and $x = 1$ exist. The singularity means that the population lies at a stable or unstable state. Positive infinity means the state with corresponding allele frequency stable, while negative infinity means the state with corresponding allele frequency unstable. Here the values of adaptive landscape for two singular points go to positive infinity. These demonstrate that there are indeed two stable states in the process. The adaptive landscape has U-shape.

Directional force actually gives the effects of driving force and random drift. Two stable states with infinite potential occur in the process. The fixed point with allele frequency $x = 1/2$ could be got through letting directional force being zero directly. It is the critical point to decide the eventual stable state of an initial state.

4.2.2 Mutation-Drift Balance

So far we have been considering situations where the change in composition of populations is caused by pure random drift. We now include the effect of mutation:

an A_1 allele may mutate with a probability μ to an A_2 allele, and an A_2 allele may mutate with a probability ν to an A_1 allele. These parameters are probabilities per generation. In the process we assume there is no difference on viability because of the absence of selection, the evolutionary process is only under the condition of mutation and random drift with unchanged environment. We focus on the evolution of the allele A_1. Previous neutral evolution regarded these evolutionary processes as neutral.

To include mutation, one chooses two alleles from the current population to die and replaces it with two types chosen. That is, when we pick alleles to be the parent of a child in the next generation, the offspring can mutate with the probability μ or ν. For example, if an allele A_1 is chosen, there is a probability $1 - \mu$ that replacement allele is also an A_1 and of μ that it is an A_2. So with mutation, in the offspring generation the frequency of allele A_1 is

$$p_{t+1} = (1 - \mu)p_t + \nu(1 - p_t). \tag{27}$$

After diffusion approximation like pure random drift process, the average change of allele frequency x: $M(x)$ and random sampling variance $V(x)$ are, respectively,

$$M(x) = -\mu x + \nu(1 - x), \tag{28}$$

$$V(x) = \frac{x(1 - x)}{2N}, \tag{29}$$

then $f(x)$, $D(x)$ in Eq. (20) are

$$\begin{aligned} f(x) &= -\mu x + \nu(1 - x) - \frac{1 - 2x}{4N} \\ &\equiv F(x - a), \end{aligned} \tag{30}$$

$$\epsilon D(x) = \frac{x(1 - x)}{4N}, \tag{31}$$

with

$$F = \frac{1 - 2N\mu - 2N\nu}{2N}, \quad a = \frac{1 - 4N\nu}{2 - 4N\mu - 4N\nu}. \tag{32}$$

Here we also take $\epsilon = 1$. So the adaptive landscape corresponding to Eq. (21) reads

$$\Phi(x) = 4N\nu \ln x + 4N\mu \ln(1 - x) - \ln x(1 - x). \tag{33}$$

There are three elements composed of adaptive landscape. The first and the second elements represent the effect due to mutation. The last element is the effect due to random drift. There exist two singular points with allele frequency $x = 0$ and $x = 1$ from the expression of adaptive landscape. Singularity indicates the state stable or unstable. We show below that the evolutionary behavior can still be explored in terms of adaptive landscape. This method is insensitive to the singular expression. Fixed points are derived by

$$\Phi'(x) = 0. \tag{34}$$

That is

$$(4N\mu - 1 + 4N\nu - 1)x - (4N\nu - 1) = 0. \tag{35}$$

Then we get only one fixed point with allele frequency $x = a$ (its expression is Eq. (32)), and it is also the zero point of $f(x)$. In another word, when directional force is expressed in linear form, the fixed point can be read from its form directly.

We could investigate the dynamics by analyzing the relative position of the fixed point and singular points. In the following we give the detailed information.

(i) $a < 0$

There are two regimes holding the same dynamical behavior. In one of the regimes parameters satisfy $\mu \in (0, 1/4N)$ and $\nu \in (1/4N, (1 - 2N\mu)/2N)$. In another regime parameters satisfy $\mu \in (1/4N, 1)$ and $\nu \in ((1 - 2N\mu)/2N, 1/4N)$. There are two unstable states and one stable state in the process. The states at the fixed point with allele frequency $x = a$ and at the singular point with allele frequency $x = 0$ are unstable while the state with allele frequency $x = 1$ is stable. This means the population tends to fix at the state with allele frequency $x = 1$. Alleles composed of the population are likely to be allele A_1.

(ii) $a = 0$

There are two regimes having the same dynamical behavior. In one of the regime mutation rates satisfy $\mu \in (0, 1/4N)$ and $\nu = 1/4N$. In another regime mutation rates satisfy $\mu \in (1/4N, 1)$ and $\nu = 1/4N$. Among these regimes the fixed point mixed the singular point with allele frequency $x = 0$. Among these cases the factor which can dominate the process is mutation rate. If parameters satisfy $\mu \in (0, 1/4N)$ and $\nu = 1/4N$, the probability of mutation from allele A_2 to allele A_1 is greater. So there is only one stable state with allele frequency $x = 1$ and one unstable state with allele frequency $x = 0$ in the process. If parameters satisfy $\mu \in (1/4N, 1)$ and $\nu = 1/4N$, the probability of mutation from allele A_1 to allele A_2 is greater. So there is one stable state with allele frequency $x = 0$ and one unstable state with allele frequency $x = 1$ in the process.

(iii) $0 < a < 1$

Two regimes have the same dynamical behavior. In one of the regimes mutation rates satisfy $\mu \in (0, 1/4N]$ and $\nu \in (0, 1/4N)$. In another regime mutation rates satisfy $\mu \in [1/4N, 1)$ and $\nu \in (1/4N, 1)$. The fixed point in the two regimes lies at the interval $(0, 1)$. Among these cases the factor which can determine stability is mutation rate. If parameters satisfy $\mu \in (0, 1/4N]$ and $\nu \in (0, 1/4N)$, the probability of mutation from allele A_2 to allele A_1 and the probability of mutation from allele A_1 to allele A_2 are very small. There is only one unstable state with allele frequency $x = a$ and two stable states with allele frequency $x = 1$ and $x = 0$ in the process. Ultimately which state the population tends to fix is determined by the initial state. If its initial state with allele frequency is less than $x = a$, the population would move to the state with allele frequency $x = 0$. If parameters satisfy $\mu \in [1/4N, 1)$ and $\nu \in (1/4N, 1)$, the probability of mutation from allele A_2 to allele A_1 and the probability of mutation from allele A_1 to allele A_2 are very great. In one generation mutation number $2N(\mu + \nu)$ is greater than one. None of the two mutation directions can dominate the process. This results in individuals with genotype of two different genes becoming much. There is one stable state with allele frequency $x = a$ and two unstable states with allele frequency $x = 0$ and $x = 1$ in the process. Ultimately the population tends to fix at the state with allele frequency $x = a$. The population moves to the stable state with genotype A_1A_2.

(iv) $a = 1$

There is only one point with the dynamical behavior. If parameters satisfy $(\mu, \nu) = (1/4N, 1/4N)$, the potential of the whole population is plain. In fact any state has equal chance to be the ultimate state. That is, the effects of mutation and random drift have no influence on the allele change. The effect of mutation is offsetted by that of random drift. The evolution has no bias on any state. We call this process new neutral evolution. Previous neutral evolution described the dynamical processes in the absence of selection. It has bias on some special states such as the case driven by unequal effects of mutation and random drift. Its description is not completely neutral.

(v) $a > 1$

There are two regimes with the same dynamical behavior. In one of the regimes parameters satisfy $\mu \in (0, 1/4N)$ and $\nu \in (1 - 2N\mu)/2N, 1]$. In another regime parameters satisfy $\mu \in (1/4N, 1/2N)$ and $\nu \in (0, (1 - 2N\mu)/2N]$. Among these cases stronger mutation rate from one of the two directions dominates the process. If parameters satisfy $\mu \in (0, 1/4N)$ and $\nu \in [1 - 2N\mu)/2N, 1)$, the probability of mutation from allele A_2 to allele A_1 is greater. So there is one stable state with allele frequency $x = 1$ and one unstable state with allele frequency $x = 0$ in the process. Ultimately the population tends to fix at the state with allele frequency $x = 1$. If parameters satisfy $\mu \in (1/4N, 1/2N)$ and $\nu \in (0, (1 - 2N\mu)/2N]$, the probability of mutation from allele A_1 to allele A_2 is greater. So there is one stable state with allele frequency $x = 0$ and one unstable state with allele frequency $x = 1$ in the process. Ultimately the population tends to fix at the state with allele frequency $x = 0$.

We theoretically analyze the dynamical behavior of the population and give the critical point in parameters space. Furthermore we can get the cause of the evolution clearly.

4.2.3 Selection-Drift Balance

The process in the presence of selection, and random drift is much complex. We only focus on some special cases to show how the process is characterized by adaptive landscape. Natural selection works when genotypes have different fitnesses [30]. But the dynamics of selection depends on relative fitness. Suppose we denote the relative fitness of selection on genotype A_1A_1 by $1 + s$, A_1A_2 by $1 + sh$, A_2A_2 by 1. Here s is called selection coefficient, a measure of the fitness of A_1A_1 relative to that of A_2A_2 and h is called heterozygote effect [30], a measure of the fitness of the heterozygote relative to the selective difference between the two homozygotes. We here assume $0 < s < 1$, A_1A_1 is fitter than A_2A_2. Let x be the frequency of allele A_1, then we can define mean fitness as

$$\overline{\omega} = x^2(1+s) + 2x(1-x)(1+sh) + (1-x)^2. \tag{36}$$

For the case of selection-random drift, the average frequency change of allele A_1 is

$$M(x) = \frac{x(1-x)}{2\overline{\omega}}\frac{d\overline{\omega}}{dx}. \tag{37}$$

For convenience we only assume additive fitness, put differently, it depicts the important evolutionary process parameter satisfying $h = 1/2$: additive selection-random drift process. So according to Eq. (20)

$$\begin{aligned} f(x) &= M(x) - \epsilon D'(x) \\ &= \frac{sx(1-x)}{2(1+xs)} - \frac{1-2x}{4N}. \end{aligned} \tag{38}$$

Potential function, namely, adaptive landscape derived analogous to Eq. (21) ($\epsilon = 1$) reads

$$\Phi(x) = 2N\ln(1+xs) - \ln x(1-x). \tag{39}$$

As we can see, there are two singular points with allele frequency $x = 0$ and $x = 1$ from the expression of adaptive landscape. Singularity means the population stable in the process.

Analogous to analysis of mutation-random drift balance, we get the fixed point from

$$\Phi'(x) = 0. \tag{40}$$

They are

$$x_{1,2} = \frac{2 + (2N-1)s \mp \sqrt{(2N-1)^2s^2 + 4s + 4}}{4s(N-1)}. \tag{41}$$

After some trivial calculation, we can get the relative positions of fixed points and singular points. They must satisfy the relation $0 < x_1 < 1 < x_2$ under the condition $0 < s < 1$. We can get two stable states with allele frequency $x = 0$ and $x = 1$ from the expression of adaptive landscape. In fact because of the relative positions of these points, there is only one stable state with allele frequency $x = x_1$. It is the critical state that divides the effect regimes into two stable states. This case describes that many alleles with very small effect on fitness are close to additive. That is, the heterozygote fitness is the arithmetic mean of the fitness of the two homozygotes.

4.3 Escape Time from Infinite Potential

Because the metastability is such an important phenomenon, it is expected to compute the life time of metastable states. It is an open question that when the metastable states become unstable based on potential function, such as represented by Kramers' escaping rate formulae [31].

For escape time, it is defined that the object leaves from one potential basin of an stable state and never comes back. If we take the initial state as a potential peak, the exit state as a potential valley. Then the mean first passage time can approximate escape time in general bell-shaped potential. Here we estimate the average escape time by computing mean first passage time for two cases. One is of mutation, random drift, the result is the same as [32]. The other is of selection and random drift.

If we assume transition process occurs in the presence of weak mutation and random drift under the condition of $4N\mu \to 0, 4N\nu \to 0$, the corresponding potential is U-shape. When first the population lies at the basin of stable state with allele frequency $x = 0$, then it arrives at the state with allele frequency $x = a$, and bounces back and forth and eventually reaches the state with allele frequency $x = 1$. Because the potential difference is infinite and adaptive landscape has U-shape, the time it climbed over the potential is very longer. General mean first passage time about a population leaving $(0, a)$ under the circumstance of initial Dirac distribution satisfies

$$(f(x) + \epsilon D'(x))\partial_x T(x) + \epsilon D(x)\partial_x^2 T(x) = -1. \tag{42}$$

Further we set

$$T'(0) = 0, \tag{43}$$

$$T(a) = 0. \tag{44}$$

Then we get

$$T(x) = \int_x^a \frac{1}{\epsilon D(y)} \exp(-\Phi(y))dy \int_0^y \exp(\Phi(z))dz. \tag{45}$$

Here $\Phi(x) = \int_0^x f(x')/D(x')dx'(\epsilon = 1)$ is adaptive landscape.

In general, when the landscape is sharply peaked at $x = a$, the main contribution to the integral Eq. (45) is due to a small region near $x = a$. But here there is a fat potential valley rather than a thin one. This results in the contribution to the integral Eq. (45) from a big region around $x = a$, and at the same time $\Phi'(a) = 0$, the probability from $x = 0$ to $x = a$ equals to that from $x = a$ to $x = 1$. So the escape time from $x = 0$ to $x = 1$ can be approximated to two mean first passage time from $x = 0$ to $x = a$. Because escape time is when the population leaves the basin of stable state with allele frequency $x = 0$, the escape time in the presence of weak mutation and random drift is computed as

$$\begin{aligned}
T &\approx 2 \times T_{MFPT}(0 \to a) \\
&= 8N \int_0^a y^{-4N\nu}(1-y)^{-4N\mu} dy \int_0^y z^{4N\nu-1}(1-z)^{4N\mu-1} dz \\
&= \frac{2(a-0)}{\nu} + \frac{8N\mu}{\nu} \sum_{n=1}^{\infty} \frac{a^{n+1}}{n+1} \prod_{k=2}^{n} \left(\frac{k-1+4N\mu}{k} \right) \\
&\quad +8N(1-4N\mu) \sum_{n=1}^{\infty} \frac{a^{n+1}}{(n+1)(n+4N\nu)} \prod_{k=2}^{n} \left(\frac{k-4N\mu}{k} \right) \\
&= \frac{1}{\nu} + \frac{4N\mu}{\nu} \sum_{n=1}^{\infty} \frac{2^{-n}}{n+1} \prod_{k=2}^{n} \left(\frac{k-1+4N\mu}{k} \right) \\
&\quad +4N(1-4N\mu) \sum_{n=1}^{\infty} \frac{2^{-n}}{(n+1)(n+4N\nu)} \prod_{k=2}^{n} \left(\frac{k-4N\mu}{k} \right).
\end{aligned} \tag{46}$$

Another way to get the escape time is that approximated mean first passage time from $x = 0$ to $x = 1$. It is

$$\begin{aligned}
\tau_1 &\approx T_{MFPT}(0 \to 1) \\
&= 4N \int_0^1 y^{-4N\nu}(1-y)^{-4N\mu} dy \int_0^y z^{4N\nu-1}(1-z)^{4N\mu-1} dz \\
&= \frac{1}{\nu} + \frac{4N\mu}{\nu} \sum_{n=1}^{\infty} \frac{1}{n+1} \prod_{k=2}^{n} \left(\frac{k-1+4N\mu}{k} \right) \\
&\quad +4N(1-4N\mu) \sum_{n=1}^{\infty} \frac{1}{(n+1)(n+4N\nu)} \prod_{k=2}^{n} \left(\frac{k-4N\mu}{k} \right).
\end{aligned} \tag{47}$$

These results demonstrate that the stable state with infinite potential has the probability to be unstable. It gives a quantitative measure of escape with infinite potential. From Eqs. (46) and (47), we can see the results are the same if the first element $1/\nu$ approximates the escape time.

If we assume transition process occurs in the presence of weak selection and random drift under the condition of $4Ns \to 0$, we can get the escape time leaving the basin of stable state with allele frequency $x = 0$ by the same procedure. It is

$$\begin{aligned}\tau_2 &\approx T_{MFPT}(0 \to 1) \\ &= 4N \int_0^1 \frac{4N}{y(1-y)} e^{-2N\ln(1+ys)+\ln y(1-y)} dy \int_0^y e^{2N\ln(1+zs)-\ln z(1-z)} dz \\ &\to \infty. \end{aligned} \tag{48}$$

Because the integral about the variable y is not integral at $y = 0$. Here infinite potential indeed means infinite escape time. So the above two infinite potentials have different meanings. One is true infinite, the other is false infinite.

4.4 Muller Ratchet

Muller's ratchet proposed in 1964 [33] that the genome of an asexual population accumulates deleterious mutations in an irreversible manner. It is a mechanism that has been suggested as an explanation for the evolution of sex [34]. For asexually reproducing population, without recombination, chromosomes are directly passed down to offsprings. As a consequence, the deleterious mutations accumulate so that the fittest class loses. For sexually reproducing population, because of the existence of recombination between parental genomes, a parent carrying high mutational loads can have offspring with fewer deleterious mutations. The high cost of sexual reproduction is thus offsetted by the benefits of inhibiting the ratchet [35]. Muller's ratchet has been received growing attention.

Here in one-dimensional case, we consider one locus with two alleles (for example, A and a), that is, there are two classes in the haploid asexual population, one class with allele A while the other with allele a, supposed mutation from allele A to a is deleterious. We assume fixed population size of N, which means we have N alleles in all. We also assume that $N > 1$. Generations are non-overlapping. The lifecycle of the individuals in the population is from adults to juveniles, during which we consider irreversible mutation, selection, and random drift. The frequency of the allele A for generation t is p while that of allele a is $1 - p$. Let μ be the probability that an offspring of an adult with allele A is an individual with allele a, labeled by $M_{1,0}$, that is $M_{1,0} = \mu$. Analogously, $M_{0,0} = 1-\mu$, $M_{0,1} = 0$, $M_{1,1} = 1$. The relative viability of individuals with allele A is $\nu_0 = 1$ while that of individuals with allele a is $\nu_1 = 1-\sigma$ where σ can be treated as an effective selection coefficient associated with deleterious mutations. So the values of parameters for μ and σ are from 0 to 1. Then in generation $t + 1$, when selection and deleterious mutation are active, the probability that the offspring of a parent with allele A is chosen to be with allele a is $\mu p(1 - \sigma)$, the probability that the offspring with allele A is $(1 - \mu)p$, the probability that the offspring of a parent with allele a is still with allele a is $(1-\sigma)(1-p)$. So the frequency of allele A in generation $t+1$ is

$$p' = \frac{(1-\mu)p}{1-\sigma+\sigma(1-\mu)p}. \tag{49}$$

After using diffusion approximation, frequency p is treated as continuous quantities x, and this leads to the distribution of the frequency for the allele A being the probability density. Let $\rho(x,t)$ be the probability density of the frequency for the allele A being x at time t. Then it obeys the dynamical equation

$$\frac{\partial}{\partial t}\rho(x,t) = -\frac{\partial}{\partial x}[M(x)\rho(x,t)] + \frac{1}{2}\frac{\partial^2}{\partial x^2}[V(x)\rho(x,t)], \tag{50}$$

and according to the definition of $M(x)$ and $V(x)$, the explicit expressions of them are

$$\begin{aligned} M(x) &= \frac{(1-\mu)x}{1-\sigma+\sigma(1-\mu)x} - x \\ &= \frac{x[(\sigma-\mu)-\sigma(1-\mu)x]}{1-\sigma+\sigma(1-\mu)x}, \end{aligned} \tag{51}$$

$$V(x) = \frac{x(1-x)}{N}. \tag{52}$$

So according to Eq. (21) ($\epsilon = 1$), we have adaptive landscape as the following

$$\begin{aligned} \Phi(x) = &\frac{2N\mu(1-\sigma)}{1-\sigma\mu}\ln(1-x) - \ln x(1-x) \\ &+\frac{2N(1-\mu)}{1-\sigma\mu}\ln(1-\sigma+x\sigma(1-\mu)). \end{aligned} \tag{53}$$

From the expression of adaptive landscape $\Phi(x)$, we may find there are two singular points with allele frequency $x = 0$ and $x = 1$ of adaptive landscape, characterized by infinite value. Infinity means the state relative stable or unstable in the system.

4.5 Dynamics of Muller Ratchet

To understand the mechanism of Muller's ratchet, a full characterization of dynamical process is a prerequisite for obtaining more accurate decaying time. Here we study the dynamical behavior by investigating the positions and relative stabilities of all fixed points as [36]. We further derive the parameter regions for all possible cases.

According to general analysis of a dynamical system, letting

$$\Phi'(x) = 0, \tag{54}$$

we get

$$2\sigma(1-\mu)(N-1)x^2 + (2N(\mu-\sigma) + 3\sigma - \sigma\mu - 2)x + (1-\sigma) = 0. \tag{55}$$

We solved Eq. (55) and found two fixed points. If we denote

$$\alpha = 2 - 3\sigma + \sigma\mu + 2N\sigma - 2N\mu, \tag{56}$$

$$\beta = 8\sigma(1-\mu)(N-1)(1-\sigma). \tag{57}$$

They are

$$x_{1,2} = \frac{\alpha \mp \sqrt{\alpha^2 - \beta}}{4\sigma(1-\mu)(N-1)}. \tag{58}$$

For two singular points with allele frequency $x = 0$ and $x = 1$, if allele frequency x goes to 1, and parameter satisfies $\sigma \in (\mu, (2N\mu - 1)/(2N\mu - \mu))$, then the value of adaptive landscape $\Phi(1)$ goes to $-\infty$, so the population is unstable at the state with allele frequency $x = 1$. When allele frequency x moves to 1, and selection satisfies $\sigma \in ((2N\mu - 1/(2N\mu - \mu), 1)$, the value of adaptive landscape $\Phi(1)$ tends to $+\infty$, so the population is stable at the state with allele frequency $x = 1$. For allele frequency x goes to 0, the value of adaptive landscape $\Phi(0)$ goes to $+\infty$ in almost parameters regimes except $\sigma = 1$, so the population is always stable at the state with allele frequency $x = 0$. When selection satisfies $\sigma = 1$, the viability of the suboptimal class is zero, so the population stays at the initial state.

Here we address dynamical behaviors by the positions of two real inequivalent fixed points with allele frequency $x_1 < x_2$ first.

For convenience, we denote

$$\sigma_1 = \frac{(2+2\mu-10N\mu+4N^2\mu+2N\mu^2+4(1-\mu)\sqrt{N(N-1)((2N-1)\mu-1)}}{(\mu-2N+1)^2}.$$

1. We find two different real fixed points in two regimes. When parameters satisfy $\mu \in (0, 2/(2N-1+2\sqrt{N(N-1)}))$ and $\sigma \in (\mu, 1)$, parameters satisfy $\mu \in (2/(2N-1+2\sqrt{N(N-1)}), 1)$ and $\sigma \in (\sigma_1, 1)$ two different real fixed points occur. Among them parameters regimes do not include $\mu \in ((2N-1)/4N(N-1), 1)$ and $\sigma = (2N\mu - 1)/(2N\mu - \mu)$.

We discuss the relative positions for the fixed points and the singular points with allele frequency $x = 0$, $x = 1$ and stabilities of them in the following.

(i) $1 < x_1 < x_2$

In one of the regimes parameters satisfy $\mu \in (1/(2N-1), 2/(2N-1+2\sqrt{N(N-1)})]$ and $\sigma \in (\mu, (2N\mu-1)/(2N\mu-\mu))$, in another regime parameters satisfy $\mu \in (2/(2N-1+2\sqrt{N(N-1)}), (2N-1)/4N(N-1))$ and $\sigma \in (\sigma_1, (2N\mu-1)/(2N\mu-\mu))$, the fixed points satisfy $1 < x_1 < x_2$. At the same time the state with singular point $x = 1$ is unstable. There is one stable state with allele frequency $x = 0$ in the system. The population tends to evolve to the stable state.

(ii) $1 = x_1 < x_2$

In the regions parameters satisfy $\mu \in (1/(2N-1), (2N-1)/4N(N-1))$ and $\sigma = (2N\mu-1)/(2N\mu-\mu)$, the two fixed points satisfy allele frequency $x_1 = 1$, $1 < x_2$. The state with allele frequency $x = 1$ is unstable. There is only one stable state with allele frequency $x = 0$ in the system.

(iii) $0 < x_1 < 1 < x_2$

In one of the regimes parameters satisfy $\mu \in (0, 1/(2N-1))$ and $\sigma \in (\mu, 1)$, in another regime parameters satisfy $\mu \in (1/(2N-1), 1)$ and $\sigma \in ((2N\mu-1)/(2N\mu-\mu), 1)$, the fixed points satisfy $0 < x_1 < 1 < x_2$. There is only one unstable state with allele frequency $x = x_1$ in the system, and two stable states with allele frequency $x = 1$ and $x = 0$ exist in the system. The population tends to evolve to which stable state dependent on the position of the initial state. If the initial state with allele frequency is greater than x_1, the population tends to evolve to the stable state with allele frequency $x = 1$.

(iv) $0 = x_1 < 1 < x_2$

In the regime parameters satisfy $\mu \in (0, 1)$ and $\sigma = 1$, the fixed points satisfy allele frequency $x_1 = 0$, $1 < x_2$. When selection rate satisfies $\sigma = 1$, the process stays at the initial state. Because for this case the viability of the sub-fittest class is zero.

(v) $0 = x_1 < x_2 < 1$

The case $0 = x_1 < x_2 < 1$ is impossible. For $x_1 = 0$, the other parameter must satisfy $\sigma = 1$, at the same time x_2 must be greater than one.

(vi) $0 < x_1 < x_2 < 1$

In one of the regimes parameters satisfy $\mu \in ((2N-1)/4N(N-1), (2N-1)/(4N-3))$ and $\sigma \in (\sigma_1, (2N\mu-1)/(2N\mu-\mu))$, in another regime parameters satisfy $\mu \in ((2N-1)/(4N-3), 1)$ and $\sigma \in (\sigma_1, (2N\mu-1)/(2N\mu-\mu))$, the fixed points satisfy $0 < x_1 < x_2 < 1$. The state with allele frequency x_1 is unstable while that with allele frequency x_2 is stable. There are two stable states with allele frequency $x = 0$ and $x = x_2$ and two unstable states with allele frequency $x = 1$ and $x = x_1$ in the system. The population evolves to which stable states dependent on the initial position.

(vii) $x_1 < 0$ or $x_2 < 0$

The case $x_1 < 0$ is impossible, and the case $x_2 < 0$ is impossible.

2. Then we discuss the case of two equivalent real fixed points with allele frequency $x_2 = x_1$.

In the regimes of parameters satisfying $\mu \in (2/(2N - 1 + 2\sqrt{N(N-1)}), 1)$ and $\sigma = \sigma_1$, we find two same fixed points with allele frequency

$$x_{1,2} = \frac{\alpha}{4\sigma(1-\mu)(N-1)}. \tag{59}$$

(i) $1 < x_{1,2}$

In the regimes of parameters satisfying $\mu \in (2/(2N-1+2\sqrt{N(N-1)}), (2N-1)/4N(N-1))$ and $\sigma = \sigma_1$, there are two same fixed points satisfying $1 < x_{1,2}$, and they are unstable. There is one stable state with allele frequency $x = 0$ in the process.

(ii) $1 = x_{1,2}$

At the two points of $((2N-1)/4N(N-1), 2N/(2N-1)^2)$ and $((2N-1)/(4N-3), (4(N-1)(3-6N+4N^2)+8(N-1)(4N-3)\sqrt{N(N-1)/(4N-3)})/(2N-1)^2)$, there are two same fixed points satisfying $x_{1,2} = 1$, and they are unstable. There is one stable state with allele frequency $x = 0$ in the process.

(iii) $0 < x_{1,2} < 1$

In one of the regimes parameters satisfy $\mu \in ((2N-1)/4N(N-1), (2N-1)/(4N-3))$, $\sigma = \sigma_1$, in another regime parameters satisfy $\mu \in ((2N-1)/(4N-3), 1)$ and $\sigma = \sigma_1$, there are two same fixed points satisfying $0 < x_{1,2} < 1$, and they are unstable. There is one stable state with allele frequency $x = 0$ in the process.

3. Finally we consider two imaginary fixed points $|x_1| = |x_2|$ where the $|.|$ denotes the length for an imaginary points.

In the regime of parameters satisfying $\mu \in (2/(2N-1+2\sqrt{N(N-1)}), 1)$ and $\sigma \in (\mu, \sigma_1)$, there are two imaginary fixed points in the system. There is only one stable state with allele frequency $x = 0$. The population always evolves to the stable state.

4.6 *Estimation of Single Click Time*

To evaluate the single click time and show the further power of adaptive landscape, in the following we will demonstrate how the single click time from one relative stable state to another is derived for Muller ratchet. We also start from backward

Fokker–Planck equation as general treatment [37, 38], backward Fokker–Planck equation corresponding to Eq. (50) under natural boundary condition can be expressed in the following form

$$\partial_t \rho(x,t) = (f(x) + \epsilon D'(x))\partial_x \rho(x,t) + \epsilon D(x)\partial_x^2 \rho(x,t). \tag{60}$$

General single click time dependent on initial Dirac function satisfies

$$(f(x) + \epsilon D'(x))\partial_x T(x) + \epsilon D(x)\partial_x^2 T(x) = -1. \tag{61}$$

The general solution corresponding to Eq. (61) is

$$T(x) = \int_0^x dy \frac{1}{\epsilon D(y)} \exp(-\Phi(y)) \int_y^1 dz \exp(\Phi(z)), \tag{62}$$

here $\Phi(x) = \int^x dx'(f(x')/D(x'))(\epsilon = 1)$.

Here the evolutionary process occurs when $x \in [0, 1]$. We are more interested in the escape time between the two stable states with allele frequency $x = 0$ and $x = 1$. Here the adaptive landscape has rich structure which contains escaping from finite and infinite potential. In the escaping from finite potential process, there are two important states x^*, x_0^*. Interval $(0, 1)$ contains a potential well at x^* and a potential barrier at x_0^*. The single click time is composed of two elements, one denotes forming process of fittest class, the other describes losing process of fittest class. In general, the time spent on forming process is much smaller than that spent on losing process. So the escape time approximates to the time spent on losing process. Because we assume that near x_0^* we can write

$$\Phi(x) \approx \Phi(x_0^*) - \frac{1}{2}\left(\frac{x - x_0^*}{\alpha'}\right)^2. \tag{63}$$

and near x^*

$$\Phi(x) \approx \Phi(x^*) + \frac{1}{2}\left(\frac{x - x^*}{\beta'}\right)^2. \tag{64}$$

At the same time, if the central maximum of $\Phi(x)$ is large compared with $1/N$, then $\exp(\Phi(z))$ is sharply peaked at x_0^*, while $\exp(-\Phi(y))/D(y)$ is very small near $y = x^*$. Equation (62) is evaluated as

$$\begin{aligned} T_{1\to 0} &\approx \int_{x^*}^{0} dy \frac{1}{D(y)} \exp(-\Phi(y)) \int_1^{x_0^*} dz \exp(\Phi(z)) \\ &\approx \frac{2\pi\alpha'\beta' \exp(\Phi(x_0^*) - \Phi(x^*))}{D(x^*)} \\ &\propto \frac{1}{D(x^*)} \exp(\Phi(x_0^*) - \Phi(x^*)). \end{aligned} \tag{65}$$

From the expression of Eq. (65), the single click time is not sensitive to the boundary assumption with $x = 0$ and $x = 1$, it is closely related to the potential difference $\Phi(x_0^*) - \Phi(x^*)$.

In the higher mutation rates regime, where x_0^* equals to a stable state x_2, x^* corresponds to the unstable state x_1 that the population lies between the adaptive states 0 and x_2. The potential barrier $\Phi(x_2) - \Phi(x_1)$ is finite. The difference of potential is

$$
\begin{aligned}
\Phi(x_2) - \Phi(x_1) &= \frac{2N\mu(1-\sigma) - 1 + \sigma\mu}{1-\sigma\mu} \ln\left(1 - \frac{x_2 - x_1}{1 - x_1}\right) - \ln\left(1 + \frac{x_2 - x_1}{x_1}\right) \\
&\quad + \frac{2N(1-\mu)}{1-\sigma\mu} \ln\left(1 + \frac{\sigma(1-\mu)(x_2 - x_1)}{1 - \sigma + x_1\sigma(1-\mu)}\right) \\
&= \frac{2N\mu(1-\sigma)}{1-\sigma\mu} \ln\left(1 - \frac{2\sqrt{\alpha^2 - \beta}}{\alpha - \sqrt{\alpha^2 - \beta}}\right) \\
&\quad - \ln\left(1 - \frac{4(\alpha^2 - \beta)}{(\alpha - \sqrt{\alpha^2 - \beta})^2}\right) + \frac{2N(1-\mu)}{1-\sigma\mu} \\
&\quad \times \ln\left(1 + \frac{2\sqrt{\alpha^2 - \beta}}{4N - 6 - 6N\sigma + 7\sigma - \sigma\mu + 2N\mu + \sqrt{\alpha^2 - \beta}}\right),
\end{aligned}
\tag{66}
$$

where α and β are the same as Eqs. (56) and (57). According to classical derivation, corresponding to Eq. (62) the single click time approximates to

$$
\begin{aligned}
T_{1\to 0} &= \lim_{x\to 1} \int_0^x dy \frac{1}{\epsilon D(y)} \exp(-\Phi(y)) \int_y^1 dz \exp(\Phi(z)) \\
&\approx 2N \int_0^{x^*} \frac{(1-y)^{2N\mu(\sigma-1)/(1-\sigma\mu)}}{(1 - \sigma + y\sigma(1-\mu))^{2N(1-\mu)/(1-\sigma\mu)}} dy \\
&\quad \times \int_{x_0^*}^1 z^{-1}(1-z)^{(\sigma\mu - 1 - 2N\mu(\sigma-1))/(1-\sigma\mu)} \\
&\quad \times (1 - \sigma + \sigma z(1-\mu))^{2N(1-\mu)/(1-\sigma\mu)} dz \\
&\approx \frac{1}{D(x^*)} \exp(\Phi(x_2) - \Phi(x_1)) \\
&\approx \frac{N(N-1)^2\sigma^2(1-\mu)^2}{(\alpha - \sqrt{\alpha^2 - \beta})(4\sigma(N-1)(1-\mu) - \alpha + \sqrt{\alpha^2 - \beta})}.
\end{aligned}
\tag{67}
$$

The approximated single click time varies with mutation rates. The single click time $T_{1\to 0}$ increases with population size N in certain regime, decreases with mutation rates μ and selection rates σ in the parameters regime $\mu \in (2N/4N(N-1), 1)$ and

$\sigma \in (\sigma_1, (2N\mu - 1)/(2N\mu - \mu))$. When mutation rate holds constant, with selection rates increasing, the difference of potential between two fixed points decreases, the viability of suboptimal class decreases, the population evolves to the fittest class. On the other hand, when selection holds constant, with deleterious mutation increasing, the population of suboptimal class increases, the difference of potential between two fixed points decreases.

For the lower mutation rates regime, where the potential barrier is infinite. The landscape has U-shape. The single click time can also be estimated under the condition of weak mutation and weak selection, x^* corresponds to the fixed point x_1 that the population lies at the lowest potential.

$$\begin{aligned} T_{Escape}(1 \to 0) &\approx T_{MFPT}(1 \to 0) \\ &\approx 2N \int_0^1 \frac{(1-y)^{2N\mu(\sigma-1)/(1-\sigma\mu)}}{(1-\sigma+y\sigma(1-\mu))^{2N(1-\mu)/(1-\sigma\mu)}} dy \\ &\quad \times \int_{x_0^*}^1 z^{-1}(1-z)^{(\sigma\mu-1-2N\mu(\sigma-1))/(1-\sigma\mu)} \\ &\quad \times (1-\sigma+\sigma z(1-\mu))^{2N(1-\mu)/(1-\sigma\mu)} dz \\ &\approx \frac{1-\sigma\mu}{\mu(1-\sigma)}. \end{aligned} \tag{68}$$

From expression of Eq. (68), the single click time goes to infinity with mutation rates tending to zero in the parameters regimes of $\mu \in (0, 1/(2N-1))$ and $\sigma \in (\mu, 1)$. The results of the single click time are not sensitive to the population size.

Analogous to the derivation of $T_{Escape}(1 \to 0)$, we can calculate

$$\begin{aligned} T_{0\to 1} &= \lim_{x\to 0} \int_x^1 dy \frac{1}{\epsilon D(y)} \exp(-\Phi(y)) \int_0^y dz \exp(\Phi(z)) \\ &\approx \lim_{x\to 0} 2N \int_x^1 \frac{(1-y)^{2N\mu(\sigma-1)/(1-\sigma\mu)}}{(1-\sigma+y\sigma(1-\mu))^{2N(1-\mu)/(1-\sigma\mu)}} dy \\ &\quad \times \int_0^y z^{-1}(1-z)^{(\sigma\mu-1-2N\mu(\sigma-1))/(1-\sigma\mu)} \\ &\quad \times (1-\sigma+\sigma z(1-\mu))^{2N(1-\mu)/(1-\sigma\mu)} dz \\ &\approx 2N \int_0^1 \frac{(1-y)^{2N\mu(\sigma-1)/(1-\sigma\mu)}}{(1-\sigma+y\sigma(1-\mu))^{2N(1-\mu)/(1-\sigma\mu)}} dy \\ &\quad \times \int_0^y (z^{-1} + \ldots) dz \\ &\to \infty. \end{aligned} \tag{69}$$

From expression of Eq. (69), the single click time goes to infinity with mutation rates being zero.

Our results show that the expected time to loss of the fittest class can be predicted for a wide range of parameter values that appear to be biologically relevant. Furthermore our results show that the time is insensitive to population size in the presence of weak mutation and weak selection. This tells us that the ratchet process may occur in finite but not small populations.

5 Discussion

Adaptive landscape introduced by Sewall Wright, a re-emerging powerful concept in systems biology, is used as a tool to describe complex biological processes. In the past Prugel-Bennett and Shapiro, Rattray took analogies between population evolution and statistical physics and proposed maximum polymorphism to approximate polygenic system. Their method does not ensure that the arbitrary entropy measure converges to the correct stationary distribution, though it makes accurate prediction. Barton group uses information entropy measure to analyze the process of population evolution analogous to statistical thermodynamics. It points out that the method works only for high mutation rates and breaks down for low mutation rates [15], though this method ensures the convergence to the correct stationary distribution by maximizing introduced entropy. In a word, past methods of constructing adaptive landscape must need the normalization of stationary distribution. The present chapter manifests the nonlinear and complex dynamics of the evolutionary system by adaptive landscape analytically. Especially it can describe the dynamical behavior under the action of random drift. Waxman and Loewe [26] studied the same process, but they put Dirac function at boundary. This results in the solution not satisfying the equation. The theoretical results suggest our constructive potential function may be a reasonable candidate to quantify the adaptive landscape and investigate the complex biological processes.

Recently, Zhou and Qian [39] also constructs landscape function to study the complex dynamical property by discrete and diffusion Moran process. They also meet the dilemma diffusion approximation and mismatch between fixed points and extrema of adaptive landscape. But this mismatch can be explained in our constructive method. Our method does not need the existence and normalization of the stationary distribution. Our constructive method is independent of the stationary distribution. Our method investigates the long-term dynamical property of the system and reduces the complexity of calculating stochastic differential equations. Our adaptive landscape can give a new definition of neutral evolutionary, that is, the population is under the equal action of mutation and random drift. And the analytical results of mutation-random drift balance are consistent with the results in general textbook [21]. Whatever its shape and its singularity, the adaptive landscape characterizes globally the complex dynamical behaviors for a system.

Acknowledgments The critical discussion with Prof. Zhu Xiaomei is appreciated. We also thank Jiang Pengyao, Wang Yanbo, and other members in the lab for their constructive comments.

References

1. Wright S (1932) The role of mutation, inbreeding, crossbreeding and selection in evolution. Proc Int Congr Genet 1:356–366
2. Lande R (1976) Natural selection and random genetic drift in phenotypic evolution. Evolution 30:314–334
3. Arnold SJ, Pfrender ME, Jones AG (2001) The adaptive landscape as a conceptual bridge between micro and macroevolution. Genetica 112–113: 9–32
4. Wright S (1988) Surfaces of selective value revisited. Am Nat 131:115–123
5. Wright S (1967) Surfaces of selective value. Proc Natl Acad Sci 131:165–172
6. Provine WB (1986) Sewall Wright and evolutionary biology. University of Chicago Press, Chicago, IL
7. Weinreich DM, Sindi S, Watson RA (2013) Finding the boundary between evolutionary basins of attraction, and implications for wrights fitness landscape analogy. J Stat Mech Theor Exp 01:P01001
8. Gyllenberg M, Metz JAJ, Service R (2011) When do optimisation arguments make evolutionary sense? In: Fabio ACC Chalub, Jos Francisco Rodrigues (eds) The mathematics of Darwin's legacy. Springer, Basel, pp 233–288
9. Ao P (2009) Global view of bionetwork dynamics: adaptive landscape. J Genet Genom 36: 63–73
10. Kaplan J (2008) The end of the adaptive landscape metaphor? Biol Philos 23:625–638
11. Gavrilets S (1997) Evolution and speciation on holey adaptive landscapes. Trends Ecol Evol 12:307–312
12. Ao P (2008) Emerging of stochastic dynamical equalities and steady state thermodynamics from Darwinian dynamics. Comm Theor Phys 49:1073–1090
13. Ao P (2005) Laws in Darwinian evolutionary theory. Phys Life Rev 2:117–156
14. Pigliucci M, Kaplan J (2006) Making sense of evolution: the conceptual foundations of evolutionary thoery. University of Chicago Press, Chicago, IL
15. de Vladar HP, Barton NH (2009) Statistical mechanics and the evolution of polygenic quantitative traits. Genetics 181:997–1011
16. Coe JB, Barton NH (2009) On the application of statistical physics to evolutionary biology. J Theor Biol 259:317–324
17. Assaf M, Mobilia M (2011) Fixation of deleterious allele under mutation pressure and finite selection intensity. J Theor Biol 275:93–103
18. Bharucha-Reid AT (1960) Elements of the theory of Markov processes and their applications. McGraw-Hill, New York
19. Feller W (1954) Diffusion processes in one dimension. Trans Am Math Soc 77:1–31
20. Kimura M (1964) Diffusion models in population genetics. J Appl Prob 1:177–232
21. Ewens WJ (2004) Mathematical population genetics. Springer, New York
22. Ao P (2004) Potential in stochatic differential equation: novel construction. J Phys Math Gen 37:25–30
23. Fisher RA (1930) The genetical theory of natural selection. Clarendon Press, Oxford
24. Wright S (1931) Evolution in Mendelian populations. Genetics 16:97–159
25. Blythe RA, McKane AJ (2007) Stochastic models of evolution in genetics, ecology and linguistics. J Stat Mech Theor Exp 2007:P07018
26. Waxman D, Loewe L (2010) A stochastic model for a single click of Muller's ratchet. J Theor Biol 264:1120–1132
27. Van Nimwegen E, Crutchfield JP, Huynen M (1999) Neutral evolution of mutational robustness. Proc Natl Acad Sci 96:9716–9720
28. Krakauer DC, Plotkin JB (2002) Redundancy, antiredundancy, and the robustness of genomes. Proc Natl Acad Sci 99:1405–1409
29. Waxman D (2007) Singular solutions of the diffusion equation of population genetics. J Theor Biol 247:849–858

30. Gillespie JH (2004) Population genetics: a concise guide. The Johns Hopkins University Press, Baltimore
31. Kramers HA (1940) Brownian motion in a field of force and the diffusion model of chemical reactions. Physica 7:284–304
32. Xu S, Jiao SY, Jiang PY, Yuan B, Ao P (2012) Escape from infinite adaptive peak. In: Proceedings of Sixth International Conference on System Biology, 268–273, Xi'an, 18–20 August (2012)
33. Muller HJ (1964) The relation of recombination to mutational advance. Mutat Res Fund Mol Mech Mutagen 1:2–9
34. Maynard Smith J (1978) The evolution of sex. Cambridge University Press, England
35. Etheridge A, Pfaffelhuber P, Wakolbinger A (2009) How often does the ratchet click? Facts, heuristics, asymptotics. In: Bath J, Mörters P, Scheutzow M (eds) Trends in stochastic analysis. Springer, Basel, pp 233–288
36. Jiao SY, Ao P (2012) Absorbing phenomena and escaping time for Muller's ratchet in adaptive landscape. BMC Syst Biol S1:S10
37. Van Kampen NG (1992) Stochatic processes in physics and chemistry. North Holland, Amsterdam
38. Ø ksendal B (2003) Stochatic differential equations: an introduction with applications. Springer, Berlin
39. Zhou D, Qian H (2011) Redundancy, antiredundancy, and the robustness of genomes. J Theor Biol 99:1405–1409

Scaffolding of Complex Systems Data

Philippe Blanchard and Dimitri Volchenkov

Abstract Complex systems, in many different scientific sectors, show coarse-grain properties at different levels of magnification. Discrete data sequences generated by such systems call for the relevant tools for their classification and analysis. We show that discrete time scale-dependent random walks on the graph models of relational databases can be generated by a variety of equivalence relations imposed between walks (e.g., composite functions, inheritance, property relations, ancestor–descendant relations, data queries, address allocation and assignment polices). The Green function of diffusion process induced by the random walks allows to define scale-dependent geometry. Geometric relations on databases can guide the data interpretation. In particular, first passage times in a urban spatial network help to evaluate the tax assessment value of land. We also discuss a classification scheme of growth laws which includes human aging, tumor (and/or tissue) growth, logistic and generalized logistic growth, and the aging of technical devices. The proposed classification permits to evaluate the aging/failure of combined new bio-technical "manufactured products," where part of the system evolves in time according to biological-mortality laws and part according to technical device behaviors. Moreover it suggests a direct relation between the mortality leveling-off for humans and technical devices and the observed small cure probability for large tumors.

1 Introduction

There is an impressive number of experimental verifications, in many different scientific sectors, that coarse-grain properties of systems, with simple laws with respect to fundamental microscopic algorithms, emerge at different levels of magnification

P. Blanchard (✉) • D. Volchenkov
Faculty of Physics, Bielefeld University, Universitaetsstr. 25, 33615 Bielefeld, Germany
e-mail: philippe.blanchard@uni-bielefeld.de; volchenk@physik.uni-bielefeld.de

V. Afraimovich et al. (eds.), *Nonlinear Dynamics and Complexity*, Nonlinear Systems and Complexity 8, DOI 10.1007/978-3-319-02353-3_7,

providing important tools for explaining and predicting new phenomena. Discrete data sequences obtained from observations of complex systems consisting of many interacting elementary parts are ubiquitous in the real world. The most of such systems are often considered computationally irreducible [52, 53] which means that the only way to decide about their evolution is to let them evolve in time. By capturing how the individual data units in the observed data sequences are related to each other, we can convert it into a relational database subjected to further investigations.

In our work, we, first, propose a method of feature extraction for relational databases by introducing different equivalence relations on the set of walks over its graph model (Sect. 2) following [49], and, second, describe the generalization of the classification scheme of growth and aging of complex systems (Sect. 3) following [13]. In Sect. 2.1, we discuss a general process of data interpretation based on the implicit introduction of certain equivalence relation on the set of walks defined on the database. In Sect. 2.2, we show that every equivalence relation specifies a stochastic matrix describing transitions of a discrete time scale-dependent random walk over the database such that equivalent walks correspond to equiprobable random walks. The main motivation for introducing a scale-dependent random walks originates from the basic observation that databases can include information about processes evolving in different spatial and temporal scales, and they can be structurally composed of different segments at different scales. This implies that real-world systems, in contrast to idealized mathematical models, may appear in different ways depending on the scale of observation. While analyzing the data from unknown sources, there is no way to know a priori what scales are appropriate for describing the interesting structures in that. Hence, the only reasonable approach is to consider descriptions at multiple scales in order to be able to capture the unknown scale variations that may occur. Scale-dependent random walks can play the role of suitable guidance structures, or "scaffolds" of the real world data. Given a scale-dependent random walk defined by a transition matrix, we define the geometry of self-avoiding diffusion on the graph model (see Sect. 2.3). In Sect. 2.4, we give an example of application of the diffusion geometry in estimation of tax assessment value of land in Manhattan.

Further, in Sect. 3 we focus our attention on the problem of growth and aging statistics arising in many *a priori* unrelated systems showing similar emergent properties. The understanding of aging and of late-life mortality is still an open problem and many interesting models have been proposed to explain the similar behavior in metabolic systems and in technical devices [22]. Moreover, a unifying language for the description of performance of metabolic and technical production and distribution has been recently suggested [4] to implement the idea that the robustness of metabolic systems with respect to environmental changes could represent a useful model for technical systems. A very important example is the Gompertz law (GL) [25] which applies to human mortality tables (i.e. aging) and tumor growth [39, 45, 50]. A general classification scheme for phenomenological universalities in growth laws has been given in [14]. The generalization of the classification scheme of growth laws, including human aging, tumor (and/or tissue) growth, logistic and generalized logistic growth, and the aging of technical devices, has been addressed in [13]. In Sect. 3, we consider two applications of the proposed approach: (a) a method to evaluate the aging/failure of combined new bio-technical "manufactured product," where part of the system evolves in time according to biological-mortality laws and part is a technical device; (b) an interpretation of the "tumor size effect," i.e. the small cure probability for large tumor[5, 34, 44], in analogy with the late-life mortality in aging. The generalized growth problem is discussed in Sect. 3.1. The general classification scheme of growth laws is considered in Sect. 3.2. We conclude in the last section (Sect. 4).

2 Geometry of Equivalence Partitions of Walks on a Relational Database

2.1 *Equivalence Partitions of Walks on a Relational Database*

The analysis of data collected over the real-world systems starts with the abstraction of independent entities capable of representing some data aspect distinguishable from other aspects over the complexity of a domain. By capturing how the entities are allied to one another by the binary relationships, one converts the data into a relational database. In a graph model of the relational database, entities and binary relationships can be thought of as *vertices* (nodes or states) $\mathscr{V}$ and *edges* (arcs or relationships) $\mathscr{E} \subseteq \mathscr{V} \times \mathscr{V}$, respectively. We suppose that the database is finite that is $|\mathscr{V}| = N$ and $|\mathscr{E}| = E$. The relationships between the vertices $\mathscr{V}$ are characterized by the set of edge utility functions,

$$\mathscr{S}^{\mathscr{E}} = \{A\,|A : \mathscr{E} \to \mathscr{S}\},$$

with the domain $\mathscr{E}$ and the codomain $\mathscr{S} \subset \mathbb{R}^+$ that associate with each relation a positive, real measure of its performance. For example, a *characteristic* function $A : \mathscr{E} \to \{0, 1\}$ defined as $A(e) = 1$, if $e \in \mathscr{E}$, and $A(e) = 0$ otherwise, represents the Boolean valued relationships between vertices, the *adjacency matrix* of the graph model.

We call a succession of n adjacent edges $e_1 \times e_2 \times \ldots e_{n-1}$ connecting a series of vertices in the graph model as a *walk* of length $n \geq 1$. The edge utility product function

$$A^n : \mathscr{E} \times \mathscr{E} \times \ldots \times \mathscr{E} \to \mathscr{S}$$

characterizes the total aggregated utility of all $n-$walks $\mathscr{W}^n$ over the database. Walks on a relational database can correspond to a composite function acting from the source to the destination, a variety of inheritance, property-subproperty, and ancestor–descendant relationships between vertices, data queries (in the process of information retrieval), the address allocation and assignment policies (in the process of data storage), a coding function on a space of genetic algorithms into a space of chromosomes, etc.

The process of data interpretation (or *classification*) is always based on the implicit introduction of certain *equivalence relations* (i.e., reflexive, symmetric, and transitive) on $\mathscr{W}^n$. Different equivalence relations over sequences of identification characteristics lead to different concepts of taxonomic categories. For example, the definition of "species" as a group of organisms capable of interbreeding and producing fertile offspring is a well-known wicked problem in biology. The decision upon which species an organism belongs to crucially depends upon a pragmatic choice based on the particularities of the species of concern and is by no means immutable, as new data may indicate that one previously described species actually includes two or more separately evolving groups, each of which could justifiably be recognized as a separate species. Moreover, the standard definition (of reproductive isolation) that works well for some organisms (mostly eukaryotes) can be useless for others (prokaryotes). Species appear to us as typical natural kinds (for example, the species that we call "giraffe" is a category of things that people have recognized have a lot in common with each other); however, when biologists turn to understand species evolutionarily they are revealed as changeable and without sharp boundaries [30].

Given an equivalence relation R is given, we define the *equivalence class* of a walk $w \in \mathscr{W}^n$ under R to be the set

$$\mathscr{W}_R^n(w) \;=\; \{v| \in \mathscr{W}^n \text{ and } wRv\}\,.$$

Then the set of all possible equivalence classes (the *quotient set*) of $\mathscr{W}^n$ by R is

$$\mathscr{W}^n/R \;=\; \{\mathscr{W}_R^n(w)|\,\mathscr{W}_R^n(w) \subset \mathscr{W}^n \text{ and } w \in \mathscr{W}^n\}\,.$$

The *projection*

$$p_R(w) : \mathscr{W}^n \;\to\; \mathscr{W}^n/R$$

assigning an equivalence class $\mathscr{W}_R^n(w)$ to each $w \in \mathscr{W}^n$ describes the *partition* of $\mathscr{W}^n$ into $\mathscr{W}^n/R$ by R. It is worth a mention that a function $T_R : \mathscr{W}^n \to \mathscr{S}$ such that wRv implies

$$T_R(w) \;=\; T_R(v)$$

can be chosen to be a *probability* to follow the walk $w \in \mathscr{W}_n$, so that all walks equivalent to w by R are *equiprobable*.

It is known from the theorem on projections [8] that there is exactly one function $g_R : \mathscr{W}^n/R \to \mathscr{S}$, for which $T_R = g_R \circ p_R$, so that the following diagram commute:

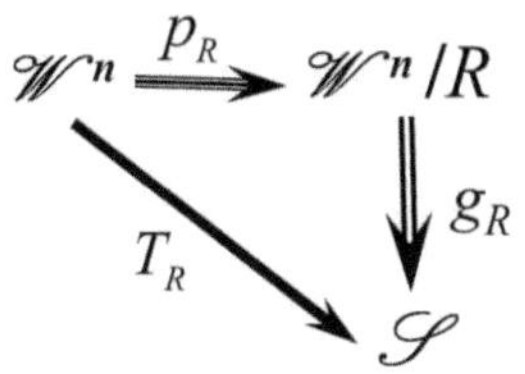

Let us denote the set of all functions $T_R : \mathscr{W}^n \to \mathscr{S}$ which respect R for each set $\mathscr{S}$ as

$$\mathscr{F}_{\mathscr{E}}(\mathscr{S}) = \{T_R |\, T_R : \mathscr{W}^n \to \mathscr{S} \text{ and } wRv \Rightarrow T_R(w) = T_R(v)\}\,.$$

Then the projection $p_R \in \mathscr{F}_{\mathscr{E}}(\mathscr{S})$, and any function $T_R \in \mathscr{F}_{\mathscr{E}}(\mathscr{S})$ constant on each equivalence class factors *uniquely* through p_R [8]. If T_R is the probability to follow a walk among all walks $\mathscr{W}^n$, the projection p_R equals the probability to choose the walk over its equivalence class $\mathscr{W}^n_R(w)$, and $g^{(n)}_R$ is the total utility of walks in the equivalence class.

2.2 *Equivalence Partitions and Scale-Dependent Random Walks*

Given a probability measure on the set $\mathscr{W}^n$, every equivalence relation over walks specifies a stochastic matrix describing transitions of a discrete time scale-dependent *random walk* between vertices $\mathscr{V}$ such that equivalent walks correspond to equiprobable random walks.

For example, two walks, w and v, can be regarded as equivalent by wR_xv if they start at the same vertex $x \in V$, or as equivalent by wR_yv if they end at the same vertex $y \in V$. The utility functions for the equivalence classes $\mathscr{W}^n_{R_x}$ and $\mathscr{W}^n_{R_y}$ are

$$g^{(n)}_{R_x} = \delta_{xi} A^n_{i,j}, \quad g^{(n)}_{R_y} = \delta_{j,y} \left(A^\top\right)^n_{i,j}$$

where $\delta_{a,b}$ is the Kronecker symbol. Then the projections describing the partitions of the set $\mathscr{W}_n$ into the quotient sets, $\mathscr{W}^n/R_x$ and $\mathscr{W}^n/R_y$ are given by

$$p^{(n)}_{R_x} = D^{-1}_n, \quad p^{(n)}_{R_y} = D^{-1}_n$$

where

$$D_n = \operatorname{diag}\left(\sum_{i=1}^{N} A^n_{1,i}, \ldots, \sum_{i=1}^{N} A^n_{N,i}\right)$$

and

$$D'_n = \operatorname{diag}\left(\sum_{i=1}^{N} \left(A^\top\right)^n_{i,1}, \ldots, \sum_{i=1}^{N} \left(A^\top\right)^n_{i,N}\right)$$

are the diagonal matrices of aggregated utilities, for each equivalence class. The transition probabilities while following the walks that belongs to the equivalence classes $\mathscr{W}^n_{R_x}$ and $\mathscr{W}^n_{R_y}$ are given by the following stochastic matrices,

$$T^{(n)}_{R_x} = D_n^{-1} A^n, \quad \text{and} \quad T^{(n)}_{R_y} = \left(A^\top\right)^n D'^{-1}_n, \tag{1}$$

respectively. The matrices defined in Eq. (1) coincide if the matrix A is symmetric. The elements of the left eigenvectors belonging to the biggest eigenvalue of $T^{(n)}_{R_x}$ and $T^{(n)}_{R_y}$ are nothing else but the *centrality measures* defined as the fractions of n-walks starting at the vertex x, or ending at y,

$$\pi^{(n)}_{R_x}(x) = \frac{\sum_{j\in\mathscr{V}} A^n_{x,j}}{\sum_{i,j\in\mathscr{V}} A^n_{i,j}}, \quad \pi^{(n)}_{R_y}(y) = \frac{\sum_{j\in\mathscr{V}} \left(A^\top\right)^n_{j,y}}{\sum_{i,j\in\mathscr{V}} \left(A^\top\right)^n_{i,j}}. \tag{2}$$

In the case of an undirected graph, $\mathscr{W}^n_{R_x} \equiv \mathscr{W}^n_{R_y}$, the first centrality measure,

$$\pi^{(1)}_{R_x} = \pi^{(1)}_{R_y} = \frac{\deg(x)}{2E},$$

is the well-known stationary distribution of the nearest-neighbor random walks.

To illustrate a dramatic dissimilarity of random walks for different values of n, we performed a simple simulation of two different random walks spread from the lower left corner of the square 30×30 containing a number of obstacles along the main diagonal. In Fig. 1a, we have presented a contour density plot of random walkers after 100 iterations, in accordance with the usual nearest neighbor random walk transition $T^{(1)}_{R_x}$. The contour density plot of random walkers after 100 iterations for $T^{(10)}_{R_x}$ is shown in Fig. 1b.

While the random walks $T^{(10)}_{R_x}$ (Fig. 1b) have already reached the stationary distribution, and its density stopped changing by $t = 100$, the random walks $T^{(1)}_{R_x}$ (Fig. 1a) still undergo a transient process, and their density is far from fixed. Random walks $T^{(1)}_{R_x}$ cover all available space uniformly, including the spaces between obstacles. In contrast to them, the random walks $T^{(10)}_{R_x}$ are "repelled from obstacles," as their density between the obstacles is minimal, and they do not penetrate beyond the obstacle at the upper right corner of the square.

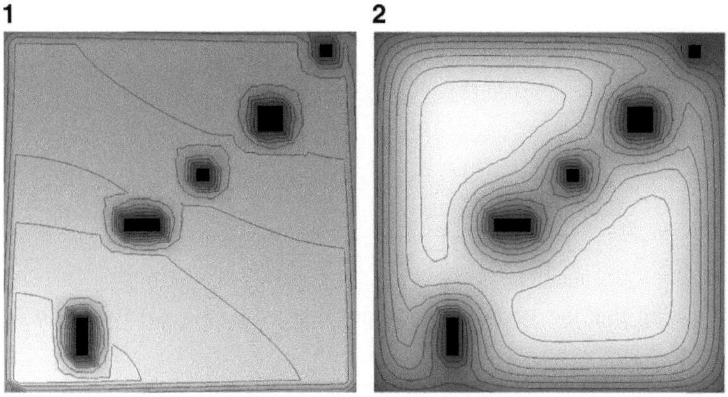

Fig. 1 Two different random walks spread from the lower left corner of the square 30 × 30 containing a number of obstacles along the main diagonal. (**a**) The contour density plot of random walkers after 100 iterations, in accordance with the usual nearest neighbor random walk transition $T^{(1)}_{R_x}$. (**b**) The contour density plot of random walkers after 100 iterations for $T^{(10)}_{R_x}$

Provided the matrix A is irreducible, the Perron–Frobenius theorem (see, for example, [26]) states that its dominant eigenvalue $\alpha > 0$ is simple, and the corresponding left and right eigenvectors can be chosen positive,

$$A\phi = \alpha\phi, \quad \phi_i > 0, \quad \text{and} \quad \varphi^\top A = \alpha\varphi^T, \quad \varphi_i > 0.$$

As $n \to \infty$, the elements of the transition matrices (Eq. 1) tend to the constant row matrices,

$$\begin{aligned} \lim_{n\to\infty}\left(T^{(n)}_{R_x}\right)_{x,y} &= \left[\frac{\varphi_x}{\sum_{x\in\mathscr{V}}\varphi_x}\right]_{x=1,\ldots,N} = \lim_{n\to\infty}\left(\pi^{(n)}_{R_x}(x)\right), \\ \lim_{n\to\infty}\left(T^{(n)}_{R_y}\right)_{x,y} &= \left[\frac{\phi_y}{\sum_{y\in\mathscr{V}}\phi_y}\right]_{y=1,\ldots,N} = \lim_{n\to\infty}\left(\pi^{(n)}_{R_y}(y)\right). \end{aligned} \tag{3}$$

The intersection of the equivalence relations, $R_x \cap R_y$, is also an equivalence relation on $\mathscr{W}^n$. The total utility function for the equivalence class $\mathscr{W}^n_{R_x\cap R_y}$ is

$$g^{(n)}_{R_x\cap R_y} = \delta_{x,i}\delta_{y,j}A^n_{i,j}.$$

Then the uniform probability to follow a walk $w \in \mathscr{W}^n$ among all equivalent walks (from x to y) equals

$$p^{(n)}_{R_x\cap R_y} = \frac{1}{\alpha^n}\frac{\phi_y}{\phi_x},$$

and therefore the transition probability to follow such a walk is

$$T^{(n)}_{R_x\cap R_y} = \frac{A^n_{x,y}\phi_y}{\alpha^n\phi_x}. \tag{4}$$

It is clear that since

$$\sum_{y\in\mathscr{V}} A^n_{x,y}\phi_y \;=\; \alpha^n\phi_x,$$

it is always

$$\sum_{y\in\mathscr{V}} T^{(n)}_{R_x\cap R_y} \;=\; 1.$$

It is interesting that the stochastic matrices defined by Eq. (4) for any n share the same Perron eigenvector,

$$\pi_{R_x\cap R_y}(i) \;=\; \varphi_i\phi_i,$$

$i \in \mathscr{V}$. As $n \to \infty$, the elements of the transition matrix (Eq. 4) tend to the constant row matrix,

$$\lim_{n\to\infty}\left(T^{(n)}_{R_x\cap R_y}\right)_{x,y} \;=\; \lim_{n\to\infty}\frac{\alpha^n\phi_x\varphi_y\phi_y}{\alpha^n\phi_x} \;=\; \left[\varphi_y\phi_y\right]_{y=1,\dots,N} \;=\; \pi_{R_x\cap R_y}(y). \quad (5)$$

The stochastic matrices defined by Eqs. (1) and (4) satisfy the Chapman–Kolmogorov equation

$$P^{(n)}_{x,y} \;=\; \sum_{v\in\mathscr{V}} P^{(\tau)}_{x,v}P^{(t-\tau)}_{v,y}, \quad 0<\tau<t,$$

which relates the probabilities

$$P^{(t)}_{i,j} \;=\; \left(T^{(n)}\right)^t_{i,j}$$

of the $t > 0$ transitions between the states i and j.

2.3 Self-Avoiding Diffusion Geometry of Undirected Graphs

Given a random walk defined by a transition matrix T on a finite connected undirected weighted graph $G(\mathscr{V},\mathscr{E})$, all vertices and their subsets can be characterized by certain probability distributions and characteristic times [10]. The stationary distribution of random walks (the left eigenvector of the transition matrix T belonging to the maximal eigenvalue $\mu = 1$) determines a unique measure on $\mathscr{V}$ with respect to which the transition operator T becomes self-adjoint and is represented by a symmetric transition matrix $\hat{T}$. The use of self-adjoint operators (such as the normalized graph Laplacian) becomes now standard in spectral graph theory, [16] and in studies devoted to random walks on graphs, [36]. Diagonalizing the symmetric matrix $\hat{T}$, we obtain

$$\hat{T} \;=\; \Psi M \Psi^{\top},$$

where Ψ is an orthonormal matrix,

$$\Psi^{\top} = \Psi^{-1},$$

and M is a diagonal matrix with entries

$$1 = \mu_1 > \mu_2 \geq \ldots \geq \mu_N > -1$$

(here, we do not consider bipartite graphs, for which $\mu_N = -1$). The rows

$$\psi_k = \{\psi_{k,1}, \ldots, \psi_{k,N}\}$$

of the orthonormal matrix Ψ are the real eigenvectors of $\hat{T}$ that forms an orthonormal basis in Hilbert space $\mathscr{H}(\mathscr{V})$, $\psi_k : \mathscr{V} \to S_1^{N-1}$, $k = 1, \ldots N$, where S_1^{N-1} is the $N - 1$-dimensional unit sphere. We consider the eigenvectors ψ_k ordered in accordance with the eigenvalues they belong to. For eigenvalues of algebraic multiplicity $m > 1$, a number of linearly independent orthonormal ordered eigenvectors can be chosen to span the associated eigenspace. The first eigenvector ψ_1 belonging to the largest eigenvalue $\mu_1 = 1$ (which is simple) is the Perron–Frobenius eigenvector that determines the stationary distribution of random walks over the graph nodes,

$$\psi_{1,i}^2 = \pi_i.$$

The diffusion process is described by the irreducible Laplace operator

$$L = 1 - T$$

which has the one-dimensional null space spanned by the vector π. As being a member of the multiplicative group under the ordinary matrix multiplication [20, 37], the Laplace operator possesses a *group inverse* (a special case of *Drazin inverse*, [6, 17, 37]) with respect to this group, $L^{\sharp}$, which satisfies the following conditions [20]:

$$LL^{\sharp}L = L, \quad L^{\sharp}LL^{\sharp} = L^{\sharp}, \quad \text{and} \quad \left[L, L^{\sharp}\right] = 0,$$

where $[A, B] = AB - BA$ denotes the commutator of matrices. The last condition implies that $L^{\sharp}$ shares the same set of symmetries as the Laplace operator, being the Green function of the diffusion equation. The methods for computing the group generalized inverse for matrices of rank$(L) = N - 1$ have been developed in [12,43] and by many other authors. Perhaps, the most elegant way is by considering the eigenprojection of the matrix L corresponding to the eigenvalue $\lambda_1 = 1 - \mu_1 = 0$ developed in [1, 12, 29],

$$L^{\sharp} = (L + Z)^{-1} - Z, \quad Z = \prod_{\lambda_i \neq 0} (1 - L/\lambda_i), \quad \lambda_i = 1 - \mu_i \tag{6}$$

where the product in the idempotent matrix Z is taken over all nonzero eigenvalues $\lambda_{i>1}$ of L. The eigenprojection (Eq. 6) can be considered as a stereographic projection that projects the points ψ_k on the sphere S_1^{N-1} to a projective manifold such that all vectors collinear to the vector ψ_1 (corresponding to the stationary distribution of random walks) are projected onto a common image point. Since

$$\psi_{1,i} \equiv \sqrt{\pi_i} \; > \; 0$$

for any $i \in \mathscr{V}$, we can define the new basis vectors $\left\{\psi'_k\right\}_{k=1}^N$, spanning the projection space $\mathrm{P}\mathbb{R}_\pi^{(N-1)}$, such that

$$\psi'_k \;=\; (1, \psi_{2,i}/\psi_{1,i}, \ldots, \psi_{N,i}/\psi_{1,i})\,.$$

We define the inner product between any two vectors $\xi, \zeta \in \mathbb{R}^N$ by

$$(\xi,\zeta)_T \;=\; \left(\xi, L^\sharp\zeta\right). \tag{7}$$

The inner product (Eq. 7) is a symmetric real valued scalar function that allows us to define the (squared) norm of a vector ξ by

$$\parallel \xi \parallel_T^2 \;=\; \left(\xi, L^\sharp\xi\right) \tag{8}$$

and an angle $\theta \in [0, 180^\mathrm{o}]$ between two vectors,

$$\theta \;=\; \arccos\left(\frac{(\xi,\zeta)_T}{\parallel \xi \parallel_T \parallel \zeta \parallel_T}\right). \tag{9}$$

The Euclidean distance between two vectors is given by

$$\|\xi - \zeta\|_T^2 \;=\; \parallel \xi \parallel_T^2 + \parallel \zeta \parallel_T^2 - 2\,(\xi,\zeta)_T\,. \tag{10}$$

For instance, let us consider the vector $\mathbf{e}_i = \{0, \ldots 1_i, \ldots 0\}$ pointing at the vertex i of the graph G in the canonical basis. The spectral representation of the generalized inverse for undirected graphs [10] is given by

$$L^\sharp_{i,j} \;=\; \sum_{k=2}^N \frac{1}{\lambda_k}\frac{\psi_{i,k}}{\psi_{i,1}}\frac{\psi_{j,k}}{\psi_{j,1}}, \qquad i,j = 1,\ldots,N, \tag{11}$$

in which λ_k are all nontrivial eigenvalues of the Laplace operator. The matrix (Eq. 11) is real symmetric semi-positive, as its smallest eigenvalue $\Lambda_1 \;=\; 0$. Accordingly Eq. (8), the spectral representation for the squared norm of e_i equals

$$\| \mathrm{e}_i \|_T^2 = \frac{1}{\pi_i} \sum_{s=2}^{N} \frac{\psi_{s,i}^2}{\lambda_s}. \tag{12}$$

In the theory of random walks on undirected graphs [36], the latter result is known as the spectral representation of the *first passage time* to the node $i \in \mathscr{V}$, the expected number of steps required to reach the node i for the first time (i.e., without visiting any node twice) starting from a node randomly chosen among all nodes of the graph according to the stationary distribution π. The Euclidean distance between any two nodes of the graph induced by the random walk,

$$K_{i,j} = \| \mathrm{e}_i - \mathrm{e}_j \|_T^2 = \sum_{s=2}^{N} \frac{1}{\lambda_s} \left(\frac{\psi_{s,i}}{\sqrt{\pi_i}} - \frac{\psi_{s,j}}{\sqrt{\pi_j}} \right)^2, \tag{13}$$

is nothing else but the *commute time*, the expected number of steps required for a random walker starting at i to visit j and then to return to i, without visiting any node twice [36]. The commute time can be represented as a sum, $K_{i,j} = H_{i,j} + H_{j,i}$, in which

$$H_{i,j} = \| \mathrm{e}_i \|_T^2 - (\mathrm{e}_i, \mathrm{e}_j)_T \tag{14}$$

is the *first-hitting time* which is the expected number of steps a random walker starting from the node i needs to reach j for the first time, [36]. The zero-diagonal matrix of first-hitting times is not symmetric, $H_{i,j} \neq H_{j,i}$, even for a regular graph. Its average with respect to the first index equals the first-passage time to the node,

$$\| \mathrm{e}_j \|_T^2 = \sum_{i \in \mathscr{V}} \pi_i H_{i,j}. \tag{15}$$

It is worth a mention that matrix (Eq. 11) can be considered as the Gram matrix, $L^\sharp = (\hat{\mathbf{e}}_i, \hat{\mathbf{e}}_j)$, with respect to the usual dot product of vectors in the projection space $P\mathbb{R}^{(N-1)}$. The vector

$$\hat{\mathbf{e}}_i = \left(\frac{1}{\sqrt{\lambda_2}} \frac{\psi_{2,i}}{\psi_{1,i}}, \ldots, \frac{1}{\sqrt{\lambda_N}} \frac{\psi_{N,i}}{\psi_{1,i}} \right) = \left(\frac{\psi'_{2,i}}{\sqrt{\lambda_2}}, \ldots, \frac{\psi'_{N,i}}{\sqrt{\lambda_N}} \right) \tag{16}$$

represents an image of the vertex $i \in \mathscr{V}$ in the projection space $P\mathbb{R}^{(N-1)}$. The image of the graph $G(\mathscr{V}, \mathscr{E})$ in the projection space $P\mathbb{R}^{(N-1)}$ constitutes a diffusion manifold of self-avoiding random walks (in which nodes cannot be visited twice) in affine subspace, as we can subtract vertices (by componentwise subtracting of their images (Eq. 16)) to get vectors, or add a vector to a vertex to get another vertex, but we cannot add new vertices. It seems natural to describe the structural properties of the graph using the topology of the manifold of self-avoiding diffusion in the projection space $P\mathbb{R}^{(N-1)}$. The differential geometric structure on the manifold can be defined using the topology of the manifold of symmetric positive-definite tensors specified at every vertex of the graph. A Riemannian structure on the manifold can be introduced with respect to the different connections on the tangent bundle of the space of symmetric matrices. First, the Hessian Riemannian structure [18] can be introduced at each point $\mathbf{x} \in P\mathbb{R}^{(N-1)}$, for every function $f(\mathbf{x})$ such that the Hessian $\mathfrak{g}_{yz}(\mathbf{x}) = \nabla_y \otimes \nabla_z(\mathbf{x}) f(\mathbf{x})$ is positive definite, where the finite difference operator $\nabla_y(\mathbf{x}) = (\mathbf{y} - \mathbf{x})/\|\mathbf{y} - \mathbf{x}\|_{P\mathbb{R}^{(N-1)}}$. Second, a Riemannian structure of non-positive curvature can be introduced with the help of the affine invariant Riemannian metric [7, 41] invariant with respect to inversions and similarity transformations of the symmetric positive-definite (SPD) diffusion tensors defined at each point $\mathbf{x} \in P\mathbb{R}^{(N-1)}$,

$$\mathfrak{D}_{y,z}(\mathbf{x}) = (\mathbf{y} - \mathbf{x}) \otimes (\mathbf{z} - \mathbf{x}), \quad \mathbf{x}, \mathbf{y}, \mathbf{z} \in P\mathbb{R}^{(N-1)}. \tag{17}$$

The $(N-1) \times (N-1)$ matrix (Eq. 17) is always diagonalizable, with strictly real positive eigenvalues.

2.4 *Example: First-passage Times Estimate Land Value in Manhattan*

In space syntax theory (see [31, 32]), built environments are treated as systems of spaces of vision subjected to a configuration analysis. Being irrelevant to the physical distances, spatial graphs representing the urban environments are removed from the physical space. It has been demonstrated in multiple experiments that spatial perception shapes peoples understanding of how a place is organized and eventually determines the pattern of local movement, [32]. The aim of the space syntax study is to estimate the relative proximity between different locations and to associate these distances with the densities of human activity along the links connecting them [3, 28, 51]. In the present section, we take a "named-streets"-oriented point of view on the decomposition of urban spatial networks into the complete sets of intersecting open spaces following our previous works [47, 48]. Being interested in the statistics of random walks defined on spatial networks of urban patterns, we assign an individual street ID code to each continuous segment of a street. The spatial graph of urban environment is then constructed by mapping all edges (segments of streets) of the city map shared the same street ID into nodes and all intersections among each pair of edges of the primary graph into the edges of the secondary graph connecting the corresponding nodes.

The notion of isolation acquires the statistical interpretation by means of random walks. The first-passage times in the city vary strongly from location to location. Those places characterized by the shortest first-passage times are easy to reach while very many random steps would be required in order to get into a statistically isolated site.

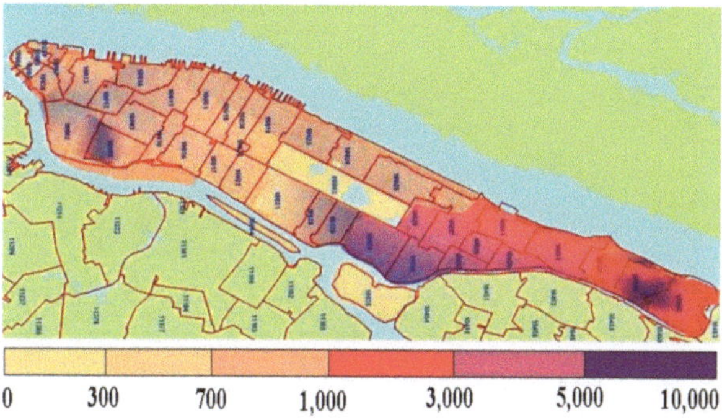

Fig. 2 Isolation map of Manhattan. Isolation is measured by first-passage times to the places. *Darker color* corresponds to longer first-passage times

Being a global characteristic of a node in the graph, the first-passage time assigns absolute scores to all nodes based on the probability of paths they provide for random walkers. The first-passage time can therefore be considered as a natural statistical centrality measure of the node within the graph, [9].

A visual pattern displayed in Fig. 2 represents the pattern of structural isolation (quantified by the first-passage times) in Manhattan (darker color corresponds to longer first-passage times). It is interesting to note that the spatial distribution of isolation in the urban pattern of Manhattan (Fig. 2) shows a qualitative agreement with the map of the tax assessment value of the land in Manhattan reported by B. Rankin (2006) in the framework of the RADICAL CARTOGRAPHY project being practically a negative image of that.

Recently, we have discussed in [9] that distributions of various social variables (such as the mean household income and prison expenditures in different zip code areas) may demonstrate the striking spatial patterns which can be analyzed by means of random walks. In the present work, we analyze the spatial distribution of the tax assessment rate (TAR) in Manhattan.

The assessment tax relies upon a special enhancement made up of the land or site value and differs from the market value estimating a relative wealth of the place within the city commonly referred to as the "unearned" increment of land use [11]. The rate of appreciation in value of land is affected by a variety of conditions, for example it may depend upon other property in the same locality, will be due to a legitimate demand for a site, and for occupancy and height of a building upon it.

The current tax assessment system enacted in 1981 in the city of New York classifies all real estate parcels into four classes subjected to the different tax rates set by the legislature: (1) primarily residential condominiums; (2) other residential property; (3) real estate of utility corporations and special franchise properties; (4) all other properties, such as stores, warehouses, and hotels. However, the scarcity of physical space in the compact urban pattern on the island of Manhattan will naturally set some increase of value on all desirably located land as being a restricted commodity. Furthermore, regulatory constraints on housing supply

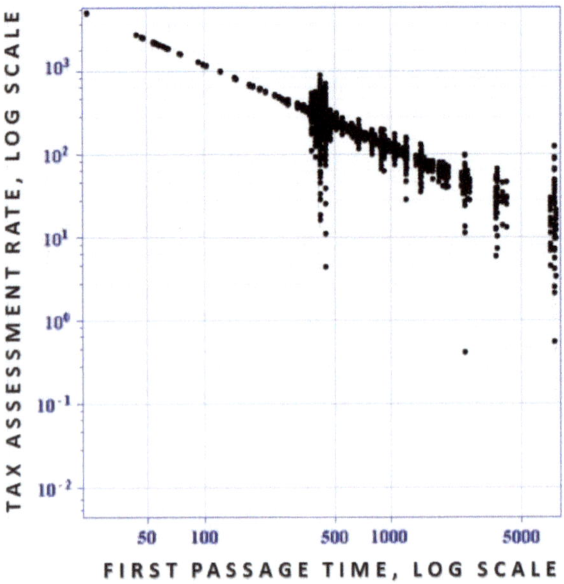

Fig. 3 Tax assessment rate (TAR) of places in Manhattan (the vertical axes, in $/fit^2) is shown in the logarithmic scale vs. the first-passage times (FPT) to them (the horizontal axes)

exerted on housing prices by the state and the city in the form of "zoning taxes" are responsible for converting the property tax system in a complicated mess of interlocking influences and for much of the high cost of housing in Manhattan, [24].

Being intrigued with the likeness of the tax assessment map and the map of isolation in Manhattan, we have mapped the TAR figures publicly available through the Office of the Surveyor at the Manhattan Business Center onto the data on first-passage times to the corresponding places. The resulting plot is shown in Fig. 3, in the logarithmic scale. The data presented in Fig. 3 positively relates the geographic accessibility of places in Manhattan to their "unearned increments" estimated by means of the increasing burden of taxation. The inverse linear pattern dominating the data is best fitted by the simple hyperbolic relation between the tax assessment rate (TAR) and the value of first-passage time (FPT),

$$\mathrm{TAR} \propto \frac{c}{\mathrm{FPT}}, \tag{18}$$

in which $c \simeq 120{,}000\ \$ \times \mathrm{Step}/\mathrm{fit}^2$ is a fitting constant.

3 Unified Approach to Growth and Aging in Biological, Technical, and Biotechnical Systems

3.1 Generalized Growth Problem

In general, a growth problem is characterized by a function $f(t)$, which describes the time evolution of some macroscopic quantity, and by the specific rate, α, defined as

$$\frac{1}{f}\frac{df}{dt} = \alpha(t).$$

In the GL α has an exponential dependence on time:

$$\frac{1}{f}\frac{df}{dt} = \alpha(t) = a\,e^{bt}, \tag{19}$$

where a and b are constants. In aging $f(t)$ indicates the survival probability; while with regard to tumor growth it corresponds to the number of cells $N(t)$ (depending on the specific case a and b can be positive or negative).

For technical devices the specific rate of the survival probability has a power-law time behavior

$$\frac{1}{f}\frac{df}{dt} = \alpha(t) = a\,t^{n}, \tag{20}$$

with $n > 1$, called Weibull law (WL) [2, 42]. The analogy with the biological systems is intriguing (for clarity, as necessary, one defines the specific rate $\alpha_h(t)$ for the human mortality, $\alpha_f(t)$ for the technical systems and $\alpha_c(t)$ for tumor growth) and deeper than the similarity between Eqs. (19) and (20).

Indeed, many independent analyses of experimental data on humans and animals suggest that at advanced ages (more than 85–90 years for humans) there is a deceleration in mortality [21, 40, 46]: in the large range 20–85 years for humans the mortality rate is well described by the Gompertz law and then there is a late-life mortality (although a definite conclusion has yet to be reached [23]). A similar trend is observed for technical devices [19], confirming the analogy between biological and technical systems.

3.2 General Classification Scheme

Let us start with the general classification scheme. It turns out that a classification of the growth laws according to the simple equation

$$\frac{1}{f}\frac{df}{dt} = \alpha(t)$$

is obtained by considering the power expansion in α of the function (see [14] for details)

$$\Phi(\alpha) \;=\; \frac{d\alpha}{dt} \;=\; \Sigma_i b_i \alpha^i, \qquad i = 0, 1, 2 \ldots \tag{21}$$

which for $b_0 = 0$ and $b_i = 0$ for $i > 1$ gives a time independent specific rate α_0 and therefore an exponential growth; for $b_0 \neq 0$ and $b_i = 0$ for $i > 1$ describes a linear time-dependent specific rate and again an exponential growth; at the first order in α, for $b_0 = 0, b_1 \neq 0$ and $b_i = 0$ for $i > 1$, reproduces an exponential time behavior of the specific growth and therefore the GL; the second order term, $O(\alpha^2)$, for $b_0 = 0, b_1, b_2 \neq 0$ and $b_i = 0$ for $i > 2$ generates the logistic and generalized logistic growth.

The feedback effect, that is the dependence of the specific growth rate α on the function $f(t)$, can be easily derived by the temporal behavior of the specific rate. For the GL for a growing number of cells, $N(t)$, one has the well-known logarithmic nonlinearity,

$$\frac{1}{N(t)}\frac{dN(t)}{dt} = a - b\ln\frac{N(t)}{N_0} = b\ln\frac{N_\infty}{N(t)} \quad \text{Gompertz}, \tag{22}$$

and for the (generalized) logistic law one gets the typical power-law behavior

$$\frac{1}{N(t)}\frac{dN(t)}{dt} = c\left[1 - \left(\frac{N(t)}{N_\infty}\right)^{\gamma}\right] \quad \text{gen. logistic}, \tag{23}$$

where a, b, c, γ are constants and the carrying capacity, N_∞, corresponds to $\alpha = 0$.

In order to describe technical devices, the previous classification scheme has to be generalized since the specific growth rate of Weibull law has a power law dependence on time which is not reproduced by Eq. (21). The behavior

$$\alpha_f(t) \;\simeq\; t^n,$$

with n positive integer, corresponds to terms $O(\alpha^{(n-1)/n})$ in the expansion of $\Phi(\alpha)$ and therefore for a general classification scheme of the specific growth/aging/failure rate of biological and technical systems one has to consider:

$$\Phi(\alpha) \;=\; \Sigma_{n>2}^{\infty} c_n \alpha^{(n-1)/n} + \Sigma_{n\geq 1} b_n \alpha^n. \tag{24}$$

Note that: (a) $0 < (n-1)/n < 1$ and the nth term in the power series in $\alpha^{(n-1)/n}$ tends for large n to α, i.e. to the Gompertz law; (b) the term $b_0 \neq 0$, i.e. the exponential growth, has been neglected because one considers the GL, the generalized logistic or more complex growth laws for the biological systems (there is no problem to include this term in the expansion); (c) the first sum in the expansion has fractional powers that recall a Puiseux expansion.

As a by-product of the proposed classification scheme one can easily evaluate the aging/failure of combined new bio-technical "manufactured products" by taking explicitly into account the mutual "interference" between the aging behavior of the biological part and the failure of the technical one. The "interference" effect strongly depends on the typical time scales in the coefficients c_n and b_n in the previous expansion: if the life-time of the technical device is much larger than the life-time of the biological part (or vice versa) there is essentially no effect [38].

Let us first consider aging/failure of a combined bio-technological "manufactured product," where part of the system evolves in time according to GL, i.e. the term $O(\alpha)$, and the behavior of technical part is described by a single term $O(\alpha^{n-1/n})$, i.e.

$$\Phi(\alpha) \;=\; c_n\alpha^{(n-1)/n} + b_1\alpha \tag{25}$$

By introducing dimensionless variables in time unit $1/b_1$, i.e.

$$\tau \;=\; b_1 t, \quad \bar{\alpha} \;=\; \frac{\alpha}{b_1}, \quad \text{and} \quad \bar{c}_n \;=\; c_n b_1^{-1-1/n},$$

after simple calculations the time dependence of the specific rate is given by:

$$e^{\tau} \;=\; \frac{\bar{\alpha}}{\bar{\alpha}_0}\frac{\left[1+\frac{\bar{c}_n}{\sqrt[n]{\bar{\alpha}}}\right]^n}{\left[1+\frac{\bar{c}_n}{\sqrt[n]{\bar{\alpha}_0}}\right]^n}, \tag{26}$$

where $\bar{\alpha}_0 = \bar{\alpha}(\tau = 0)$. Of course in the limit $c_n \to 0$ one recovers the GL and for $b_1 \to 0$ the Weibull one. By previous equation, for $\bar{\alpha}_0 = 1$, one obtains:

$$\ln\bar{\alpha} \;=\; n\ln\left[(1+\bar{c}_n)e^{\tau/n} - \bar{c}_n\right] \tag{27}$$

which describes the combined effect of the two growth laws. The quantitative effect is depicted in Figs. 4 and 5 where the previous function is plotted for different values of n at fixed $\bar{c}_n$ and for various values of $\bar{c}_n$ at fixed n.

The next step is to include the term $b_2\alpha^2$ in the expansion of $\Phi(\alpha)$ (b_2 is dimensionless) which corresponds to a generalized logistic evolution. As we shall see this term is crucial in understanding the late-life mortality effect.

By repeating analogous calculations it turns out that

$$\tau = ln(\bar{\alpha}/\bar{\alpha}_0) - \int_{\bar{\alpha}_0}^{\bar{\alpha}} dx \frac{b_2 + \bar{c}_n x^{-(1-n)/n}}{1 + b_2 x + \bar{c}_n x^{-1/n}} \tag{28}$$

In Fig. 6 is shown that the term $b_2\alpha^2$ completely changes the time evolution with respect to GL and/or WL producing a leveling-off of the specific rate.

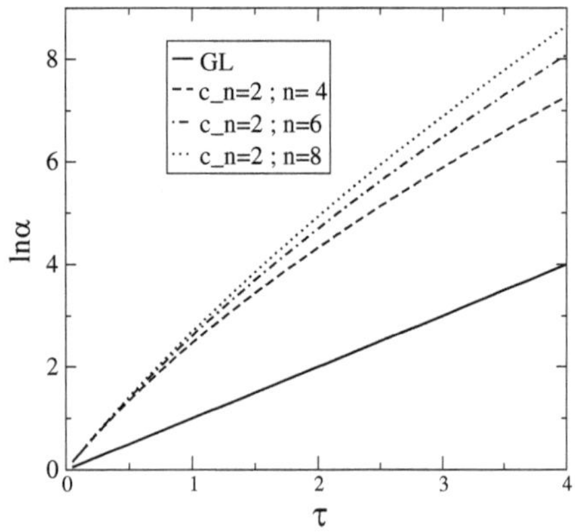

Fig. 4 Comparison of the GL, the WL, and the combined effect for a biotechnical device for $\ln \bar{\alpha}$. $\tau = b_1 t$ and the curves are for a fixed value of the coefficient $\bar{c}_n = 2$ and different values of $n = 4, 6, 8$

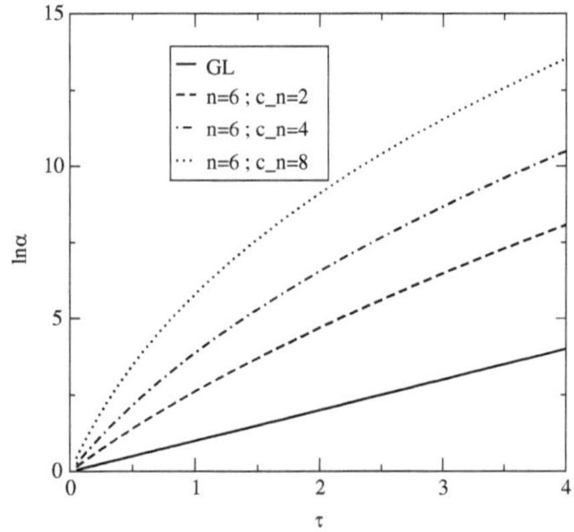

Fig. 5 Comparison of the GL, the WL, and the combined effect for a biotechnical device for $\ln \bar{\alpha}$. $\tau = b_1 t$ and the curves are for $n = 6$ and the coefficient $\bar{c}_n = 2, 4, 8$

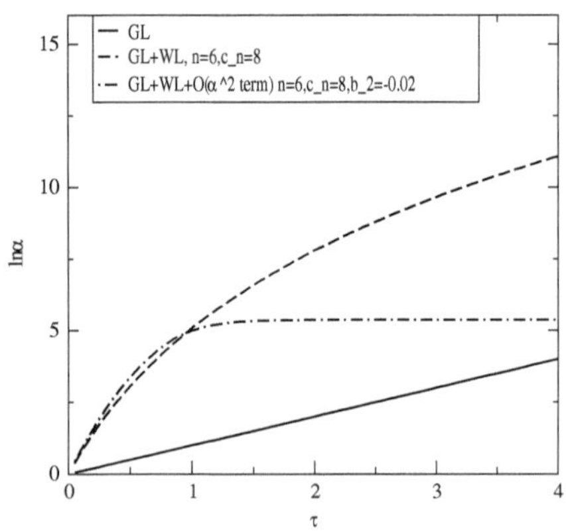

Fig. 6 Comparison for $\ln \bar{\alpha}$ of the GL, the WL, and the effects of $O(\alpha^2)$ term for $n = 6, \bar{c}_n = 8$ and $b_2 = -0.02$

Therefore the general expansion of $\Phi(\alpha)$ in Eq. (24) can describe the aging/failure of any biological and technical system including the leveling-off at late mortality which is obtained by taking into account the term $O(\alpha^2)$ in $\Phi(\alpha)$, i.e. by the transition from the GL or WL to a logistic type law [33].

The proposed unification scheme suggests a practical method to understand growth patterns. Given a set of data on some growth process, the first step of the analysis is a fit in power of α of the derivative of the specific growth rate, i.e. of the function $\Phi(\alpha)$. Therefore: (a) if the best fit is linear, the growth is a Gompertzian one; (b) if the best fit is quadratic, look at the sign of the coefficients of the expansion. For $b_1 > 0$ and $b_2 < 0$ the growth is logistic (or generalized logistic) corresponding to a competitive dynamics; (c) if the best fit indicates a fractional power the growth follows the WL. Of course, it is always possible to obtain a better agreement with data by increasing the number of coefficients. However, should increasing the number of parameters indicate only a marginal improvement in the description of data one concludes that the added terms in the expansion are irrelevant.

4 Discussion and Conclusions

We have discussed the methods of structural analysis for graphs and databases. The process of data classification is based on the implicit introduction of certain equivalence relations on the set of walks over the database. Given a probability measure on the set of walks, every equivalence relation specifies a stochastic matrix describing transitions of a discrete time scale-dependent random walk on the graph model of the database such that equivalent walks correspond to equiprobable random walks. The Green function of the relevant diffusion process shares all the symmetries of the graph and can be used in purpose of data geometrization. In particular, the first passage times, the expected number of steps required for a random walker starting from a node chosen randomly among all nodes of the graph to reach the node for the first time, without visiting any node twice, can be considered as a measure of accessibility to the node in the graph.

Sociologists think that isolation worsens an area's economic prospects by reducing opportunities for commerce and engenders a sense of isolation in inhabitants, both of which can fuel poverty and crime. It is well known that many social variables demonstrate striking spatial distribution patterns, and therefore may be detected and predicted by a structural analysis. In particular, we demonstrated that random walks and diffusions defined on spatial city graphs might spot hidden areas of geographical isolation in the urban landscape going downhill. First passage time to a place correlates with assessed value of land in that. The method accounting the average number of random turns at junctions on the way to reach any particular place in the city from various starting points could be used to identify isolated neighborhoods in big cities with a complex web of roads, walkways, and public transport systems.

We have also considered the generalization of the classification scheme of growth laws including human aging, tumor growth, logistic and generalized logistic growth, and the aging of technical devices.

As previously discussed, there is a deceleration of mortality in aging at late time which is described as a "transition" from a Gompertz law to a generalized logistic behavior. On the other hand, tumors evolve in time according to the GL. The obvious indication is to verify if a phenomenon corresponding to the deceleration of mortality, i.e. a transition from the GL to a power law, exists for cancer growth at a later time. As we shall see, this aspect has strong consequences on the therapy.

For tumor growth the $b_1\alpha$ term gives the GL in Eq. (22) and the introduction of the $O(\alpha^2)$ term corresponds to the power law nonlinear feedback in Eq. (23). Therefore one has to investigate if at late-life of a tumor growth there is such a modification in the dependence of the specific growth rate on the cell number $N(t)$. Since direct information in vivo are almost impossible, the question has to be addressed in an indirect way by considering radiotherapy.

The radiotherapic tumor treatment consists in series of radiation doses at fixed time intervals. However tumors start to re-grow in the interval between two treatments: the re-growth during radiotherapy is therefore an important clinical parameter [35] and the probability of treatment benefit critically depends on the tumor re-growth pattern.

The so-called tumor size effect is a reduction of radiotherapeutic results for large tumors (which, presumably, has grown since long time). The dependence of the surviving fraction on the tumor volume was already observed by Stanley et al. in 1977 in lung tumors [44] and re-emphasized by Bentzen et al. and Huchet et al. in [5, 34].

The effect of re-growth rate on radiotherapy has been quantitatively investigated in [15] and the results clearly indicate that to understand the tumor size effect the re-growth rate for large tumor has to follow a power law [27] rather than the GL.

From this point of view the "tumor size effect" is a phenomenon which indicates that in late-time tumor growth there is a change from a GL specific rate to a power-law behavior, corresponding to the deceleration in mortality at advanced age.

One should conclude that such a common feature in aging and in failure in biological and/or technical systems should be considered as a "bifurcation" or a "phase transition" in the specific growth rate at large time from GL or WL to a logistic or generalized logistic behavior.

In closing, the general expansion of $\Phi(\alpha)$ in Eq. (24) can describe the growth/aging/failure of biological and technical systems and the transition to a different ("phase") specific growth rate at late-life could be a common feature of those systems independently of the microscopic dynamics.

Acknowledgments Financial support by the project *MatheMACS* ("Mathematics of Multilevel Anticipatory Complex Systems"), grant agreement no. 318723, funded by the EC Seventh Framework Programme FP7-ICT-2011-8 is gratefully acknowledged.

References

1. Agaev RP, Chebotarev PYu (2002) On determining the eigenprojection and components of a matrix. Autom Rem Contr 63(10):1537
2. Barlow RE, Proschan F (1975) Statistical theory of reliability and testing. Probability models. Wiley, New York

3. Batty M (2004) A new theory of space syntax. Working paper, 75. UCL Centre For Advanced Spatial Analysis Publications, CASA, London
4. Becker T et al (2011) Flow control by periodic devices: a unifying language for the description of traffic, production and metabolic systems. J Stat Mech 2011: P05004
5. Bentzen SM, Thomas HD (1996) Tumor volume and local control probability: clinical data and radiobiological interpretations. Int J Radiat Oncol Biol Phys 36:247–251
6. Ben-Israel A, Greville ThNE (2003) Generalized inverses: theory and applications, 2nd edn. Springer, New York
7. Bio Th, Fillard P, Ayache N, Pennec X (2004) A Riemannian framework for tensor computing. Int J Comput Vis 66:41
8. Birkhoff G, Mac Lane S (1979) Algebra, 2nd edn. Macmillan, New York, p 35, Th. 6
9. Blanchard Ph, Volchenkov D (2009) Probabilistic embedding of discrete sets as continuous metric spaces. Stochastics: Int J Prob Stoch Proc (formerly: Stochast Stochast Rep) 81(3):259
10. Blanchard Ph, Volchenkov D (2011) Introduction to random walks on graphs and databases. Springer series in synergetics, vol 10. Springer, Berlin/Heidelberg, ISBN 978-3-642-19591-4
11. Bolton RP (1922) Building For Profit. Reginald Pelham Bolton, New York
12. Campbell SL, Meyer CD, Rose NJ (1976) Applications of the Drazin Inverse to Linear Systems of Differential Equations with Singular Constant Coefficients. SIAM J Appl Math 31(3):411
13. Castorina P, Blanchard Ph (2012) Unified approach to growth and aging in biological, technical and biotechnical systems. SpringerPlus 1(7). doi:10.1186/2193-1801-1-7
14. Castorina P, Delsanto PP, Guiot C (2006) Classification scheme for phenomenological universalities in growth problems in physics and other sciences. Phys Rev Lett 96:188701
15. Castorina P, Deisboeck TS, Gabriele P, Guiot C (2007) Growth laws in cancer: implications for radiotherapy. Rad Res 169:349
16. Chung FRK (1997) Lecture notes on spectral graph theory. AMS Publications, Providence
17. Drazin, MP (1958) Pseudo-inverses in associative rings and semigroups. Am Math Mon 65:506
18. Duistermaat JJ (2001) On the boundary behaviour of the Riemannian structure of a self-concordant barrier function. Asymptotic Anal 27(1):9
19. Economos AC (1979) A non-gompertzian paradigm for mortality kinetics of metazoan animals and failure kinetics of manufactured products. Age 2:74–76
20. Erdélyi I (1967) On the matrix equation $Ax = \lambda Bx$. J Math Anal Appl 17:119
21. Gavrilov LA, Gavrilova NS (1991) The Biology of life span: a quantitative approach. Harwood Academic Publisher, New York
22. Gavrilov LA, Gavrilova NS (2001) The reliability theory of aging and longevity. J Theor Biol 213(4):527–545
23. Gavrilov LA, Gavrilova NS (2011) Mortality measurement and modeling beyond age 100. In: Living to 100 Symposium, Orlando, FL
24. Glaeser EL, Gyourko J (2003) *Why is Manhattan So Expensive?* Manhattan Institute for Policy Research, *Civic Report*, No. 39
25. Gompertz B (1825) On the nature of the function expressive of the law of human mortality and a new mode of determining life contingencies. Phil Trans R Soc 115:513
26. Graham A (1987) Nonnegative matrices and applicable topics in linear algebra. Wiley, New York
27. Guiot C, Degiorgis PG, Delsanto PP, Gabriele P, Deisboeck TS (2003) Does tumor growth follow a universal law? J Theor Biol 225:147–151
28. Hansen WG (1959) How accessibility shapes land use. J Am Inst Planners 25:73
29. Hartwig RE (1976) More on the Souriau-Frame algorithm and the Drazin inverse. SIAM J Appl Math 31(1):42
30. Hey J (2001) The mind of the species problem. Trends Ecol Evol 16(7):326
31. Hillier B, Hanson J (1984) The social logic of space. Cambridge University Press, Cambridge, ISBN 0-521-36784-0
32. Hillier B (1999) Space is the Machine: A Configurational Theory of Architecture. Cambridge University Press, Cambridge, ISBN 0-521-64528-X (1999)

33. Horiuchi S, Wilmoth JR (1998) Deceleration in tha age pattern of mortality at older ages. Demography 35:391
34. Huchet A, Candry H, BelkaceniY (2003) L'effet volume en radiotherapie. Premiere parie: effect volume et tumeur. Canc Radiother 7:79–89
35. Kim JJ, Tannock IF (2005) Repopultaion of cancer cells during therapy: an important cause of treatment failure. Nat Rev Canc 5:516–525
36. Lovász L (1993) Random walks on graphs: a survey. *Bolyai Society Mathematical Studies* **2**: *Combinatorics, Paul Erdös is Eighty*, 1, Keszthely (Hungary)
37. Meyer CD (1975) The role of the group generalized inverse in the theory of finite Markov chains. SIAM Rev 17:443
38. Muller Ch, et al (1988) The survival rate of patient with pace-maker is essentially the same of the normal population. Eur Heart J 9:1003
39. Norton LA (1988) Gompertzian model of human breast cancer growth. Canc Res 48:7067–7071
40. Olshansky SJ (1998) On the biodemography of aging: a review essay. Popul Dev Rev 24:381–393
41. Pennec X (2004) Intrinsic statistics on Riemannian manifolds: Basic tools for geometric measurements. J Math Imag Vis. 25(1):127
42. Rigdon SE, Basu AP (2000) Statistical methods for the reliability of repairable systems. Wiley, New York
43. Robert P (1968) On the group inverse of a linear transformation. J Math Anal Appl 22:658
44. Stanley JA, Shipley WU, Steel GG (1977) Influence of tumor size on the hypoxic fraction and therapeutic sensitivity of Lewis lung tumour. Br J Canc 36:105–13
45. Steel GG (1977) Growth kinetics of tumours. Clarendon Press, Oxford
46. Vaupel JW, Carey JR, Christensen K, Johnson T, Yashin AI, Holm NV, Iachine IA, Kannisto V, Khazaeli AA, Liedo P, Longo VD, Zeng Y, Manton K, Curtsinger JW (1998) Biodemographic trajectories of longevity. Science 280:855–860
47. Volchenkov D, Blanchard Ph (2007) Random walks along the streets and channels in compact cities: spectral analysis, dynamical modularity, information, and statistical mechanics. Phys Rev E 75:026104
48. Volchenkov D, Blanchard Ph (2008) Scaling and universality in city space syntax: between Zipf and Matthew. Physica A 387(10):2353
49. Volchenkov D (2013) Markov chain scaffolding of real world data, Discontinuity, Nonlinearity, and Complexity 2(3):289–299
50. Wheldon TE (1988) Mathematical models in cancer research. Adam Hilger Publisher, Bristol
51. Wilson AG (1970) Entropy in urban and regional modeling. Pion Press, London
52. Wolfram S (1984) Cellular automata as models of complexity. Nature 311:419
53. Wolfram S (2002) A new kind of science. Wolfram media, Champaign, IL

Treasure Hunting in Virtual Environments: Scaling Laws of Human Motions and Mathematical Models of Human Actions in Uncertainty

Dimitri Volchenkov, Jonathan Helbach, Marko Tscherepanow, and Sina Kühnel

Abstract Searching experiments conducted in different virtual environments over a gender balanced group of people revealed a gender irrelevant scale-free spread of searching activity on large spatiotemporal scales. The better performance of men in virtual environments can be associated with the regularly renewed computer game experience, essentially in games played through a first-person perspective. We suggested a simple self-organized critical model of search, in which the experimentally observed scale-free behavior can be interpreted as a trade-off between the value of exploitation versus exploration amid uncertainty.

1 Introduction

Virtual environments (VE) provide a simplified way to see and experience the real world, supporting the sense of spatial presence via virtual locomotion, rendering a clear sense of navigation, and allowing for interactions with objects through a user interface [40]. Not surprisingly, VE gained widespread use in recent years as a tool for studying human behavior, maintaining the capacity to create unique experimental scenarios under tightly controlled stimulus conditions. A major problem for users of VE is maintaining knowledge of their location and orientation

D. Volchenkov
Faculty of Physics, Bielefeld University, Universitaetsstr. 25, 33615 Bielefeld, Germany
e-mail: volchenk@physik.uni-bielefeld.de

J. Helbach • M. Tscherepanow
Technical Faculty, Bielefeld University, Universitaetsstr. 25, 33615 Bielefeld, Germany.
e-mail: jonathan.helbach@gmail.com; marko@techfak.uni-bielefeld.de

S. Kühnel (✉)
Physiological Psychology, Bielefeld University, Post office box 100131, 33501 Bielefeld, Germany
e-mail: skuehnel@uni-bielefeld.de

V. Afraimovich et al. (eds.), *Nonlinear Dynamics and Complexity*, Nonlinear Systems and Complexity 8, DOI 10.1007/978-3-319-02353-3_8,

while they move through the space, essentially when the whole path cannot be viewed at once but occluded by objects in the environment [16]. Most of the human spatial abilities (such as navigating a large-scale space and identifying a place) have evolved in natural environments over a very long time, using properties present in nature as cues for spatial orientation and wayfinding [41]. However, many of the natural body-based self-motion cues are absent in VE, causing systematic spatial orientation problems in subjects and calling for the new adaptive strategies to move through VE under reduced multisensory conditions.

In order to understand the adaptive movement strategy in VE and clarify the role of environmental structure in searching and browsing, we conducted a treasure hunting experiment (Sect. 2.1) with a gender balanced group of participants (Sect. 2.3), in the different office VE (Sect. 2.2). Based on the results of the statistical data analysis, we discuss on the gender disparity in video- and computer game experience as a factor potentially influencing the better performance of men in VE (Sect. 3.1), on the role of scanning and reorientations in a compensation of information deficiency while moving through VE (Sect. 3.2), and on the experimentally observed scale-free spread of searching activity on large spatiotemporal scales reminiscent of Lévy flights that become ubiquitous in modern movement ecology (Sect. 3.3).

Our results shed new light on the well-known exploration-exploitation dilemma [13]. Subjects participated in the study had to decide how to proceed amid uncertainty. On the one hand, there was the option to continue searching in the nearest neighborhood (exploitation), in the hope to get a reward beyond the next door. Alternatively, subject could explore the parts of environment she never been to. Therefore, the actual searching trajectory of a participant in the VE resulted from a permanent balance between exploitation and exploration confronted at all levels of behavior across all time-scales. In general, how agents should and do respond to the trade-off between exploration and exploitation is poorly understood [13]. It is important to note that the setting of our experiments had nothing to do with the assumptions of [21] presented an optimal strategy for trading off exploration and exploitation. The probability of delivering the reward was not fixed in our case, subjects did not discount the value of each reward exponentially as a function of when it is acquired, and eventually the time of experiment was essentially limited (in contrast to over an infinite horizon in the Gittins approach). Subjects in our study acted amid uncertainty, so that any calculation of an optimal strategy was impossible for them. Appearance of Levy statistics in searching behavior can help us to understand how humans manage the balance between exploration and exploitation in non-stationary environments.

In Sect. 4.1, we give arguments in favor of that the observed statistical properties of search patterns can be explained in the framework of a self-organized critical model of search amid uncertainty, which can help generate biologically relevant hypothesis about fundamental process of a decision making when no precise information on a possibility of rewards is available. The mathematical model of decision making in random search is formulated in Sect. 4.2 and solved analytically in some practically important cases in Sect. 4.3. We conclude in the last section.

2 Methods

2.1 *Experimental Design and Procedure*

In our treasure searching experiment, every participant was asked to browse an office VE searching for collectable objects. For each time frame, the position and heading orientation of the participant were tracked and subsequently analyzed by considering the collections of displacements and turns as a series of random events whose spatial and temporal distributions are assumed to possess certain statistical regularities.

To motivate searching, each found object were rewarded with an extra 50 cents coin, in addition to the basis remuneration for participation in the study. None of the treasure hunters neither visited a real prototype of the VE model nor foresaw its floor plan. The objects of search were big enough, contrast colored, clearly visible toys: teddy bears and locomotives. At the beginning of each trial, a number of toys (10 toys, for the smaller environment, and 15 toys, for the bigger one; see the Sect. 2.2 for details) were allocated in randomly chosen offices, beyond the closed doors, one toy per room. Objects could be found immediately, as soon as subject opens the door and enters the room. In order to focus subjects on the tasks no communication between experimenter and subject was performed during the experiment.

Before entering the main exploration areas, every participant was trained in a virtual tutorial room, in order to get used to stereoscopic imaging of computer-simulated environment (two slightly different images accounting for the interpupillary distance paired with stereo glasses providing a three-dimensional display of the environment), to get a good command of a *Nintendo* Wii remote controller, and to judge their perceived motions via button presses. Although the time of search was not limited, we have restricted the total number of doors subjects could open during the experiment (by 10 doors, in the smaller environment, and by 15 doors, in the bigger one), in order to prevent a sequential search at each office and stimulate exploration activity in subjects. All participants were informed about that some rooms can be interconnected, at a benefit for the search, as only one door should be opened but two rooms can be searched. The experiment ended when the subject opened 10(15) doors.

Two AVI video fragments showing the records of actual searching experiment from a first-person perspective can be found under the URLs [35].

2.2 *Virtual Environments*

Virtual models of two actual office environments of the University of Bielefeld were rendered with the *Autodesk ® 3ds Max ® Design 2010* software and then projected for any user viewpoint onto a wall-wide laboratory screen (4×2 m) with the use of the *Barco Galaxy NH-12* active stereoscopic 3D stereo projector. Natural color reproduction, extended gray levels, and high brightness support the sense of spatial presence in subjects. The control of user viewpoint motion through the VE was implemented via the *Bluetooth* connected *Wiimote*, the primary controller for *Nintendo's* Wii console featured with motion sensing capability, which allows the user to manipulate items on screen via gesture recognition and pointing through the use of accelerometer and optical sensor technology.

The VE model A (Fig. 1a) exactly reproduces the second floor of a temporary building (2012) belonging to the Cognitive Interaction Technology—Center of Excellence (CITEC, Bielefeld University), and the VE model B (Fig. 1b) has as a prototype the ground floor of the future Interactive Intelligent Systems Institute (Bielefeld University) presently under construction. The both environments consist of the standard adjacent offices, meeting rooms, hallways and corridors providing space where people can move, meet, and discuss. Emergency exits and elevators that exist in the actual prototype buildings were not taken into account in our experiments. The VE model A consists of 48 interconnected individual spaces of movement (the nodes of the spatial graph shown in Fig. 1a), and the VE model B includes 68 interconnected individual spaces of movement (the nodes of the spatial graph shown in Fig. 1b).

The spatial structure is important because of its effect on proximity: greater connectedness of a built environment generally means more direct routes and thus shorter distances between possible destinations. Connectedness also affects walking by expanding the choice of routes, thereby enabling some variety in routes within the environment. Discovering important spaces of movement and quantifying differences between them in a spatial graph of the environment is not easy since any two spaces can be related by means of many paths. In [7, 8], we suggested using the properties of random walks, in order to analyze the structure of spatial graphs and to spot structural isolation in urban environments. In contrast to classical graph theory paying attention only to the shortest paths (of least costs) all existing paths are taken into account in the approach related to random walks, although some paths are more preferable than others, as random walks respecting graph symmetries assign the probability to be traversed by a random walker to each path in the graph [9]. Each node of the graph can be characterized with respect to the entire graph structure by the *first-passage time*, the expected number of steps required to reach the node for the first time (without revisiting any intermediate node) starting from any node of the graph chosen randomly, in accordance with the stationary distribution of random walks. In built environments, the first-passage time to a place can be understood as an average number of elementary wayfinding instructions (such as "turn left/right," "pass by the door," and "walk on the corner") required to navigate a wanderer to the place from elsewhere within the environment. The values of first passage times to the nodes in the spatial graphs A and B are color indicated in Fig. 1: the central places characterized with the minimal first passage times are red colored, and the secluded places with the maximal first passage times are shown in violet.

It is worth a mention that the spatial structure of the VE model A analyzed by means of the spatial graph Fig. 1a is essentially simpler than the structure of

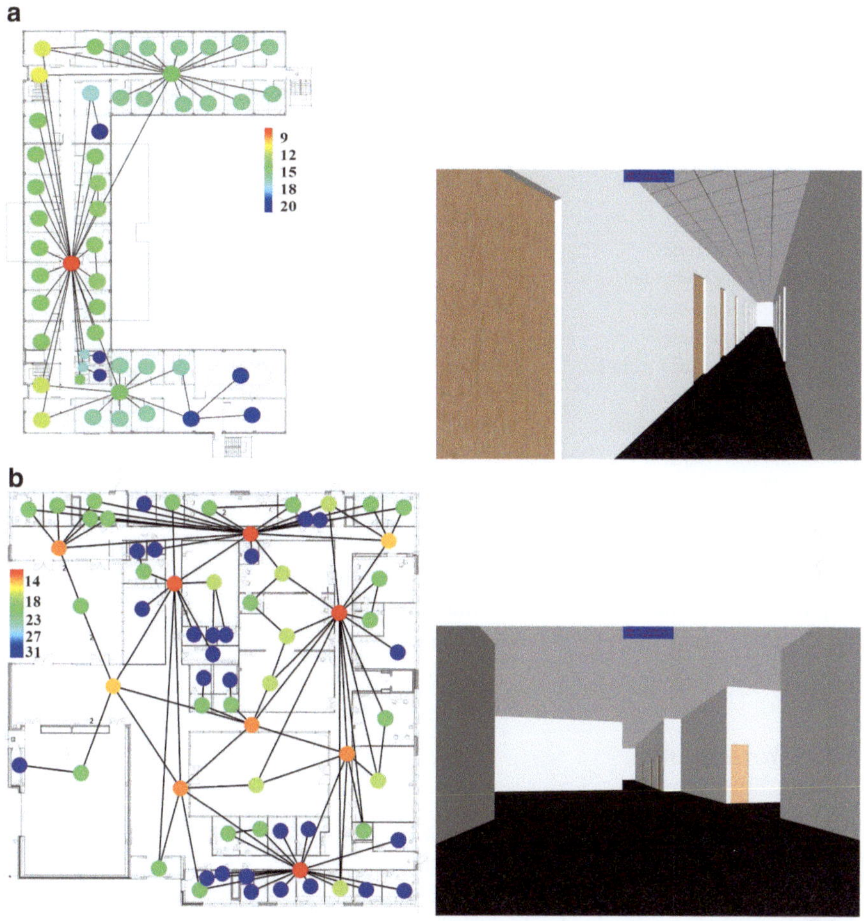

Fig. 1 (*Colored*) Two models of the office VE were used in the treasure search experiments. The individual spaces of movement are identified with the nodes in the spatial graphs. The centrality/isolation of a node with respect to the entire structure of a spatial graph is characterized by the first-passage time of random walks indicated by *color*. (**a**) The VE model A consists of 48 interconnected spaces of movement, with a single central place (the central corridor denoted by the *red node*) that locates nine steps apart from any randomly chosen node in the spatial graph. (**b**) The VE model B consists of 68 interconnected spaces of movement and contains a network of well-connected central places (halls and connecting corridors) characterized by the first passage times ranging from 10 to 14 steps

the model B. Contrasted to the model A, where offices are located along three sequentially connected corridors providing a single path between most of the possible destinations, the spatial structure of the model B allows for many cyclic trips due to a network of well-connected places and includes a number of vantage points, from which a walker could observe the substantial parts of the environment from different perspectives. The single central node (the central corridor) in the VE

Table 1 Summary table on the experiment participants

	Office environment A		Office environment B	
	Women	Men	Women	Men
Number of volunteers	20	20	21	21
Average age of a volunteer	26.2 Min: 21 Max: 51	24.5 Min: 21 Max: 31	23.6 Min: 20 Max: 37	23.4 Min: 11 Max: 29
Average time playing computer (*hours per week*)	4.1 Min: 0 Max: 35	5.4 Min: 0 Max: 35	0.9 Min: 0 Max: 10	6.2 Min: 0 Max: 30
Average time playing video games (*hours per week*)	0.42 Min: 0 Max: 10	1.4 Min: 0 Max: 5	0.09 Min: 0 Max: 2	1.2 Min: 0 Max: 10
Played first-person perspective games	1	18	10	20
Wear glasses	12	18	13	9

model A is located at nine steps apart from any randomly chosen node in the spatial graph (Fig. 1a). The VE model B contains a network of well connected spaces of movement characterized by the first passage times ranging from 10 to 14 steps.

2.3 *Participants*

Two gender balanced groups of volunteers (82 participants in total) took part in the controlled searching experiments conducted in the office VE shown in Fig. 1a, b. Although participants (mostly university students) were recruited personally and through advertisements at the University of Bielefeld, none of them have ever been familiar with the actual building prototyping the virtual model A and have ever seen the plan of the building prototyping the model B. Prior to testing, all adult participants and parents of children younger than 16 years old gave their informed written consent for participation in the study. Participation in the study was voluntary, and a participant could revoke her participation consent and quit at any time and for any reason. The standard provisions for data protection were adhered: all test results were kept confidential. All individual data were managed and processed anonymously that eliminated any possibility of identification of participants.

Before starting the experiment, volunteers were asked to estimate time spent playing computer and video game consoles (in hours per week) and to answer the question of whether or not they played games with the experience of a first-person perspective. The data summary on subjects joined the study is given in Table 1.

3 Results and Discussions

3.1 On Gender Disparity in Game Experience and Mobility in VE

Perhaps, the main feature of experiments on human behavior in VE is that male participants usually enter the study with more prior experience in computer and video games than females that may have to do with well-known video game gender disparity which receives extensive academic, corporate, and social attention regularly [18, 22]. Unfortunately, the factor of gender disparity in prior virtual game experience is systematically disregarded in studies concerning spatial cognition in VE. With some noticeable exceptions, female gamers participated in our study reported on substantially less time playing computer and video games (in hours per week) than male participants. In Fig. 2, we have presented the box plots summarizing the data on time playing virtual games reported by men and women participated in our experiments. In each box presented in Fig. 2, a central line indicates the median, a lower line is showing the first quartile, and an upper line is for the third quartile. Two lines of maximal length extending from the central box indicate points that lie outside the extent of the previously mentioned statistics.

Contrasted to female participants, almost all male subjects in the present study reported that they played games centered on projectile weapon-based combat through a first-person perspective, as shown in Fig. 3. It is important to note that first-person shooters often focus on action gameplay, being featured by large built environments, which can be explored freely often requiring interaction with the environment to varying degrees (e.g., by opening doors). Playing computer games through a first-person perspective was reported to be associated with superior mental flexibility [14], possibly because players of such games are required to develop a more responsive mindset to rapidly react to fast-moving visual and auditory stimuli. Women have been shown to prefer role playing games to first-person shooters favoring in-game communication [18] and interpersonal relationships over fast-paced first person actions preferred by men [22].

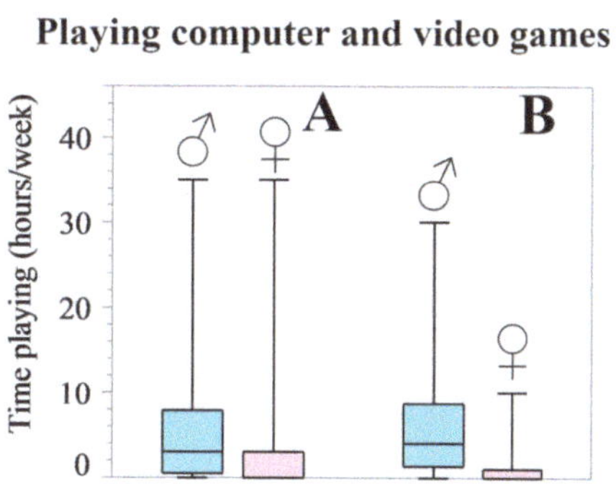

Fig. 2 The box plots summarizing the data on time playing virtual games (in hours per week) reported by men and women participated in our experiments

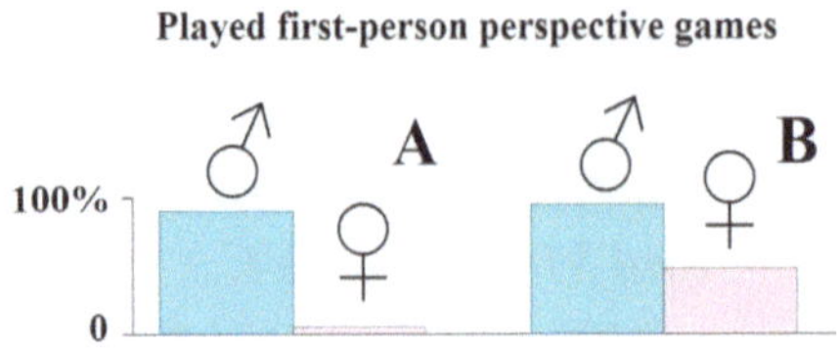

Fig. 3 The proportions of male and female subjects reported that they played computer and video games from a first-person perspective

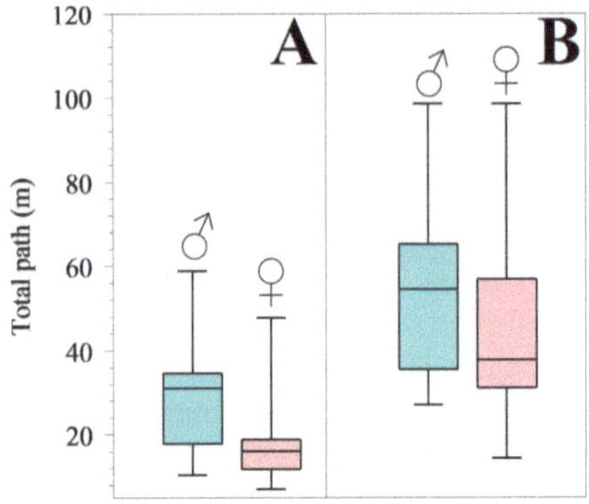

Fig. 4 The total path (in meters) walked by male and female subjects during the experiments

We must conclude that male subjects were certainly better prepared for participation in the study than females: men not only were more experienced in navigation and action in VE but also regularly renewed their experience. Proficiency in visual-spatial cognition is associated with better performance in VE [15]. It is worth a mention that male and female subjects who played computer games through a first-person perspective always started the experiment with a complete exploration of the environment, before opening the first door and starting the active treasure search.

Gender is often reported as a decisive factor in spatial cognition research [17,26]: women rated driving in unfamiliar areas as being a more serious problem than did men [19], males learn a virtual maze more rapidly than females [29]. A review of gender differences in spatial ability in real world situations can be found in [39]. Gender is also found as a factor influencing navigation in VE: males were reported to acquire route knowledge from landmarks faster than females [15].

Contrasted to locomotion in real environments, mobility in VE depends upon self-motion perception in virtual space and convenience of locomotion interface, rather than on physiological factors of an individual such as height, weight, age, or fitness. The instantaneous translation velocity in the VE is kept constant for any individual, as long as she presses a button. In order to quantify mobility of subjects in a VE, we measured the total path (in meters) travelled by subject in the experiment (summarized in the form of box plots in Fig. 4) and the mean velocity of motion calculated as a ratio of the total path and the total traveling time (see Fig. 5). Our results show that men stride longer distances in average and move faster than women. It is not a surprise that the amount of walking people do in the model B substantially exceeds the total distances travelled in the model A, as the latter model was smaller, provided less opportunity for exploration, and the searching activity

Fig. 5 The mean velocity of motion (m/s) in VE during the treasure hunting experiments

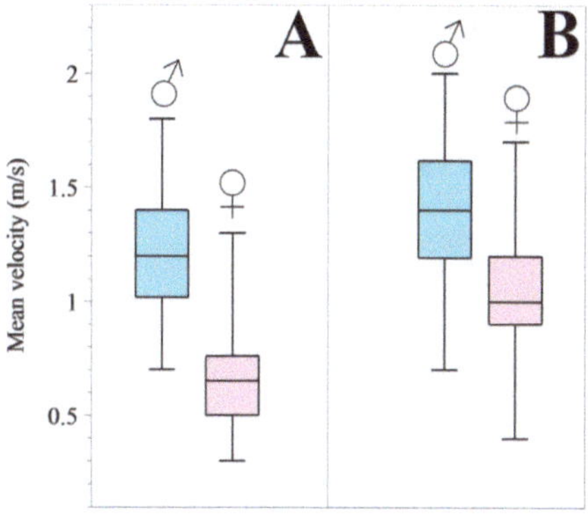

in that was limited to only 10 doors. However, it is remarkable that the surplus of mean total paths walked in the VE model B was disproportionate in men and women. Namely, the average path walked by men in B was 75 % longer than in A, in comparison with the 135 % increase of average path in women (Fig. 4). Although in average men moved faster than women in the both VE models (Fig. 5), the velocity gain in B was higher in women (54.5 %) than in men (9.5 %).

Our results indicate that human mobility can increase in the VE affording more opportunity for exploration. With some noticeable exceptions female subjects were in average less mobile than men. The better performance of men in VE can be associated with the prior computer game experience, essentially in games played through a first-person perspective, being a key to acquiring proficiency in visual-spatial cognition. It is possible that the extraordinary increase of mobility in women explored the VE model B is due to the fact that more than half of them played first-person perspective games before participating in our experiments. Further investigation is needed to clarify the role of a prior computer game experience in studies concerning spatial cognition in VE.

3.2 *Scanning Turns, Reorientations, and Explorative Rotations in a VE*

Self-motion through a VE suffers from a lack of many natural body-based cues. Natural methods of visual exploration are also restricted in VE to that experienced through the display representing only a limited field of view, suffering from the degradation of sensory cues due to device latencies and blocking out all surrounding visual input. It was concluded from various experiments that the optic flow without proprioception, at least for the limited field of view of the virtual display system, appears to be not effective for the updating of heading direction [24], and even when physical motion cues from free walking are included, this is not necessarily sufficient to enable good spatial orientation in VE [31].

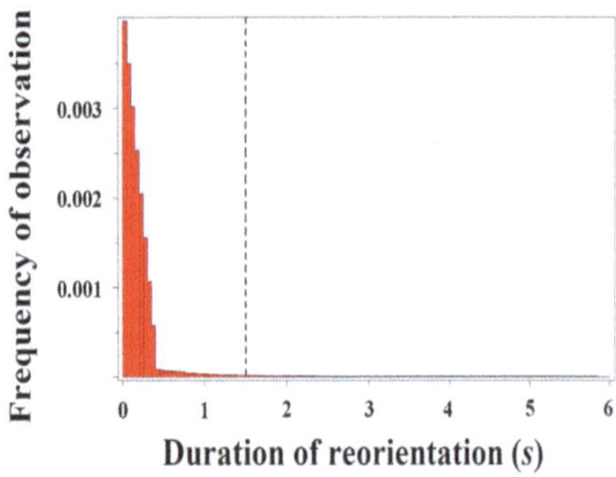

Fig. 6 The typical distribution of scanning turns, reorientations, and explorative rotation durations while exploring a VE. The *vertical dashed line* indicates the duration of turn 1.5 s

Perhaps due to systematic spatial orientation problems occurring in VE compared to real-world situations, the most of subjects participated in our study permanently performed fast scanning turns (of 200–300 ms) by the quick altering pressing on the left and right buttons of the remote controller, each time causing them to turn a greater or lesser angle. Probably, such a movement routine played an important role for the proper self-motion perception, as compensating information deficiency experienced by subjects while moving through VE under reduced multisensory conditions. Longer turns (usually, taking about half a second) were observed when subjects redirected their walk or avoided obstacles. Eventually, the very long, explorative rotations often including several complete revolutions (each time lasting up to a few seconds) occurred after far relocations, at vantage points, along the borders of two or several vista spaces, and at intersections of corridors that afford a broader view of the environment.

The typical distribution of scanning turns, reorientations, and explorative rotation durations while exploring a VE is presented in Fig. 6. The areas of adjacent rectangles in the histogram are equal to the relative frequency of observations in the duration interval. The total area of the histograms is normalized to the number of data. The data show that the vast majority of all reorientations performed by subjects were the quick scanning turns, being the essential part of the walking routine in VE. The vertical dashed line in Fig. 6 stands for the rotation duration of 1.5 s; the locations of the correspondent points, at which participants performed longer turns (without translations) are displayed on the floor plans by the circles (see Fig. 7a and b). The diameter of a circle is proportionate to a number of long turns recorded at its central point over all subjects.

Several studies conducted on small animals [4,5,25,32] suggested that the switch between scanning and reorientation behavior in movement patterns of animals that search emerges from complex mechanic-sensorial responses of animals to the local environment and could infer the effects of limited perception and/or a patchy environmental structure. When exploring patchy resources, animals could adjust turning angle distributions, selecting a preferred turning angle value that would allow organisms to stay within the patch for a proper amount of time, maximizing the energetic gain. For example, the zigzag motion of *Daphnia* appears to be an

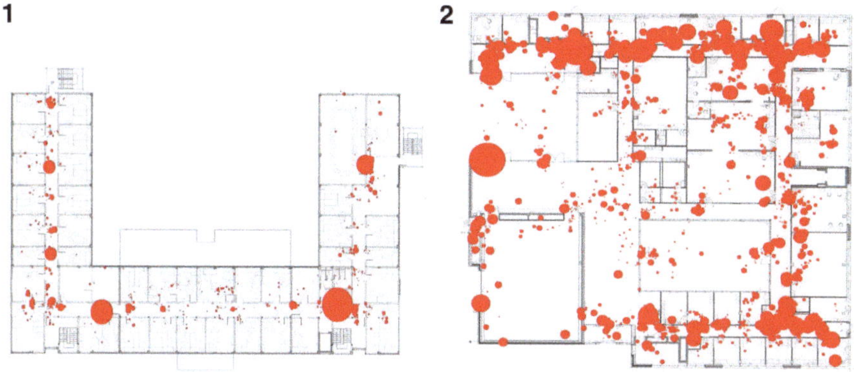

Fig. 7 The floor plans of the models A and B with the locations of vantage points (shown by *circles*), where participants performed reorientations and explorative turns lasting longer than 1.5 s. The *diameter of a circle* is proportionate to a number of long turns recorded at its central point

optimal strategy for patch exploitation [32]. Therefore, the distinction between quick scanning turns and a longer reorientation behavior is crucial to understand the statistical patterns of search [3].

Being a part of the travelling routine in a VE, quick scanning turns performed by subjects during the walk can produce correlations between turning durations and displacements, on relatively small spatiotemporal scales sensitive to the local structural features of the environment. After reaching the natural limits of available space of motion or entering a new movement zone, the subject performs a longer reorientation, perhaps in order to explore the new environment visually that breaks the correlations. We analyzed the data on durations of reorientations of subjects travelling through the both VE models with the use of the *root mean square fluctuations* (RMSF) suitable for detecting correlations [23, 36]. The RMSF of displacements is calculated by

$$F_s(n) = \sqrt{\left\langle (\Delta\mathfrak{S}(n))^2 \right\rangle - \langle \Delta\mathfrak{S}(n) \rangle^2}, \quad (1)$$

in which the net displacement of the walker by the n-th reorientation is $\mathfrak{S}(n) = \sum_{k=1}^{n} \|\mathbf{r}_{k+1} - \mathbf{r}_k\|$, $\mathbf{r}_k$ is the recorded position of the k-th reorientation, $\Delta\mathfrak{S}(n) = \mathfrak{S}(n + n_0) - \mathfrak{S}(n_0)$, and the angular brackets denote averaging over all data points $n_0 = 1, \dots n_{\max}$. Similarly, the RMSF of rotation durations is calculated by

$$F_\tau(n) = \sqrt{\left\langle (\Delta\mathfrak{T}(n))^2 \right\rangle - \langle \Delta\mathfrak{T}(n) \rangle^2}, \quad (2)$$

in which the total rotation duration of the subject by the n-th reorientation is $\mathfrak{T}(n) = \sum_{k=1}^{n} \tau_k$, τ_k is the duration of the k-th reorientation, $\Delta\mathfrak{T}(n) = \mathfrak{T}(n + n_0) - \mathfrak{T}(n_0)$, and the angular brackets again denote the average over all data points. In Fig. 8, we

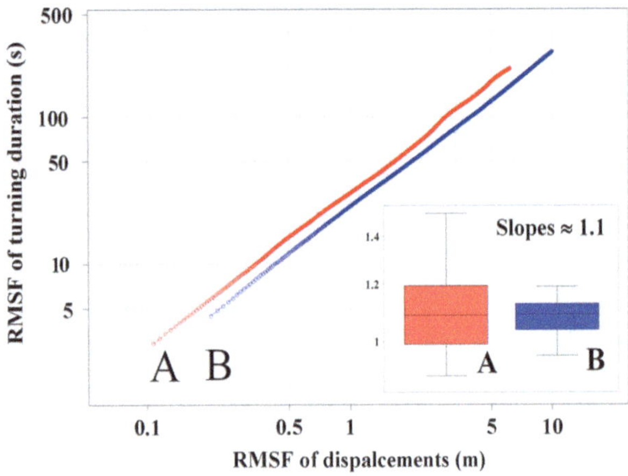

Fig. 8 The RMSF of the total durations of turning is shown via the RMSF of the net displacement in the log–log scale, for all recorded reorientation points of all subjects, in the both VE models

juxtaposed the RMSF of the net displacement and the RMSF of the total rotation duration, in the log–log scale, for all recorded reorientations of subjects, separately for the VE models A and B. Identically, for the both VE models, the graphs show the super linear slope,

$$\frac{d\log F_\tau}{d\log F_s} \cong 1.1,$$

indicating the presence of a strong positive relation that reinforces the total duration of quick scanning turns (up to 30 s), with the increase of the net displacement of subject (by 10 m). In the long run, the correlations generated by the quick scanning behavior vanish that is typical for a correlated random walk process [5]. It is obvious that such an intensive scan is performed by subjects only within their immediate neighborhoods and principally cannot be extended neither to the entire VE nor even to any of its significant parts, as the superlinear increase of required time makes the scanning process on large spatiotemporal scales biologically unfeasible. Thus, after completing a phase of intensive search within the patch of a size depending on the structural properties of the environment, a treasure hunter moves to some other area, where the phase of intensive search is resumed. It is important to mention that changes in reorientation behavior, on large spatiotemporal scales, can generate different anomalous diffusion regimes, which in turn, can affect the search efficiency of random exploration processes [5].

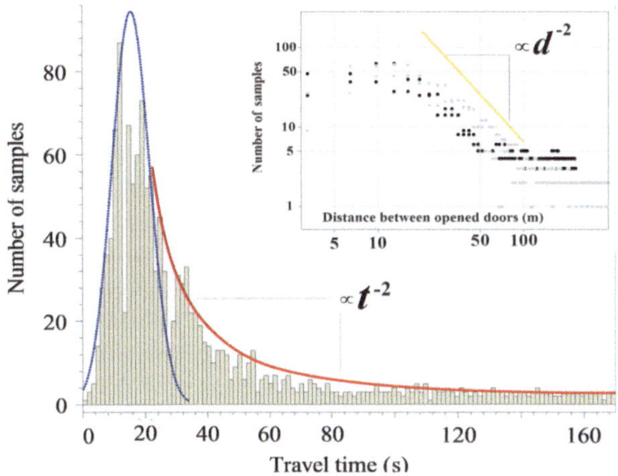

Fig. 9 The histogram shows the frequency of travelling times between the consequent door openings as a function of time. The *bell-shaped dotted curve* represents the Gaussian distribution. Occasional long travels contribute into the long right tail of the distribution dominated by a quadratic hyperbola. On the outline, the data on dispersal of the consequent searching events are shown as a function of distance in the log–log scale for the VE model A (*by black points*) and for the VE model B (*by grey points*). The *solid guide line* indicates the inverse quadratic slope

3.3 Pareto–Lévy Distributions of Human Travels in a VE

In order to identify phases of specific activity in recorded movement patterns and to reveal the underlying cognitive mechanisms from their statistical properties, we have studied the probability distributions of time intervals and distances between consequent observable searching events (door openings) as they can determine strong changes in the diffusive properties of movement and in relevant spatial properties of the trajectories. The distributions representing the data on dispersal of the treasure hunters during the experiment are shown in Fig. 9, for both VE models. The form of the distributions is not gender specific.

The data show that the most of the consequent door openings happen in the immediate neighborhood of the actual position of a participant. The dispersal statistics on the small spatiotemporal scales shown in Fig. 9 can be well approximated by uncorrelated Gaussian random walks (see the dotted bell shaped curve) insensitive to the local structure of the environment. The both distributions in Fig. 9 are remarkable for the long right tails dominated by the *quadratic hyperbolas* attenuating the superdiffusive spread of treasure hunters on large spatiotemporal scales. A power-law tail in the probability distributions of both travelling times and travelled distances could arise from processes in which neither time nor distance has a specific characteristic scale, so that rare but extremely long and far cry travels can occur, alternating between sequences of many short travels featuring local searches. The power-law tails in the dispersal data provide an evidence in favor of a strong coupling between movements and a large-scale environmental structure which cannot be viewed at once [6]. A special class of discrete-step random walks with movement displacements drawn from a probability distribution with a power-law tail (the Pareto–Lévy distribution) is known as *Lévy flights* [28]. Geometrically, they consist of walking clusters with very large displacements between them repeated over a wide range of scales [33]. Although Lévy flights are ubiquitous for representing intermittent search, cruise, and foraging strategies in living species ranging from microorganisms to big marine predators (see [1, 4–6, 10, 11, 27, 34, 36, 37] and many others), to our knowledge, this is the first report on observation of the Lévy flight search patterns in humans exploring VE. It has long been recognized that the existence of episodic behaviors resulting in random alternating of active exploration phases with phases of fast ballistic motion can strongly modify the success of a search process. In particular, it was shown that the optimal strategy for a searcher looking for sparsely and randomly distributed fixed targets is to be drawn from an inverse square power-law distribution of flight length, $P(l) \sim l^{-2}$, corresponding to Lévy flight motion [11, 36, 37]. The Lévy search hypothesis [3, 38] implies that while searching for a randomly located objects, an animal could adopt a movement strategy that takes the advantage of Lévy stochastic process, in order to minimize the mean time for target detection or mean first-passage time to a random target, as well as to maximize the energetic gain in case of sparsely and randomly distributed resources, since the probability of returning to a previously visited site is smaller for a Lévy flight than for a Gaussian distributed walk [37]. However, a particular biological mechanism that allows animals and humans to reproduce exactly the statistics of Lévy flights on each trial without demanding and tedious computations remains unknown.

4 Self-organized Critical Model of Search Amid Uncertainty

4.1 Arguments in Favor of a Self-organized Critical Model

The observed statistical properties of human search travels in a VE call for a model that could exhibit a "phase transition" to a variety of spatiotemporal scale-invariant behaviors spontaneously, for variable parameters that could be changed widely without affecting the emergence of critical behavior. In statistical physics, such a property of dynamical systems is known as *self-organized criticality* [2]. Furthermore, a plausible model has to be of a discrete nature, as the biological principle of intermittent locomotion assumes that animal behavior unavoidably produces observable punctuations, "producing pauses and speeding patterns on the move" [5]. These motivations reflect a fundamental "trade-off" confronted by a treasure hunter choosing between an *exploitation* of the scanning movement routine within the familiar environment of nearest neighborhood, possibly at no reward, and a fast relocation aiming at *exploration* of unknown but potentially more rewarding areas.

From a theoretical perspective, it is known that in a stationary setting there exists an optimal strategy for exploration [21] maximizing the reward over an infinite horizon when the value of each reward is discounted exponentially as a function of when it is acquired. However, to date, there is no generally optimal solution to the exploration versus exploitation problem [12, 13], as human and other animals are prone to dynamically update their estimates of rewards in response to diverse, mutable and perhaps discrepant factors, including elapsed time of search, annoying failures to predict the location of a searched object, and instantaneous mood swings that could change in a matter of seconds. Therefore, it seems that a stochastic model managing a balance between exploration and exploitation may be more biologically realistic [13].

There is growing evidence that the neuromodulatory system involved in assessing reward and uncertainty in humans is central to the trade-off decision [12]. The problem can be cast in terms of a distinction between *expected uncertainty*, coming from known unreliability of predictive cues and coded in the brain by a neuromodularity system with acetylcholine signals, and *unexpected uncertainty*, triggered by strongly unexpected observations promoting exploration and coded in the brain with norepinephrine signals [42]. It was suggested in [42] that an individual decides on whether to stay or to go accordingly to the current levels of acetylcholine and norepinephrine, encoding the different types of uncertainty.

Summarizing the above mentioned arguments, we are interested in a self-organized critical model driven by a discrete time random process of competing between two factors featuring the different types of uncertainty.

4.2 *Mathematical Model of Decision Making in Random Search*

The movement ecology framework [30] explicitly recognizes animal movement as a result of a continuous "dialogue" between environmental cues (external factors) and animal internal states, [5]. In our model, we rationalize the dialogue nature of a decision-making process to take on searching in highly unpredictable situation when no precise information on a possibility of rewards is available. Despite its inherent simplicity, the mathematical model formulated below can help generate hypothesis about fundamental biological processes and bring the possibility to look for a variety of biological mechanisms under a common perspective.

We assume that an individual decides on whether to "exploit" an immediate neighborhood by search beyond a next door or to "explore" other parts of the environment by comparing the guessed chances, $q \in [0, 1]$ of getting a reward beyond the next door and $p \in [0, 1]$ of finding a treasure elsewhere afar. It is not necessary that $p + q = 1$. We suppose that at each time click subject updates one or both estimates and decides to proceed to a part of the environment yet to be explored if $q < p$. Otherwise, if $q \geq p$, she picks a next door randomly among those not yet opened and searches in the room behind. We consider p and q to be the random variables distributed over the interval $[0, 1]$ with respect to the probability distribution functions (pdf) $\Pr\{p < u\} = G(u)$ and $\Pr\{q < u\} = F(u)$, respectively. In general, F and G are two arbitrary left-continuous increasing functions satisfying the normalization conditions $F(0) = G(0) = 0$, $F(1) = G(1) = 1$.

We model the intermittent search patterns by a discrete time random process in the following way. At time $t = 0$, the variable q is chosen with respect to pdf F, and p can chosen with respect to pdf G. If $q < p$, subject relocates by pressing a button on the controller and goes to time $t = 1$. Given a fixed real number $\eta \in [0, 1]$, at time $t \geq 1$, the following events happen:

(i) with probability η, the chance to find a treasure in the immediate neighborhood (q) is estimated with pdf F, and the chance to get a reward elsewhere (p) is chosen with pdf G.

Otherwise,

(ii) with probability $1 - \eta$, the chance to find a treasure in the immediate neighborhood (q) is estimated with pdf F, but the chance to get a reward elsewhere (p) keeps the value it had at time $t - 1$.

If $q \geq p$, the local search phase continues; however if $q < p$, subject presses the controller button and moves further, going to time $t + 1$. Eventually, at some time step t, when the estimated chance q exceeds the value p, subject stops and resumes searching within the immediate neighborhood. The integer value $t = T$ acquired in this random process limits the time interval (and travelled distance) between sequential phases of searching activity.

4.3 Analytical Solutions for Decision Making Amid Uncertainty

While studying the above model, we are interested in the distribution of durations of the relocation phases $P_\eta(T; F, G)$ provided the probability distributions F and G are given and the control parameter η is fixed. For many distributions F and G, the model can be solved analytically. We shall denote $P_\eta(T; F, G)$ simply by $P(T)$. A straightforward computation shows directly from the definitions that

$$P(0) = \int_0^1 dG(p)\,(1 - F(p))\,.$$

For $T \geq 1$, the individual can either depart elsewhere ("D") or stay in the neighborhood ("S"). Both events can take place either in the "correlated" way (with probability η; see (i)) (we denote them D_c and S_c), or in the "uncorrelated" way (with probability $1 - \eta$; see (ii)) (D_u and S_u). For $T = 1$, we have for example

$$\begin{aligned} P(1) &= P[DS_c]+P[DS_u] \\ &= \int_0^1 dG(p)F(p)\eta(1-F(p))+\int_0^1 dG(p)F(p)(1-\eta)\int_0^1 dG(z)(1-F(z)) \\ &= \eta B(1)+(1-\eta)A(1)B(0). \end{aligned}$$

Similarly,

$$P(2) \;=\; \eta^2 B(2)+\eta(1-\eta)A(1)B(1)+\eta(1-\eta)A(2)B(0)+(1-\eta)^2 A(1)^2 B(0)$$

where we define, for $n = 0, 1, 2, \ldots,$

$$A(n) \;=\; \int_0^1 dG(p)F(p)^n$$

and

$$B(n) \;=\; \int_0^1 dG(p)F(p)^n\,(1-F(p)) \;=\; A(n)-A(n+1).$$

It is useful to introduce the generating function of $P(T)$:

$$\hat{P}(s) \;=\; \sum_{T=0}^{\infty} s^T P(T).$$

The generating property of the function $\hat{P}(s)$ is such that

$$P(T) \;=\; \frac{1}{T!}\left.\frac{d^T\hat{P}(s)}{ds^T}\right|_{s=0}. \tag{3}$$

Defining the following auxiliary functions

$$\begin{aligned} x(l) &= \eta^l A(l+1), && \text{for} \quad l \geq 1, \quad x(0)=0, \\ y(l) &= (1-\eta)^l A(1)^{l-1}, && \text{for} \quad l \geq 1, \quad y(0)=0, \\ z(l) &= \eta^l\left[\eta B(l+1)+(1-\eta)A(l+1)B(0)\right], && \text{for} \quad l \geq 1, \quad z(0)=0, \\ \rho &= \eta B(1)+(1-\eta)A(1)B(0), \end{aligned} \tag{4}$$

we find

$$\hat{P}(s) \;=\; B(0)+\rho s+\frac{s}{1-\hat{x}(s)\hat{y}(s)}\left[\hat{z}(s)+\rho\hat{x}(s)\hat{y}(s)+\rho A(1)\hat{y}(s)+A(1)\hat{y}(s)\hat{z}(s)\right] \tag{5}$$

where $\hat{x}(s)$, $\hat{y}(s)$, and $\hat{z}(s)$ are the generating functions of $x(l)$, $y(l)$, and $z(l)$, respectively.

In the marginal cases $\eta = 0$ and $\eta = 1$, the probability $P(T)$ can be readily calculated. For $\eta = 0$, Eqs. (4) and (5) give

$$\hat{P}_{\eta=0}(s) = \frac{B(0)}{1 - sA(1)}. \tag{6}$$

From Eq. (6), one gets

$$P_{\eta=0}(T) = A(1)^T B(0) = \left[\int_0^1 dG(p)F(p)\right]^T \int_0^1 dG(p)(1 - F(p)). \tag{7}$$

Therefore, in this case, for any choice of the pdf F and G, the probability $P(T)$ decays exponentially. For $\eta = 1$, Eqs. (4) and (5) yield $\hat{P}_{\eta=1}(s) = \hat{B}(s)$, so that

$$P_{\eta=1}(T) = B(T) = \int_0^1 dG(p)F(p)^T(1 - F(p)). \tag{8}$$

for the special case of uniform densities $dF(u) = dG(u) = du$, for all $u \in [0, 1]$ and for any $\eta \in [0, 1]$. In this case, simpler and explicit expressions can be given for $\hat{P}(s)$ and $P(T)$. Namely, from Eq. (5), we get

$$\hat{P}(s) = \frac{1}{1 + (1 - \eta)\gamma(s)}\left[\frac{1 + \gamma(s)}{s} - \eta\gamma(s)\right], \quad \gamma(s) \equiv \frac{\ln(1 - \eta s)}{\eta s}. \tag{9}$$

The asymptotic behavior of $P(T)$ as $T \to \infty$ is determined by the singularity of the generating function $\hat{P}(s)$ that is closest to the origin. For $\eta = 0$, the generating function $\hat{P}(s) = (2 - s)^{-1}$ has a simple pole, and therefore $P(T)$ decays exponentially that agrees with the result (Eq. 7).

For the intermediate values $0 < \eta < 1$, the generating function $\hat{P}(s)$ has two singularities. The first pole, $s = s_0$, corresponds to the vanishing denominator $1 + (1 - \eta)\gamma(s)$, where $s_0 = s_0(\eta)$ is the unique nontrivial solution of the equation

$$-\ln(1 - \eta s) = \frac{\eta s}{1 - \eta}.$$

The second singularity, $s = s_1 = \eta^{-1}$, corresponds to the vanishing argument of the logarithm. It is easy to see that $1 < s_0 < s_1$, so that the dominant singularity of $\hat{P}(s)$ is of the polar type, and for times much larger than the crossover time $T_c(\eta) \sim \ln(s_0(\eta))^{-1}$ the corresponding decay of $P(T)$ is exponential, with the rate $\ln(s_0(\eta))$. Eventually, when η tends to 1, the two singularities, s_0 and s_1 merge. More precisely, we have

$$\hat{P}_{\eta=1}(s) = \frac{s + (1 - s)\ln(1 - s)}{s^2}. \tag{10}$$

The corresponding dominant term in Eq. (10) is of order $O(T^{-2})$ [20]. This obviously agrees with the exact result one can get from Eq. (8), with $dF(u) = dG(u) = du$,

$$P_{\eta=1}(T) = \frac{1}{(T+1)(T+2)}. \tag{11}$$

Let us note that in the case of uniform densities it is possible to get an expression of $P(T)$ for all times, and for any value of η,

$$P(T) = \frac{\eta^T}{(T+1)(T+2)} + \sum_{k=1}^{T} \frac{\eta^T}{k(T-k+1)(T-k+2)} \sum_{m=1}^{k} \left(\frac{1-\eta}{\eta}\right)^m c_{m,k}, \tag{12}$$

in which

$$c_{m,k} = m! \sum_{l_1+\ldots+l_m=k} \frac{l_1 l_2 \ldots l_{m-1} l_m}{(l_1+1)(k-l_1)\ldots(l_{m-1}+1)(k-l_1-\ldots-l_{m-1})(l_m+1)}.$$

When $\eta \neq 0$, there is an alternative way of writing the previous expression:

$$P(T) = \frac{\eta^T}{(T+1)(T+2)} + \sum_{k=1}^{T} \frac{\eta^{T+1}}{(T-k+1)(T-k+2)} \sum_{i_1+\ldots+i_l=k} \sum_{l=1}^{\infty} \frac{(1-\eta)^l}{(i_1+1)\ldots(i_l+1)}.$$

In Fig. 10, we have presented the probability distributions of the searching durations for increasing values of η. The proposed mathematical model suggests that the algebraic tail dominated by a quadratic hyperbola observed in the distributions of time intervals and travelling distances between sequential phases of searching activity (see Fig. 9) can arise due to a trade-off between the value of exploitation versus exploration amid uncertainty. We have shown that when balancing the chances to be rewarded in the immediate neighborhood ("now and here") and later elsewhere ("then and there") subject estimates them uniformly at random, the inverse quadratic tail always dominates the distributions on large spatiotemporal scales.

5 Conclusions

We report on a solid quantitative assessment of human travelling statistics in VE by analyzing a comprehensive data set over tens of thousands individual reorientations and displacements on a variety of spatiotemporal scales performed by a gender

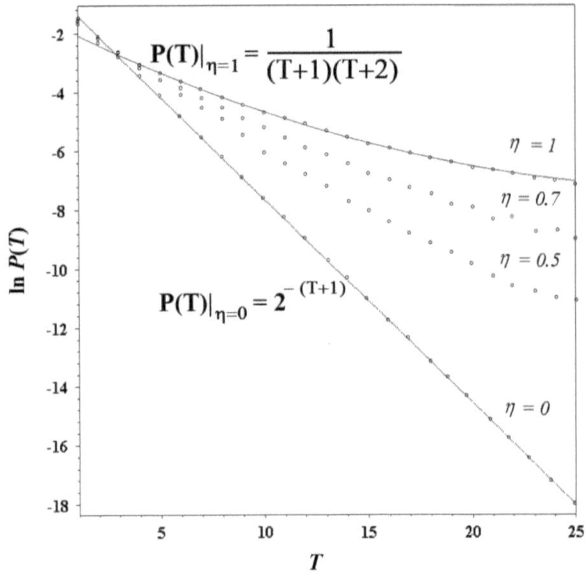

Fig. 10 The distributions of searching durations, in the case of the uniform densities $dF(u) = dG(u) = du$ at different values of η. $P_\eta(T)$ decays exponentially for $\eta = 0$, consistently with the analytical result, $P(T) = 2^{-(T+1)}$ (given by the *solid line*). $P_\eta(T)$ exhibits a power-law decay for $\eta \to 1$; the *solid line* is for $P(T) = 1/(T+1)(T+2)$

balanced group of 82 subjects participated in a virtual treasure hunting experiment. The collections of displacements and turns were considered as a series of random events subjected to a thorough statistical analysis in search of certain statistical regularities.

Our results indicate that human locomotion in a VE is sensitive to the structure affording more opportunity for exploration. A higher mobility of male subjects in a VE can be associated with the regularly renewed computer game experience, essentially in games played through a first-person perspective. The data show that the vast majority of reorientations performed by subjects in the study were the quick scanning turns, being the essential part of the adaptive movement strategy under reduced multisensory conditions in VE. The analysis of the root mean square fluctuations gives an evidence of a mechanism strongly reinforcing the total reorientation duration with the net displacement of subjects that makes the thorough scanning process biologically unfeasible on large spatiotemporal scales.

We found that dispersal of searching activity on large spatiotemporal scales in VE is anomalous. Namely, the distributions of time intervals and travelling distances between sequential phases of searching activity exhibit long tails dominated by quadratic hyperbolas, indicating that trajectories of treasure hunters were reminiscent of scale-free random walks (Lévy flights). We suggested a discrete time self-organized critical model rationalizing the dialog nature of the decision-making process in people that search. The model reproduces the experimentally observed statistics of searching patterns in case when the chances to be rewarded in the immediate neighborhood and elsewhere are estimated by subject uniformly at random, i.e. amid uncertainty.

Acknowledgments The treasure hunting experiments have been supported by the Cognitive Interaction Technology—Center of Excellence (CITEC, Bielefeld University).

D.V. gratefully acknowledges the financial support by the project *MatheMACS* ("Mathematics of Multilevel Anticipatory Complex Systems"), grant agreement no. 318723, funded by the EC Seventh Framework Programme FP7-ICT-2011-8.

References

1. Atkinson RPD, Rhodes CJ, MacDonald DW, Anderson RM (2002) Scale-free dynamics in the movement patterns of jackals. Oikos 98(1):134–140
2. Bak P, Tang C, Wiesenfeld K (1987) Self-organized criticality: an explanation of $1/f$ noise. Phys Rev Lett 59(4):381–384
3. Bartumeus F (2007) Lévy processes in animal movement: an evolutionary hypothesis. Fractals 15(2):151–162
4. Bartumeus F, Catalan J, Viswanathan GM, Raposo E, da Luz MGE (2008) The influence of turning angles on the success of non-oriented animal searches. J Theor Biol 252:43–55
5. Bartumeus F, Levin SA (2008) Fractal reorientation clocks: Linking animal behavior to statistical patterns of search. PNAS 105(49):19072–19077. doi:10.1073/pnas.0801926105
6. Bartumeus F (2009) Behavioral intermittence, levy patterns, and randomness in animal movement. Oikos 118:488–494
7. Blanchard Ph, Volchenkov D (2008) Intelligibility and first passage times in complex urban networks. Proc Roy Soc A 464:2153–2167. doi:10.1098/rspa.2007.0329
8. Blanchard Ph, Volchenkov D (2009) Mathematical analysis of urban spatial networks. Springer Series: Understanding complex systems. Springer, Berlin/Heidelberg
9. Blanchard Ph, Volchenkov D (2011) Introduction to random walks on graphs and databases. Springer Series: Synergetics. Springer, Berlin/Heidelberg
10. Brockmann D, Hufnagel L, Geisel T (2006) The scaling laws of human travel. Nature 439:462–465
11. Buldyrev SV, Havlin S, Kazakov AYa, da Luz MGE, Raposo EP, Stanley HE, Viswanathan GM (2001) Average time spent by Lévy flights and walks on an interval with absorbing boundaries. Phys Rev E 64:041108
12. Cohen JD, Aston-Jones G (2005) Cognitive neuroscience: decision amid uncertainty. Nature 436:471–472
13. Cohen JD, McClure SM, Yu AJ (2007) Should I stay or should I go? How the human brain manages the trade-off between exploitation and exploration. Phil Trans R Soc Lond B 362:933–942
14. Colzato LS, van Leeuwen PJA, van den Wildenberg WPM, Hommel B (2010) DOOM'd to switch: superior cognitive flexibility in players of first person shooter games. Front Psychol 1:8
15. Cutmore TRH, Hine TJ, Maberly KJ, Langford NM, Hawgood G (2000) Cognitive and gender factors influencing navigation in a virtual environment. Int J Hum Comput Stud 53:223–249
16. Darken RP, Sibert JL (1993) A toolset for navigation invirtual environments. Proceedings of ACM User Interface Software and Technology, Atlanta, GA, pp 157–165, 3–5 November 1993
17. Devlin AS, Bernstein J (1995) Interactive wayfinding: Use of cues by men and women. J Environ Psychol 15:23–38

18. Dillon BA (2009) Event wrap-up: Girls "N" Games 2006. Gamasutra. 18 May 2006
19. Federal Highway Administration (1997) Nationwide personal transportation study 1995: Transportation user's views of quality. U.S. Department of Transportation, Washington, DC
20. Flajolet P, Sedgewick R (2009) Analytic combinatorics. Cambridge University Press, New York
21. Gittins JC (1989) Multi-armed bandit allocation indices. Wiley - Interscience Series in Systems and Optimization. Wiley, Chichester
22. Hamilton E (2009) The girl gamer's manifesto. GameSpot. 11 February 2009
23. Hurst HE, Black RP, Simaika YM (1965) Long-term storage: an experimental study. Constable, London
24. Klatzky RL, Loomis JM, Beall AC, Chance SS, Golledge RG (1998) Spatial updating of self-position and orientation during real, imagined, and virtual locomotion. Psychol Sci 9(4): 293–298
25. Komin N, Erdmann U, Schimansky-Geier L (2004) Random walk theory applied to daphnia motion. Fluctuation Noise Lett 4:L151–L159
26. Lawton CA (1994) Gender differences in way-finding strategies: Relationship to spatial ability and spatial anxiety. Sex Roles 30(11–12):765–779
27. Levandowsky M, Klafter J, White BS (1988) Swimming behavior and chemosensory responses in the protistan microzooplankton as a function of the hydrodynamic regime. Bull Mar Sci 43:758–763
28. Levy M, Solomon S (1996) Power Laws are Logarithmic Boltzmann Laws. Int J Mod Phys C 7(4):595–601
29. Moffat SD, Hampson E, Hatzipantelis M (1998) Navigation in a "virtual" maze: sex differences and correlation with psychometric measures of spatial ability in humans. Evol Hum Behav 19:73–78
30. Nathan R, Getz WM, Revilla E, Holyoak M, Kadmon R, Saltz D, Smouse PE (2008) A movement ecology paradigm for unifying organismal movement research. PNAS 105(49): 19052–19059
31. Riecke BE, Schulte-Pelkum J, Bülthoff HH (2005) Perceiving simulated ego-motions in virtual reality—comparing large screen displays with HMDs. SPIE 2005 Conference Proceedings, San Jose, USA, 2005
32. Schimansky-Geier L, Erdmann U, Komin N (2005) Advantages of hopping on a zigzag course. Phys A 351(1):51–59
33. Shlesinger M, Zaslavsky GM, Klafter J (1993) Strange kinetics. Nature 363:31–37
34. Sims DW et al (2008) Scaling laws of marine predator search behaviour. Nature 451: 1098–1102
35. Two AVI video fragments showing the records of actual searching experiments from the first-person perspective can be found at http://youtu.be/17aNxvZFMRw (the VE model A) and http://youtu.be/_Jooi9ZXRGs (the VE model B)
36. Viswanathan GM, Afanasyev V, Buldyrev SV, Murphy EJ, Prince PA, Stanley HE (1996) Lévy flight search patterns of wandering albatrosses. Nature (London) 381:413
37. Viswanathan GM, Buldyrev S, Havlin S, da Luz MGE, Raposo EP, Stanley HE (1999) Optimizing the success of random searches. Nature (London) 401:911
38. Viswanathan GM, Raposo EP, da Luz MGE, Lévy flights and uperdiffusion in the context of biological encounters and random searches. Phys Life Rev 5(3):133–150
39. Voyer D, Voyer S, Bryden MP (1995) Magnitude of sex differences in spatial abilities: a meta-analysis and consideration of critical variables. Psychol Bull 117:250–270
40. Wann JP, Mon-Williams M (1996) What does virtual reality NEED? Human factors issues in the design of three-dimensional computer environments. Int J Hum Comput Stud 44:829–847
41. Werner S, Long P (2003) Cognition meets le Corbusier—Cognitive principles of architectural design. Spatial Cognition III. Lecture notes in artificial intelligence. Springer, Berlin, pp 112–126
42. Yu AJ, Dayan P (2005) Uncertainty, neuromodulation, and attention. Neuron 46:681–692

Vibrational Resonance in Time-Delayed Nonlinear Systems

S. Rajasekar and M.A.F. Sanjuán

Abstract Time-delay is ubiquitous in many dynamical systems. The role of single and multiple time-delay on vibrational resonance in a single Duffing oscillator and in a system of n Duffing oscillators coupled unidirectionally and driven by both a low- and a high-frequency periodic force is presented. The investigation is performed through both theoretical approach and numerical simulation. Theoretically determined values of the amplitude of the high frequency force and the delay-time at which resonance occurs are in very good agreement with the numerical simulation. A major consequence of time-delay feedback is that it gives rise to a periodic or quasiperiodic pattern of vibrational resonance profile with respect to the time-delay parameter. For the system of n-coupled oscillators with a single time-delay coupling, the response amplitudes of the oscillators are shown to be independent of the time-delay. In the case of a multi time-delayed coupling, undamped signal propagation occurs for coupling strength (δ) above a certain critical value (denoted as δ_{u}). Further, the response amplitude approaches a limiting value Q_{L} with the oscillator number i. Analytical expressions for both δ_{u} and Q_{L} are determined.

1 Introduction

Signal detection and signal amplification are very important in engineering, physics and biology. In recent years certain nonlinear phenomena are explored in this

---

S. Rajasekar
Nonlinear Dynamics, Chaos and Complex Systems Group, Departamento de Física, Universidad Rey Juan Carlos, Tulipán s/n, 28933 Móstoles, Madrid, Spain
e-mail: rajasekar@cnld.bdu.ac.in; srj.bdu@gmail.com

M.A.F. Sanjuán (✉)
Nonlinear Dynamics, Chaos and Complex Systems Group, Departamento de Física, Universidad Rey Juan Carlos, Tulipán s/n, 28933 Móstoles, Madrid, Spain
e-mail: miguel.sanjuan@urjc.es

V. Afraimovich et al. (eds.), *Nonlinear Dynamics and Complexity*, Nonlinear Systems and Complexity 8, DOI 10.1007/978-3-319-02353-3_9,

context. Particularly, certain resonance dynamics are shown to be useful for weak signal detection. Examples include stochastic resonance [1, 2], chaotic resonance [3], coherence resonance [4] and vibrational resonance (VR) [5]. In the case of stochastic resonance a bistable or an excitable system is driven by a weak periodic signal and noise. When the noise intensity is varied the signal-to-noise ratio (SNR) at the frequency of the input periodic signal becomes maximum at an optimum value. On either side of the optimum value of the noise intensity the SNR decreases. In place of noise one can use a chaotic signal of a system. The resulting resonance is called chaotic resonance. It is possible to realize a noise-induced resonance in the absence of the external periodic force and then is termed coherence resonance.

In the VR set-up, a nonlinear system is driven by a biharmonic force consisting of two frequencies ω and Ω with $\Omega \gg \omega$. In a typical VR, when the amplitude (or frequency) of the high-frequency force is varied, the response amplitude of the system at the low-frequency ω displays one or more resonance peaks. VR can occur even in overdamped single-well systems. Theoretical approaches have been developed to analyse VR [6, 7]. VR has been studied in monostable [8], bistable [5–7, 9, 10], three well [11] and spatially periodic potential [12] systems. It has also been observed in excitable systems [13], fractional-order systems [14], maps [15], small-world networks of FitzHugh–Nagumo equations [16] and ecological systems [17]. VR is found to induce undamped low-frequency signal propagation in one-way coupled [18] and globally coupled [19] bistable systems. Experimental evidence of VR has been demonstrated in analog simulations of the overdamped Duffing oscillator [9], in an excitable electronic circuit with Chua's diode [13] and in a bistable optical cavity laser [20].

It is important to study VR in different kinds of systems and explore its features with specific emphasis on the development of theoretical analysis and the role of properties of the systems on VR. This is precisely what motivates us to consider here a class of dynamical systems called time-delayed systems. When the state of a system at time t depends on its state at a shifted earlier time, say, $t - \alpha$, then a time-delayed feedback term is introduced in the equation of motion of the system. The study of time-delayed systems has received a great interest in recent years because time-delay is ubiquitous in many systems [21, 22]. The common sources of time-delay are finite propagation time of transport of information and energy, finite reaction times, memory effects and finite switching speed of amplifiers. Time-delay is easily amenable in networks [23], laser arrays [24–26], electronic circuits [27], neural systems [28–30] and optical and optoelectronic circuits. The features of vibrational resonance in the presence of a single time-delayed feedback have been analysed in the Langevin equation [31], two-coupled overdamped anharmonic oscillators [32], underdamped and overdamped Duffing oscillators [33], FitzHugh–Nagumo neuronal model [34], a genetic toggle-switch [35] and a system of n-coupled bistable oscillators [36]. There are some notable reports on the influence of two or three time-delayed feedbacks or coupling terms [37–44] on the dynamics of certain systems.

Motivated by the above ideas, here we present our investigation on the effect of single and multi time-delayed feedback (MTDF) on VR in a single nonlinear system and signal transduction in a system of unidirectionally coupled systems with multi time-delayed coupling (MTDC). We choose the paradigmatic Duffing oscillator as the reference model system. The MTDF is of the form $(\gamma/L)\sum_{l=1}^{L} x(t-l\alpha)$. $L=1$ corresponds to a single time-delay. Our main goal is to explore the enhancement of the response amplitude by the single and MTDF and MTDC.

The equation of motion of the single Duffing oscillator with an MTDF and driven by the biharmonic force is given by

$$\ddot{x} + d\dot{x} + \omega_0^2 x + \beta x^3 + \frac{\gamma}{L}\sum_{l=1}^{L} x(t-l\alpha) = f\cos\omega t + g\cos\Omega t, \tag{1}$$

where $\Omega \gg \omega$. When $d=0$, $\gamma=0$, $f=0$ and $g=0$ the potential of the Duffing oscillator is $V(x)=\frac{1}{2}\omega_0^2 x^2 + \frac{1}{4}\beta x^4$. For $\omega_0^2<0$ and $\beta>0$ the potential becomes a double-well shape, while for ω_0^2 and $\beta>0$ it becomes a single-well form. We treat the double-well and the single-well cases separately. For $f \ll 1$ and because of $\Omega \gg \omega$ it is reasonable to assume that the response of the system (Eq. 1) essentially contains only a slow component $X(t)$ with the low-frequency ω and a fast component $\psi(t,\Omega t)$ with the high-frequency Ω. Through a theoretical approach we obtain an analytical expression for the variables X and ψ. We define the ratio of the amplitude $A_{\rm L}$ of slow motion and the amplitude f of the input low-frequency force as response amplitude Q. First we analyse the occurrence of VR in the system (Eq. 1) with single time-delay ($L=1$). From the theoretical expression of Q we determine the values of g and α denoted as $g_{\rm VR}$ and $\alpha_{\rm VR}$, respectively, at which VR occurs, i.e., Q becomes a maximum. We verify the theoretical predictions with the numerical simulation. We illustrate the mechanism of the resonance and compare the change in the slow motion $X(t)$ and the actual motion $x(t)$ when the control parameters g and α are varied. Next, we analyse the effect of MTDF on VR. We determine the regions in $(\gamma-\alpha)$ parameter space for which $Q_{\max}(\gamma) > Q_{\max}(\gamma=0)$ for a few fixed values of L. It displays a band-like structure with the number of bands being the number of time-delayed terms L. We analyse the effect of L on resonance.

Then we take up the n-coupled Duffing oscillators whose equations read

$$\begin{aligned} \ddot{x}_1 + d\dot{x}_1 + \omega_0^2 x_1 + \beta x_1^3 &= f\cos\omega t + g\cos\Omega t, \\ \ddot{x}_i + d\dot{x}_i + \omega_0^2 x_i + \beta x_i^3 &= \frac{\delta}{L}\sum_{l=1}^{L} x_{i-1}(t-l\alpha), \end{aligned} \tag{2}$$

where $i=2,3,\cdots,n$. Our prime interest is on the analysis of signal propagation in the system (Eq. 2) with $n=200$. Applying the theoretical treatment used for the system (Eq. 1) we express Q_i in terms of Q_{i-1} except for the first oscillator. When $L=1$, the analytical expression of Q_i, $i>1$ is found to be independent of the time-delay parameter α. This implies that the time-delay has no influence on Q_i.

For $i \gg 1$ the theoretical Q_i deviates largely from the numerically computed Q_i. This is because of the neglect of nonlinear terms in the equation of motion of the slow variable $X_i(t)$. Inclusion of nonlinear terms leads to a set of coupled nonlinear equations for Q_i. Interestingly, the Q_i's computed by solving this set of equations are in very good agreement with the numerically calculated Q_i. The coupled system shows undamped signal propagation (that is, $Q_{200} > Q_1$) for certain range of values of the control parameters α and δ. In the undamped signal propagation, Q_i exhibits sigmoidal type of variation with i, that is, $Q_i \to Q_{\rm L}$ for sufficiently large i. We are able to obtain an analytical expression for the limiting value of Q, $Q_{\rm L}$, and the critical value of δ, $\delta_{\rm u}$, above which undamped signal propagation takes place. Interestingly, both $Q_{\rm L}$ and $\delta_{\rm u}$ are independent of the parameter g.

Before taking up the main theme, in the next two sections we briefly point out that time-delay is ubiquitous and show the absence of VR in a linear system with time-delayed feedback and driven by the biharmonic force.

2 Time-Delay Is Ubiquitous

A first-order delay differential equation is of the form

$$\dot{x} = F(t, x(t), x(t-\alpha_l)), \tag{3}$$

where $\alpha_l > 0$, $l = 1, 2, \cdots$ are delay times. The time-delays α_l can be of different types depending upon the nature of the sources of the time-delay. In the nonlinear dynamics literature the effect of the following types of time-delay is investigated in a variety of nonlinear systems [22]:

1. A single constant delay: $\alpha_l = \alpha =$ a constant, $l = 1$.
2. Multiple time-delay: $\alpha_l = l\alpha$, $\alpha =$ a constant, $l = 1, 2, \cdots, L$. An example is $F(x(t-\alpha_l)) = \dfrac{1}{L}\displaystyle\sum_{l=1}^{L} x(t - l\alpha)$.
3. Integrative time-delay or distributive delay: A delay term of this kind is represented as $\langle x \rangle_\alpha = \dfrac{1}{\alpha}\displaystyle\int_{t-\alpha}^{\alpha} x(t')\,{\rm d}t'$.
4. Time-dependent delay: $\alpha(t) = a + bt$ or $a + b\sqrt{t}$ with $\alpha < t$.
5. State-dependent delay: $\alpha = F(x(t))$. An example is $\alpha(x(t)) = |x(t)|$.

Typical examples of systems with delays are given by maturation times [45], hydrodynamic problems [46], chemical surface reactions [47] and feedback regulated voltage-controlled oscillators [48, 49]. In nonlinear optics, periodic and chaotic outputs are realized by a delayed feedback [50]. Such delay-induced dynamics are used to design practical systems including high-frequency and broadband optical chaotic oscillators for secure chaos communication [51] or high-speed random number generation [52] or to develop alternative imaging techniques [50].

It is noteworthy to mention that delayed self-communication is of great significance because of its regulatory mechanism in nature and technology [53, 54]. Examples include excitable gene regulatory systems [55], eye movements [56], human balance [57] and optically communicating semiconductor lasers [58]. In neural networks communicating between the various areas may take place in delays ranging from few milliseconds to hundreds of milliseconds due to the finite speed of the transfer of data in the axons and dendrites and because of the processing latency in the synapses [59]. In neural systems the source for a precise firing of basket cells in hippocampus during Theta and Gamma rhythm is a delayed feedback. Experimental evidences are reported for delayed recurrent excitations inducing regulation of the structure of the interspike intervals in the presence of noise [60]. Delayed self-coupling in the study of the pacemaker cells of crayfish [61] is found to give rise bursting and high-frequency discharges with relatively long quiescent intervals [62]. One can treat the time-delay as an external force.

In coupled systems, coupling transmits one or more variables to neighbours. Often the transmission time is larger than the internal time scales of the systems. In this case coupling terms consisting of time-delayed variables are realistic. Message decoding in chaos-based communication systems would require chaos synchronization between multiple time-delayed transmitter and receiver systems. In high speed chaos-based communication systems [63] external cavity semiconductor lasers form an integral part. In practical applications such lasers may be subject to more than one optical reflection. This would lead to multiple time-delays. It has been pointed out that inclusion of several external cavities could provide higher security for such communication systems [64].

3 Resonance in a Linear System with Time-Delayed Feedback

In this section we consider a linear system with linear MTDF and driven by two periodic forces. The equation of motion of the system is given by

$$\ddot{x} + d\dot{x} + \omega_0^2 x + \frac{\gamma}{L}\sum_{l=1}^{L} x(t - l\alpha) = f\cos\omega t + g\cos\Omega t, \tag{4}$$

where $\omega_0^2 > 0$ and $d > 0$. The general solution of Eq. (4) for $f = g = 0$ is not known. Equation (4) with $f = g = \gamma = 0$ is a damped linear system. Its equilibrium point $(x^*, \dot{x}^*) = (0, 0)$ is stable. When $f = g = 0$ and $\gamma \neq 0$, Eq. (4) exhibits a damped or periodic or growing oscillation depending upon the values of γ, L and α [22]. Here, we are interested in the long time behaviour of Eq. (4). In the limit $t \to \infty$ we seek the solution of Eq. (4) in the form

$$x(t) = A_\omega \cos(\omega t + \phi_\omega) + A_\Omega \cos(\Omega t + \phi_\Omega) . \tag{5}$$

The unknowns A_ω and ϕ_ω are determined as

$$A_\omega = \frac{f}{\sqrt{S_\omega}}, \quad S_\omega = \left[\omega_0^2 - \omega^2 + \frac{\gamma}{L}\sum_{l=1}^{L}\cos l\alpha\omega\right]^2 + \left[d\omega - \frac{\gamma}{L}\sum_{l=1}^{L}\sin l\alpha\omega\right]^2, \tag{6}$$

$$\phi_\omega = \tan^{-1}\left[\frac{d\omega - \frac{\gamma}{L}\sum_{l=1}^{L}\sin l\alpha\omega}{\omega^2 - \omega_0^2 - \frac{\gamma}{L}\sum_{l=1}^{L}\cos l\alpha\omega}\right]. \tag{7}$$

Replacement of ω and f by Ω and g, respectively, in Eqs. (6) and (7) gives A_Ω, S_Ω and ϕ_Ω. As $f(g)$ increases $A_\omega(A_\Omega)$ also increases. Resonance does not occur at the frequencies ω or Ω when f or g is varied from a small value. Thus, there is no vibrational resonance in the linear system with time-delayed feedback.

4 Single Duffing Oscillator: Theoretical Expression for the Response Amplitude Q

The main objective of this section is to obtain an expression for the response amplitude Q for the single Duffing oscillator system of Eq. (1) [65].

We assume the solution of the system (Eq. 1) for $\Omega \gg \omega$ as $x = X + \psi$ where $X(t)$ and $\psi(\tau = \Omega t)$ are a slow motion with period $2\pi/\omega$ in the time t and a fast motion with period 2π in the fast time τ, respectively. Further, we assume that the average value of ψ over the period 2π is $\langle\psi\rangle = \frac{1}{2\pi}\int_0^{2\pi}\psi \, d\tau = 0$. Substitution of $x = X + \psi$ in Eq. (1) gives the following equations for the variables X and ψ:

$$\ddot{X} + d\dot{X} + \left(\omega_0^2 + 3\beta\langle\psi^2\rangle\right)X + \beta\left(X^3 + \langle\psi^3\rangle\right) + 3\beta X^2\langle\psi\rangle$$
$$+\frac{\gamma}{L}\sum_{l=1}^{L}X(t - l\alpha) = f\cos\omega t, \tag{8}$$

$$\ddot{\psi} + d\dot{\psi} + \omega_0^2\psi + 3\beta X^2(\psi - \langle\psi\rangle) + 3\beta X\left(\psi^2 - \langle\psi^2\rangle\right)$$
$$+\beta\left(\psi^3 - \langle\psi^3\rangle\right) + \frac{\gamma}{L}\sum_{l=1}^{L}\psi(\Omega t - l\alpha\Omega) = g\cos\Omega t, \tag{9}$$

where $\langle\psi^n\rangle = \frac{1}{2\pi}\int_0^{2\pi}\psi^n \, d\tau$. Because ψ is a fast variable we can neglect the nonlinear terms in Eq. (9). In the limit of $t \to \infty$, referring to the solution of the linear system (Eq. 4) given by Eqs. (5)–(7), we write the solution of Eq. (9) as

$$\psi = \mu\cos(\Omega t + \phi), \tag{10}$$

where $\mu = g/k$,

$$k^2 = \left(\omega_0^2 - \Omega^2 + \frac{\gamma}{L}\sum_{l=1}^{L}\cos l\alpha\Omega\right)^2 + \left(d\Omega - \frac{\gamma}{L}\sum_{l=1}^{L}\sin l\alpha\Omega\right)^2 \tag{11}$$

and

$$\phi = \tan^{-1}\left(\frac{d\Omega - \frac{\gamma}{L}\sum_{l=1}^{L}\sin l\alpha\Omega}{\Omega^2 - \omega_0^2 - \frac{\gamma}{L}\sum_{l=1}^{L}\cos l\alpha\Omega}\right). \tag{12}$$

For sufficiently large values of Ω we can further approximate the solution (Eqs. 10–12) by dropping the terms ω_0^2 and $d\Omega$. However, in our treatment we keep these terms in the solution. From the above solution we obtain $\langle\psi\rangle = 0$, $\langle\psi^2\rangle = \mu^2/2$ and $\langle\psi^3\rangle = 0$. Then Eq. (8) becomes

$$\ddot{X} + d\dot{X} + C_1 X + \beta X^3 + \frac{\gamma}{L}\sum_{l=1}^{L} X(t - l\alpha) = f\cos\omega t, \tag{13}$$

where $C_1 = \omega_0^2 + \frac{3}{2}\beta\mu^2$. When $f = 0$, the equilibrium points of Eq. (13) are

$$X_0^* = 0, \quad X_\pm^* = \pm\sqrt{-\frac{C_1 + \gamma}{\beta}}. \tag{14}$$

Slow oscillations occur around these equilibrium points.

Substituting $X = Y + X^*$, where Y is the deviation of the slow motion from X^*, in Eq. (13), we obtain

$$\ddot{Y} + d\dot{Y} + \omega_r^2 Y + 3\beta X^* Y^2 + \beta Y^3 + \frac{\gamma}{L}\sum_{l=1}^{L} Y(t - l\alpha) = f\cos\omega t. \tag{15}$$

The solution of the linear version of Eq. (15) in the limit $t \to \infty$ and $f \ll 1$ is $Qf\cos(\omega t + \theta)$, where the response amplitude Q and the phase θ are given by

$$Q = \frac{1}{\sqrt{S}}, \quad S = \left(\omega_r^2 - \omega^2 + \frac{\gamma}{L}\sum_{l=1}^{L}\cos l\alpha\omega\right)^2 + \left(d\omega - \frac{\gamma}{L}\sum_{l=1}^{L}\sin l\alpha\omega\right)^2 \tag{16}$$

and $\theta = \phi(\Omega = \omega, \omega_0^2 = \omega_r^2)$. ω_r is the resonant frequency of the linear version of the equation of motion of the slow variable $X(t)$. In the next section we analyse the occurrence of VR in the system (Eq. 2) with a single time-delayed feedback term and then take up the system with MTDF.

5 Resonance in the System with a Single Time-Delay

In the absence of the damping term, external periodic forces and feedback term the potential of the Duffing oscillator is $V(x) = \frac{1}{2}\omega_0^2 x^2 + \frac{1}{4}\beta x^4$. $V(x)$ becomes a double-well for $\omega_0^2 < 0$, $\beta > 0$, and a single-well for ω_0^2, $\beta > 0$, respectively. Moreover, for $\omega_0^2 > 0$, $\beta < 0$ the potential has a single-well with double-hump form. These three forms of the potential are depicted in Fig. 1. For ω_0^2, $\beta < 0$ the potential has an inverted single-well form. We treat the double-well and single-well cases of the system separately.

5.1 Resonance Analysis in the Double-Well System

The equilibrium points around which slow oscillations take place are given by Eq. (14). There are three equilibrium points for $g < g_c$ where

$$g_c = \left[\frac{2k^2}{3\beta} \left(|\omega_0^2| - \gamma \right) \right]^{1/2}, \quad |\omega_0^2| - \gamma > 0, \tag{17}$$

For $g < g_c$ the system admits two slow motions, one about X_+^* and the other about X_-^*. X_0^* is unstable. For $g > g_c$, X_0^* is the only real equilibrium point and a slow orbit occurs about it. That is, the effective potential of the slow variable X undergoes a transition from the double-well to a single-well at g_c.

The possibility of occurrence of resonance when a control parameter is varied and the values of a parameter at which resonance occurs can be determined from the theoretical expression of Q. The response amplitude Q is a maximum when the function S in Eq. (16) is a minimum. When a parameter, say, g is varied then resonance occurs at a value of g_{VR}, where g_{VR} is a root of the equation $\mathrm{d}S/\mathrm{d}g = 0$. This condition requires

$$\omega_r^2 = \omega^2 - \gamma \cos \alpha\omega. \tag{18}$$

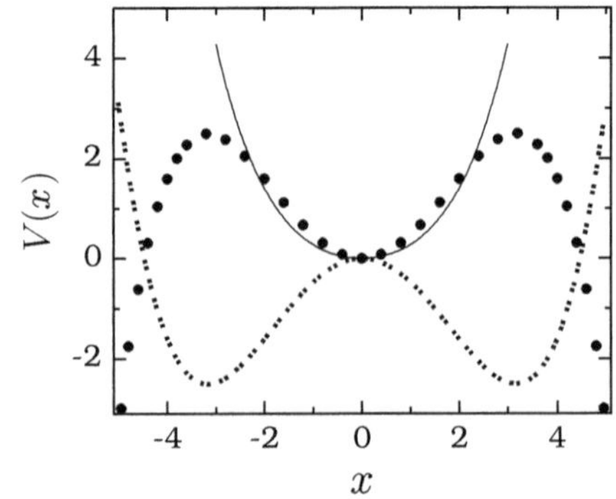

Fig. 1 Single-well (represented by a *continuous line*, $\omega_0^2 = 0.5$, $\beta = 0.1$), double-well (represented by a *dashed line*, $\omega_0^2 = -1$, $\beta = 0.1$) and single-well with double-hump (represented by *solid circles*, $\omega_0^2 = 1$, $\beta = -0.1$) forms of the potential of the Duffing oscillator

From this resonance condition one can obtain an analytical expression for g_{VR}. We obtain the following results.

Case 1: $\gamma < 0$

When

$$\gamma < 0, \quad |\gamma| < |\gamma_{\mathrm{c}<}| = \frac{\omega^2}{1 - \cos\alpha\omega} \tag{19}$$

there are two resonances. The values of g at which resonance occurs are given by

$$g_{\mathrm{VR}}^{(1)} = \left[\frac{k^2}{3\beta}\left(2|\omega_0^2| - \omega^2 - 3\gamma + \gamma\cos\alpha\omega\right)\right]^{1/2} < g_{\mathrm{c}}, \tag{20}$$

$$g_{\mathrm{VR}}^{(2)} = \left[\frac{2k^2}{3\beta}\left(|\omega_0^2| + \omega^2 - \gamma\cos\alpha\omega\right)\right]^{1/2} > g_{\mathrm{c}}. \tag{21}$$

The response amplitude is the same at these two values of g. For $\gamma < 0$ and $|\gamma| > |\gamma_{\mathrm{c}<}|$ only one resonance is possible and in this case $g_{\mathrm{VR}}^{(1)} = g_{\mathrm{c}}$. For $|\gamma| < |\gamma_{\mathrm{c}<}|$ the resonances are due to the matching of ω_{r}^2 with $\omega^2 - \gamma\cos\alpha\omega$ (refer to Eq. (18)), while the resonance at $g = g_{\mathrm{c}}$ for $|\gamma| > |\gamma_{\mathrm{c}<}|$ is due to the local minimization of ω_{r}^2.

Case 2: $\gamma > 0$

For $\gamma > 0$, one resonance occurs at $g = g_{\mathrm{VR}}^{(2)}$ given by Eq. (21) provided $|\omega_0^2| + \omega^2 > \gamma$. Another resonance occurs at $g = g_{\mathrm{VR}}^{(1)}$ given by Eq. (20) if

$$|\omega_0^2| > \gamma, \quad \gamma < \gamma_{\mathrm{c}>} = \frac{2|\omega_0^2| - \omega^2}{3 - \cos\alpha\omega}. \tag{22}$$

The two resonances are resulting from the resonance condition (Eq. 18).

To verify the theoretical predictions, we numerically compute the sine and cosine components Q_{s} and Q_{c}, respectively, at the low-frequency ω of the numerical solution $x(t)$ of the system (Eq. 1). In the calculation of Q_{s} and Q_{c} we use the solution $x(t)$ corresponding to 200 drive cycles of the input signal after leaving a sufficient transient. Then $Q = \sqrt{Q_{\mathrm{s}}^2 + Q_{\mathrm{c}}^2}/f$. We choose the values of the parameters as $d = 0.5$, $\omega_0^2 = -1$, $\beta = 0.1$, $f = 0.1$, $\omega = 1$ and $\Omega = 10$. Equation (1) is integrated numerically using the Euler method with time step 0.01. The time-delay parameter α takes always multiple values of 0.01.

Figure 2a presents both theoretical and numerical g_{VR} as a function of γ for $\alpha = 1$ and 3. We notice a very good agreement between the theory and the numerical simulation. For $\alpha = 1$ and 3 we find $\gamma_{\mathrm{c}<} = -2.17534$ and -0.50251, respectively. For $\gamma < 0$, there are two resonances for $|\gamma| < |\gamma_{\mathrm{c}<}|$ and only one for $|\gamma| > |\gamma_{\mathrm{c}<}|$. For $|\gamma| < |\gamma_{\mathrm{c}<}|$, as g increases from 0 the quantity ω_{r}^2 decreases from $2|\omega_0^2| + 3|\gamma|$ and reaches the minimum value $|\gamma|$ at $g = g_{\mathrm{c}}$. As g increases from g_{c} the value

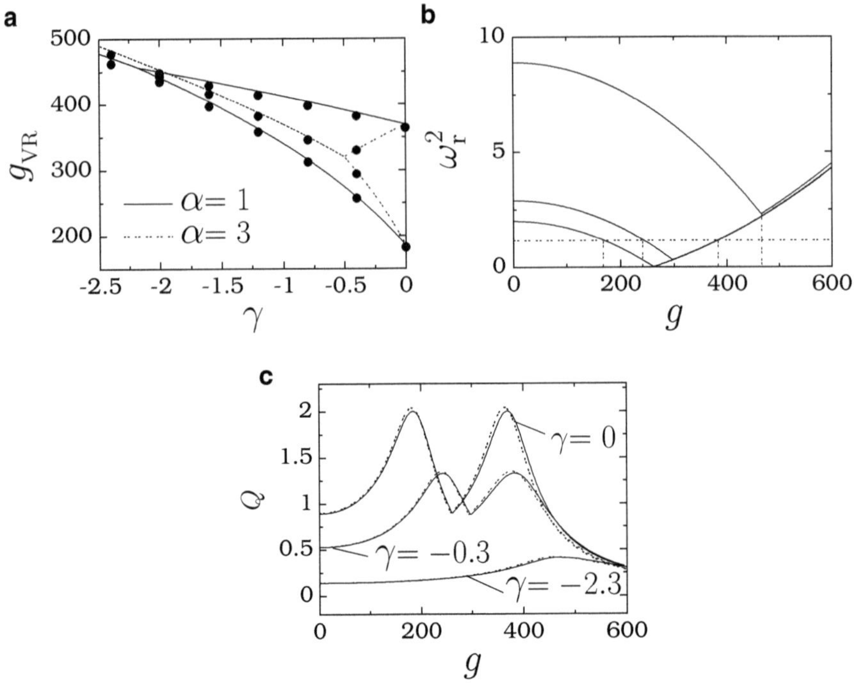

Fig. 2 (**a**) Theoretical and numerical g_{VR} versus the parameter γ for the system (Eq. 1) with the double-well potential case. The number of delay term is 1. The *solid circles* are the numerically computed g_{VR} and the *lines* are theoretical g_{VR}. (**b**) ω_{r}^2 versus g. From *bottom* to *top curves* the values of γ are 0, −0.3 and −2.3, respectively. The *horizontal dashed line* represents the value of $(\omega^2 - \gamma \cos \alpha\omega)$. The *vertical dashed lines* mark the values of g_{VR}. (**c**) Q as a function of g with $\alpha = 1$. The *continuous lines* are theoretical Q while the *dashed lines* are numerically calculated Q

of ω_{r}^2 increases from $|\gamma|$. This is shown in Fig. 2b for $\gamma = 0, -0.3$ and -2.3. For $\gamma = 0$ and -0.3 at $g_{\mathrm{VR}}^{(1)}$ and $g_{\mathrm{VR}}^{(2)}$ we have $\omega_{\mathrm{r}}^2 = \omega^2 - \gamma \cos \alpha\omega$ (indicated by the horizontal dashed line in Fig. 2b) and hence Q becomes maximum with $Q_{\mathrm{max}} = 1/|d\omega - \gamma \sin \alpha\omega|$. In Fig. 2c we observe two resonances. In the absence of time-delay feedback Q becomes maximum when $\omega_{\mathrm{r}} = \omega$ and the maximum value of Q is $1/(d\omega)$. For $\gamma = -0.3$ the theoretical values of $g_{\mathrm{VR}}^{(1)}$ and $g_{\mathrm{VR}}^{(2)}$ are 242.75 and 382.95, while the numerically computed values are 240.34 and 377.43, respectively. For $|\gamma| > |\gamma_{\mathrm{c}<}|$ the value of ω_{r}^2 is always $> (\omega^2 - \gamma \cos \alpha\omega)$. However, it has a local minimum at $g = g_{\mathrm{c}}$ and thus a resonance. These are shown in Fig. 2b,c for $\gamma = -2.3$. Note that Q is minimum at $g = g_{\mathrm{c}}$ for $|\gamma| < |\gamma_{\mathrm{c}<}|$.

In Fig. 2c the value of Q at resonance is always lower than the case $\gamma = 0$. $Q_{\mathrm{max}}(\gamma, g) = Q(\gamma, g_{\mathrm{VR}}) > Q(\gamma = 0, g_{\mathrm{VR}})$ can be realized for a range of values of α and γ. Figure 3a presents the variation of $Q(g_{\mathrm{VR}})$ in (γ, α) parameter space for $\omega = 1$ and $\gamma < 0$. We can clearly see that $Q(\gamma, g_{\mathrm{VR}}) > Q(\gamma = 0, g_{\mathrm{VR}})$ for $\alpha \in [\pi, 2\pi]$. Figure 3b is the three-dimensional plot of Q as a function of γ and g

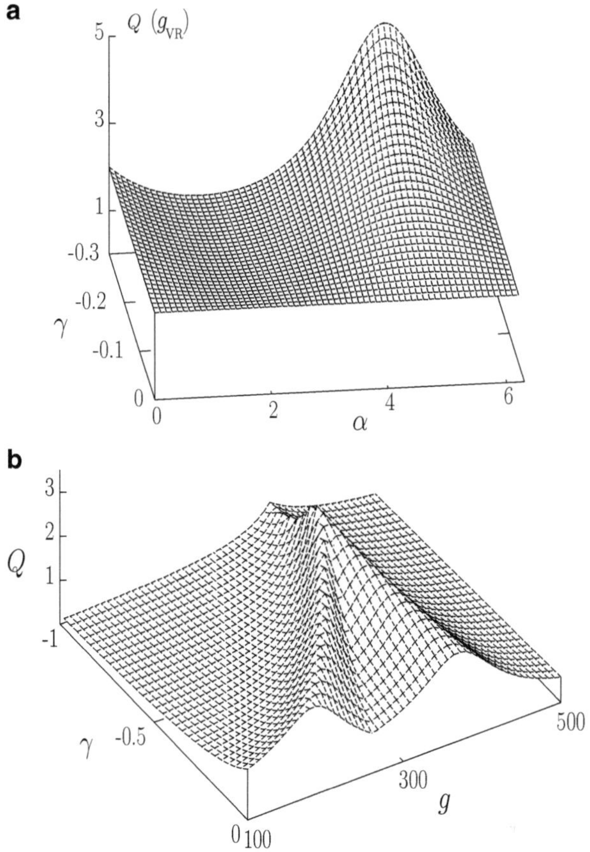

Fig. 3 (**a**) Variation of $Q(g = g_{\rm VR})$ as a function of γ and α for the double-well case of the system (Eq. 1) with single time-delay. (**b**) Three-dimensional plot of Q versus the parameters γ and g for $\alpha = 3.5$

for $\alpha = 3.5$. For $\alpha = 3.5$ the value of $|\gamma_{c<}|$ is 0.51641. In Fig. 3b for $|\gamma| < 0.51641$ there are two resonances with $Q(\gamma, g_{\rm VR}) > Q(\gamma = 0, g_{\rm VR})$. Only one resonance occurs for $|\gamma| > 0.51641$.

Now we compare the change in the slow motion $X(t)$ and the actual motion $x(t)$ when the parameter g is varied. For $\gamma = -0.3$ and $\alpha = 1$ the numerically computed values of $g_{\rm VR}$ are 240.5 and 376.85. The phase portrait of slow motion is plotted in Fig. 4 for several values of g. For $g < g_{\rm c}$ $(= 296.95)$ there are two slow motions: one around X_+^* and the other around X_-^*. As g increases from a small value the equilibrium points about which $X(t)$ and $x(t)$ occur move towards the origin. This is shown in Fig. 4 for four values of $g < g_{\rm c}$. In this figure the orbits coexisting around X_-^* are not shown for clarity. For $g > g_{\rm c}$, as noted earlier, $X_0^* = 0$ is the only equilibrium point and hence both $X(t)$ and $x(t)$ occur around the origin. This is evident in Fig. 4 for three values of $g > g_{\rm c}$. We observe that at the resonance

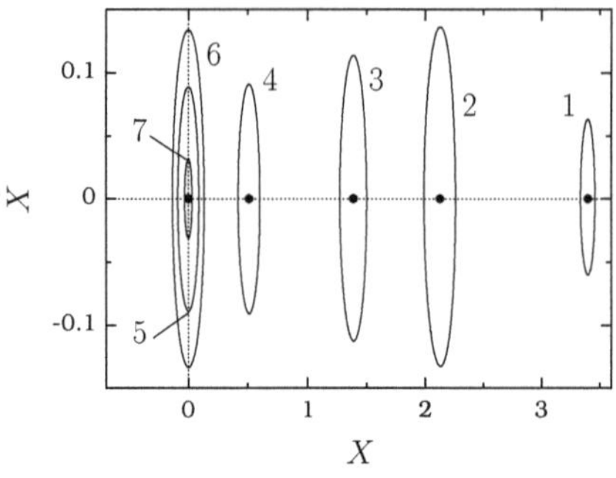

Fig. 4 Phase portraits of the slow variable X of the double-well system (Eq. 1) with single time-delay for several values of g. The values of g for the numbers $1-7$ are 100 (1), 240.5 (2), 275 (3), 295 (4), 300 (5), 376.85 (6) and 600 (7). Here $d = 0.5$, $\omega_0^2 = -1$, $\beta = 0.1$, $\gamma = -0.3$, $\alpha = 1$, $f = 0.1$, $\omega = 1$ and $\Omega = 10$. The equilibrium points X_0^* and X_+^* are marked by the *solid circles*

($g = 240.50$) both $x(t)$ and $X(t) > 0$. That is, cross-well motion and bistability are not necessary ingredients for VR. As a matter of fact, it can occur in monostable systems [8].

Next, we present the effect of the time-delay parameter α on VR. The condition for a resonance to occur when α is varied is given by (from $\mathrm{d}S/\mathrm{d}\alpha = 0$)

$$\left(\omega_{\mathrm{r}}^2 - \omega^2\right)\omega_{\mathrm{r}\alpha}^2 + \left(\omega_{\mathrm{r}\alpha}^2 - d\omega^2\right)\gamma\cos\alpha\omega - \gamma\omega\left(\omega_{\mathrm{r}}^2 - \omega^2\right)\sin\alpha\omega = 0, \qquad (23)$$

where $\omega_{\mathrm{r}\alpha}^2 = \mathrm{d}\omega_{\mathrm{r}}^2/\mathrm{d}\alpha$. Analytical expressions for the roots of the above equation are difficult to obtain. However, the roots denoted as α_{VR} can be determined numerically from Eq. (23). We compute theoretical α_{VR} (from Eq. (23)) and numerical α_{VR} (by numerically solving the Eq. (2)) for a range of values of g with $\gamma = -0.3$. In Fig. 5a $\alpha_{\mathrm{VR}} < 3 \times 2\pi/\omega$ are alone plotted. (α_{VR} are periodic with period $2\pi/\omega$). Figure 5b presents numerical Q as a function of α and g. We can clearly see the periodicity of Q with respect to α.

Figure 6 illustrates the effect of α on the slow motion $X(t)$ for $g = 250$ and 350. When α is increased from a small value with $g < g_{\mathrm{c}}$, then the shift in the locations of $X_\pm^*$ is very small. However, the amplitude of the slow orbits varies and resonance occurs at $\alpha = \alpha_{\mathrm{VR}}$. When α is varied for $g > g_{\mathrm{c}}$ then the slow orbit occurs about $X_0^* = 0$. These are shown in Fig. 6 for $g = 250 < g_{\mathrm{c}}$ and $g = 350 > g_{\mathrm{c}}$. For $g = 250 (< g_{\mathrm{c}})$ the slow motion occurs about X_+^* and X_-^*. This is shown in Fig. 6a, b for a few fixed values of α. In these figures the resonant orbits are marked by the label 2. The orbits marked by 1 and 3 correspond to the values of α on either side of $\alpha = \alpha_{\mathrm{VR}}$. In Fig. 6c,d, for $g = 350 > g_{\mathrm{c}}$ slow motion occurs about $X_0^* = 0$. Tuning time-delay is an advantage when it is desired to observe the response of a system and VR with the centre of the orbit (slow as well as the actual orbit) almost remains the same.

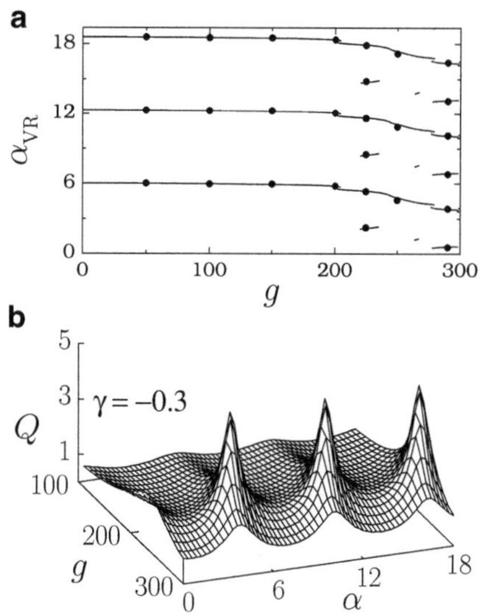

Fig. 5 (**a**) α_{VR} versus the amplitude g of the high frequency force for $\gamma = -0.3$. The system is the double-well single system. *Continuous lines* and *painted circles* represent theoretically and numerically computed values of α_{VR}, respectively. (**b**) Periodic variation of the response amplitude Q with the time-delay parameter α for various values of g in the interval $[100, 300]$ for $\gamma = -0.3$

5.2 *Resonance Analysis in the Single-Well System*

For ω_0^2, $\beta > 0$ the potential $V(x)$ of the system has a single-well shape with a local minimum at $x = 0$. Unlike the double-well system, the effective potential of X remains as a single-well when the amplitude g of the high-frequency force is varied. Consequently, slow oscillation always occurs about $X_0^* = 0$.

For the single-well case the resonance value of g is given by

$$g_{VR} = \left[\frac{2k^2}{3\beta} \left(\omega^2 - \omega_0^2 - \gamma \cos \alpha\omega \right) \right]^{1/2}. \tag{24}$$

We recall that in the double-well case a resonance is possible for all set of values of γ and α when g is varied. In contrast to this, in the single-well system a resonance is possible only for a set of values of γ and α for which $\omega^2 - \gamma \cos \alpha\omega > \omega_0^2$. Further, in the double-well system two resonances are possible while in the single-well system at most one resonance is possible.

In Fig. 7a we plot the variation of theoretical g_{VR} with γ and α for $\omega_0^2 = 0.5$ and $\beta = 0.1$. For a fixed value of γ as α increases from zero the value of g_{VR} increases and becomes maximum at $\alpha = \pi/\omega$ and then decreases. g_{VR} is periodic in α with period $2\pi/\omega$ and $Q_{\max} = 1/|d\omega - \gamma \sin \alpha\omega|$. In Fig. 7b the maximum value of Q at $g = g_{VR}$ for $\alpha = 2$ increases when γ increases. For a certain range of values of α, g_{VR} decreases when γ increases and the value of Q at resonance increases. For example, when $\alpha = 1$ the value of g_{VR} decreases when γ increases. For $\alpha = 2$ and 3, g_{VR} increases when γ increases.

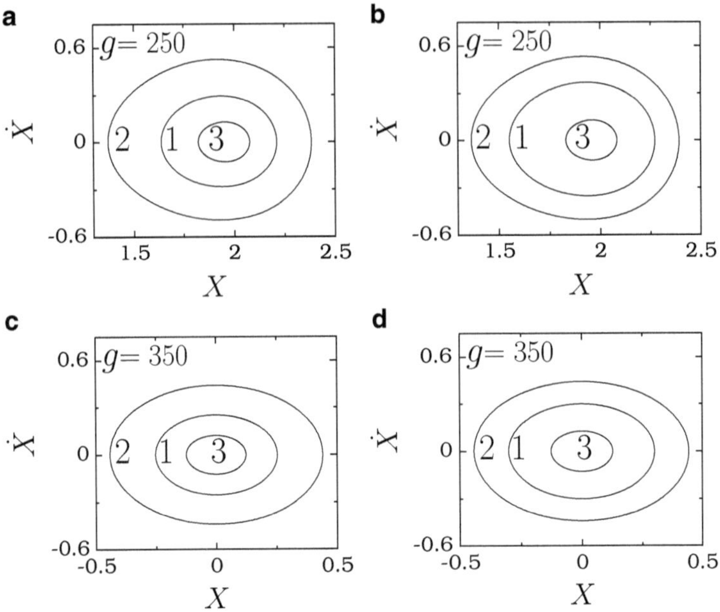

Fig. 6 X versus $\dot{X}$ as a function of $\alpha(\leq 14)$ for two fixed values of g. (**a**) $g = 250$, $\alpha = 4$ (1), 4.57 (2) and 8 (3). (**b**) $g = 250$, $\alpha = 10.5$ (1), 10.97 (2) and 14 (3). (**c**) $g = 350$, $\alpha = 3.5$ (1), 4.41 (2) and 7 (3). (**d**) $g = 350$, $\alpha = 10$ (1), 10.69 (2) and 13 (3). Resonance occurs at $\alpha = 4.57$ and 10.97 for $g = 250$ and at $\alpha = 4.41$ and 10.69 for $g = 350$

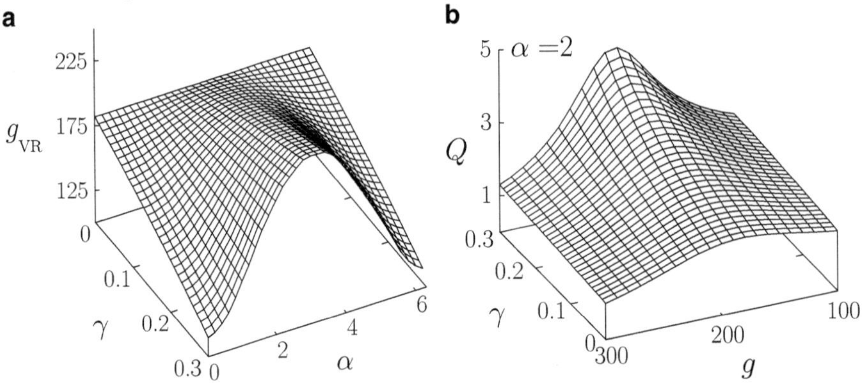

Fig. 7 (**a**) Theoretical g_{VR} versus the time-delay feedback parameters γ and α for the single-well case of the system (Eq. 1) with $L = 1$, $\omega_0^2 = 0.5$ and $\beta = 0.1$. (**b**) Q versus the parameters γ and g for $\alpha = 2$

Figure 8a presents both theoretical and numerical α_{VR} versus g for $\gamma = 0.15$. α_{VR} is periodic with period $2\pi/\omega$. For a fixed value of g resonance occurs at only one value of α for $\alpha \in [0, 2\pi/\omega]$. In Fig. 5a, corresponding to the double-well system, a double resonance is found for a certain range of fixed values of g. The presence

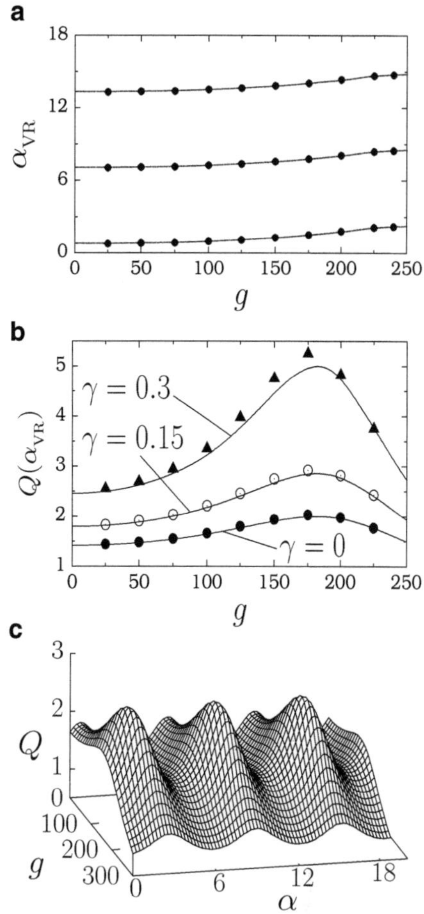

Fig. 8 (**a**) α_{VR} versus g for $\gamma = 0.15$. The system is the single-well case of Eq. (2) with single delay-time. *Continuous lines* and *solid circles* are the theoretically determined and numerically computed values of α_{VR}, respectively. (**b**) $Q(\alpha_{VR})$ versus g for three values of γ. (**c**) Periodic variation of Q with α for various values of g for $\gamma = 0.15$

of only one resonance in Fig. 8a for $\alpha \in [0, 2\pi/\omega]$ implies that the variation of k^2 due to the terms $\gamma \sin \alpha\Omega$ and $\gamma \cos \alpha\Omega$ (see Eq. (11)) is negligible and $k^2 \approx (\omega_0^2 - \Omega^2)^2 + d^2\Omega^2$. Then from Eq. (16) the expression for α_{VR} is obtained as

$$\alpha_{VR} = \frac{1}{\omega} \tan^{-1}\left(\frac{d\omega}{\omega^2 - \omega_r^2}\right), \quad \omega_r^2 = C_1. \tag{25}$$

α_{VR} is independent of γ; however, the maximum value of Q at resonance varies with γ. This is confirmed in the numerical simulation. The maximum value of Q for $\gamma \neq 0$ is always found to be greater than the value of Q for $\gamma = 0$ (see Fig. 8b). Figure 8c demonstrates the periodic variation of Q with the delay parameter α. For irrational values of the ratio Ω/ω the response amplitude exhibits a quasiperiodic pattern.

6 Single Duffing Oscillator: Effect of Multi Time-Delay

In this section we consider the system (Eq. 1) with MTDF. We restrict our analysis to the double-well case alone.

We choose $d = 0.5$, $\omega_0^2 = -1$, $\beta = 0.1$, $f = 0.1$, $\omega = 1$ and $\Omega = 10$. Figure 9 presents both theoretically and numerically computed Q as a function of the control parameter g for $L = 1, 2, 3$ and 5 for $\gamma = 0.3$ and for two values of α. The result for $\gamma = 0$ is also shown in this figure. VR is observed for all the values of L chosen. The theoretical Q value is in good agreement with the numerical Q value. In Fig. 9 two values of α are chosen in such a way that for one value $Q_{\max}(\gamma) > Q_{\max}(\gamma = 0)$ (the value of Q at resonance) while for the other value $Q_{\max}(\gamma) < Q_{\max}(\gamma = 0) = 2$.

Using the theoretical expression of Q, in $(\gamma - \alpha)$ parameter space we identify the regions where $Q_{\max}(\gamma) > Q_{\max}(\gamma = 0)$ for a few fixed values of L. The result is presented in Fig. 10. For both $\gamma < 0$ and $\gamma > 0$, Fig. 10 has L bands. In the bands filled with dots $Q_{\max}(\gamma) > Q_{\max}(\gamma = 0)$. The width of the bands is unequal.

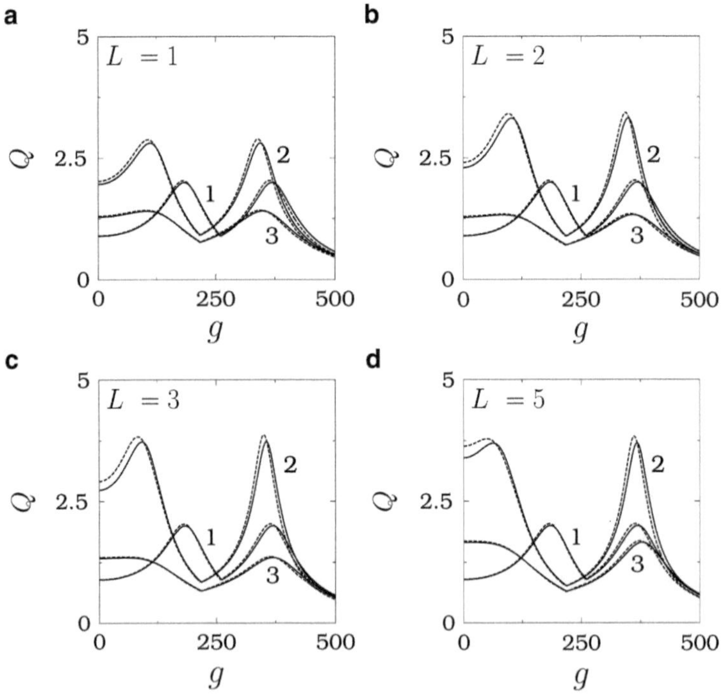

Fig. 9 Q versus g for a few fixed number of time-delayed feedback terms. We fixed $d = 0.5$, $\omega_0^2 = -1$, $\beta = 0.1$, $f = 0.1$, $\omega = 1$ and $\Omega = 10$. The *continuous and dashed lines* are the theoretically and numerically calculated values of Q, respectively. In all the subplots, $\gamma = 0$ for curve 1. For the curves 2 and 3 $\gamma = 0.3$ and $\alpha = 0.5$ and 5.5, respectively

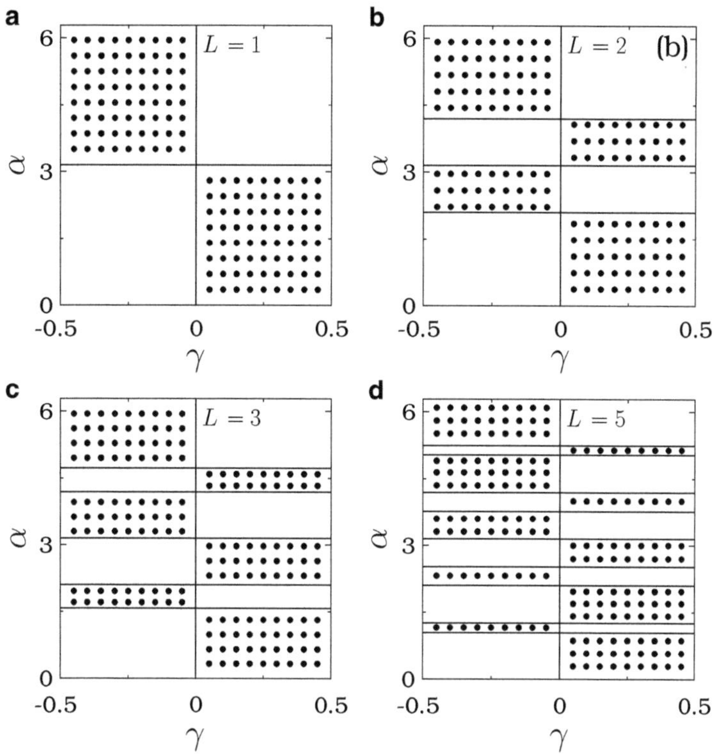

Fig. 10 Plot of regions (marked by *dots*) in the $(\gamma - \alpha)$ parameter space where $Q_{\max}(\gamma, \alpha) > Q_{\max}(\gamma = 0)$ for $\omega = 1$

The filled bands of $\gamma < 0$ become the unfilled bands of $\gamma > 0$. From the theoretical expression of Q the condition for the enhanced response at resonance due to the time-delayed feedback term is $\gamma \sum_{l=1}^{L} \sin l\alpha\omega > 0$ (refer Eq. (16)) and is realized in the regions filled with dots in Fig. 10. For each value of L the total length of α intervals where $Q_{\max}(\gamma) > Q_{\max}(\gamma = 0)$ is $\approx \pi$.

We define $G = Q_{\max}(\gamma)/Q_{\max}(\gamma = 0)$ as the gain factor. Figure 11 shows the variation of G as a function of the parameters γ and α for few values of L. In this figure data with $G > 1$ alone is plotted. For wide ranges of α and γ the gain factor is > 2, that is, the delay is able to increase the value of Q at resonance more than twice the value of Q in its absence. The addition of more and more delay terms decreases the maximum value of G. Moreover, it produces new regions with $G > 1$ in the $(\gamma - \alpha)$ parameter space and decreases the value of G to less than 1 in certain regions where $G > 1$ earlier.

The resonance condition given by Eq. (18) and the results of cases 1 $(\gamma < 0)$ and 2 $(\gamma > 0)$ presented in Sect. 5 for a single delay-time are applicable for the MTDF with $\gamma \cos\alpha\omega$ replaced by $(\gamma/L)\sum_{l=1}^{L} \cos l\alpha\omega$. We point out that in the system (Eq. 1), in absence of MTDF, there are two resonances for $2|\omega_0^2| > \omega^2$, while one for $2|\omega_0^2| < \omega^2$. With MTDF the number of resonances for $\gamma < 0$ depends on

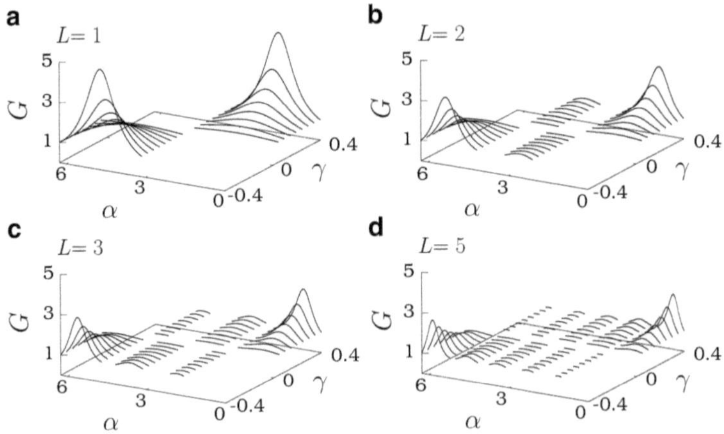

Fig. 11 $G = Q_{max}(\gamma)/Q_{max}(\gamma = 0) > 1$ versus γ and α for various values of L

the parameters ω, L and α. For $\gamma > 0$ the number of resonances depends also on the parameter ω_0^2. Thus, by suitably choosing the values of γ, α and L the system can be set to show either two resonances or one resonance by varying the control parameter g. That is, the number of resonances can be varied by means of an MTDF.

In the single oscillator, an amplification of a low-frequency signal can be achieved for a range of amplitude and frequency of the high-frequency force in the absence of time-delayed feedback. In this case the maximum value of Q is $1/(d\omega)$. When the MTDF is introduced, we find $Q_{max} = 1/|d\omega - (\gamma/L)\sum_{l=1}^{L} \sin l\alpha\omega|$. That is, Q_{max} can be further increased or decreased by appropriate choices of γ, α and L. Thus, the MTDF can be used to control the value of Q_{max}.

7 Signal Propagation in a System of n-Coupled Oscillators

In this section we focus our investigation on the signal propagation in a system of n-coupled Duffing oscillators, Eq. (2), with $n = 200$. In the system (Eq. 2) the external force is applied to the first oscillator alone. The coupling term is linear and has multiple time-delayed terms. The evolution of x_1 is independent of x_i, $i > 1$ while those of x_i, $i > 1$ depends on x_{i-1}.

7.1 Theoretical Approach

Writing $x_i = X_i + \psi_i$ where X_i's and ψ_i's are slow variables and fast variables, respectively, and applying the theoretical treatment used in Sect. 4, we obtain the following results:

$$Y_i(t) = Q_i f \cos(\omega t + \phi_i), \tag{26}$$

where

$$Q_1 = \frac{1}{\sqrt{\left(\omega_{\mathrm{r1}}^2 - \omega^2\right)^2 + d^2\omega^2}}, \qquad Q_i = P_i Q_{i-1}, \tag{27}$$

$$P_i = \frac{\delta\, r_\omega}{\sqrt{\left(\omega_{\mathrm{r}i}^2 - \omega^2\right)^2 + d^2\omega^2}}, \qquad i = 2, 3, \cdots, n \tag{28}$$

$$\omega_{\mathrm{r}i}^2 = C_i + 3\beta X_i^{*2}, \tag{29}$$

$$C_i = \omega_0^2 + \frac{3}{2}\beta\mu_i^2, \quad i = 1, 2, \cdots, n \tag{30}$$

$$X_1^*\left(X_1^{*2} + \frac{C_1}{\beta}\right) = 0, \tag{31}$$

$$X_i^{*3} + \frac{C_i}{\beta}X_i^* - \frac{\delta}{\beta}X_{i-1}^* = 0, \quad i = 2, 3, \cdots, n \tag{32}$$

$$\mu_1 = g/k, \quad \mu_i = \frac{\delta\, r_\Omega}{k}\mu_{i-1}, \quad i = 2, 3, \cdots, n \tag{33}$$

$$k = \sqrt{\left(\Omega^2 - \omega_0^2\right)^2 + d^2\Omega^2}\,, \tag{34}$$

$$r_\omega = \frac{1}{L}\left[\left(\sum_{l=1}^{L} \sin l\alpha\omega\right)^2 + \left(\sum_{l=1}^{L} \cos l\alpha\omega\right)^2\right]^{1/2} \tag{35}$$

and r_Ω is similar to r_ω with ω replaced by Ω in Eq. (35). The above theoretical treatment gives an important result. When the number of time-delayed terms in the coupling is only one ($L = 1$), then $r_\omega = r_\Omega = 1$ and hence the response amplitudes Q_i's, $i > 2$ are independent of the time-delay parameter α. This is because when $L = 1$ the coupling term $\delta x_{i-1}(t-\alpha)$ becomes $\delta X_{i-1}^* + \delta Q_{i-1} f \cos(\omega t - \alpha\omega + \phi_{i-1})$ in which $-\alpha\omega + \phi_{i-1}$ is an unimportant phase factor as far as the amplitudes of oscillation of x_i's are concerned. The above theoretical prediction is verified in the numerical simulation. Therefore, in the rest of our analysis we consider $L > 1$.

In obtaining the theoretical Q_i, we have neglected the nonlinear terms in the equations of motion of ψ and $Y(= X - X^*)$. In the systems of n-coupled oscillators, the error in the theoretical Q due to the above approximation is found to be large for $i \gg 1$. To show this we define $\Delta Q_i = Q_{i,\mathrm{T}} - Q_{i,\mathrm{N}}$, where $Q_{i,\mathrm{T}}$ and $Q_{i,\mathrm{N}}$ represent Q_i values determined theoretically and numerically. In Fig. 12 we plot ΔQ_i with i for three values of g. For first few oscillators $\Delta Q_i \approx 0$ and then it diverges with i. In obtaining μ_i given by Eq. (33) we assumed that $\ddot{\psi}_i \gg \psi_i^2$ and ψ_i^3. This can be a valid assumption for $i = 1$, where the first oscillator is driven by the high-frequency force $g\cos\Omega t$. Because the other oscillators are not driven explicitly by the high-frequency force, the assumption $\ddot{\psi}_i \gg \psi_i^2$ and ψ_i^3 is not valid for $i \gg 1$. Further, nonlinear terms in the equations of Y_i are neglected in obtaining Q_i. Moreover, the

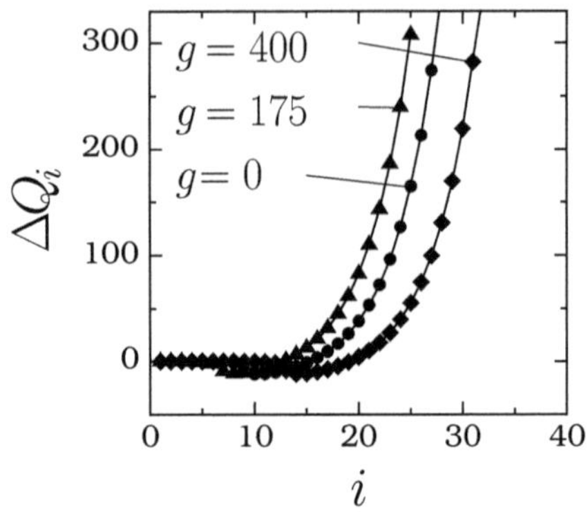

Fig. 12 Variation of ΔQ_i (difference between the theoretically calculated Q_i and the numerically computed Q_i) with i for $d = 0.5$, $\omega_0^2 = 1$, $\beta = 1$, $f = 0.1$, $\omega = 1.5$, $\Omega = 15$, $L = 2$, $\delta = 2.5$, $\alpha = 1$ and for three values of g

errors in ψ_i and Y_i propagate to the $(i+1)$th oscillator through the coupling term. Consequently, ΔQ_i is negligible for the first few oscillators and becomes large for $i \gg 1$.

In order to minimize the error in the theoretical Q_i and also to reduce the propagation of this error through the consecutive oscillators, we include nonlinear terms in the calculation of the amplitudes of oscillation of the fast and slow variables [18]. We assume

$$\psi_i = \mu_i \cos(\Omega t + \phi_i), \quad X_i = A_i \cos(\omega t + \theta_i). \tag{36}$$

Substitution of (Eq. 36) in the equations

$$\ddot{\psi}_1 + d\dot{\psi}_1 + \omega_0^2 \psi_1 + \beta \psi_1^3 = g \cos \Omega t,$$

$$\ddot{\psi}_i + d\dot{\psi}_i + \omega_0^2 \psi_i + \beta \psi_i^3 = \frac{\delta}{L} \sum_{l=1}^{L} \psi_{i-1}(t - l\alpha), \tag{37}$$

$$\ddot{X}_1 + d\dot{X}_1 + \omega_{01}^2 X_1 + \beta X_1^3 = f \cos \omega t,$$

$$\ddot{X}_i + d\dot{X}_i + \omega_{0i}^2 X_i + \beta X_i^3 = \frac{\delta}{L} \sum_{l=1}^{L} X_{i-1}(t - l\alpha), \tag{38}$$

where $i = 2, 3, \cdots, n$ and $\omega_{0j}^2 = \omega_0^2 + \frac{3}{2}\beta\mu_j^2, \quad j = 1, 2, \cdots, n$ gives [65]

$$\mu_i^6 + a_\mu \mu_i^4 + b_\mu \mu_i^2 - R_{i\mu} = 0, \tag{39}$$

$$A_i^6 + a_{iA} A_i^4 + b_{iA} A_i^2 - R_{iA} = 0, \quad i = 1, 2, \cdots, n \tag{40}$$

where

$$a_\mu = \frac{8}{3\beta}\left(\omega_0^2 - \Omega^2\right), \quad b_\mu = \frac{16}{9\beta^2}\left[\left(\omega_0^2 - \Omega^2\right)^2 + d^2\Omega^2\right], \tag{41}$$

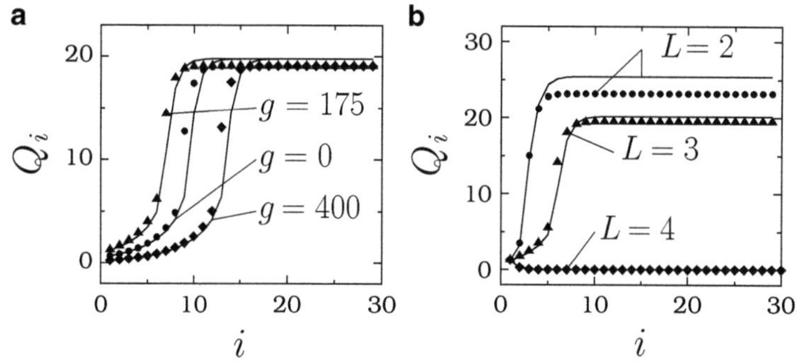

Fig. 13 Variation of Q_i with i for (**a**) three values of g with $L = 2$, $\alpha = 1$ and $\delta = 2.5$ and (**b**) three values of L with $g = 175$, $\delta = 5$ and $\alpha = 1$. The *continuous lines* and the *symbols* represent the theoretically and numerically computed values of Q_i, respectively

$$R_{1\mu} = \frac{16g^2}{9\beta^2}, \quad R_{1A} = \frac{16f^2}{9\beta^2}, \tag{42}$$

$$a_{iA} = \frac{8}{3\beta}\left(\omega_{0i}^2 - \omega^2\right), \quad b_{iA} = \frac{16}{9\beta^2}\left[\left(\omega_{0i}^2 - \omega^2\right)^2 + d^2\omega^2\right], \quad i = 1, 2, \cdots, n \tag{43}$$

$$R_{i\mu} = \frac{16\delta^2 r_\Omega^2 \mu_{i-1}^2}{9\beta^2}, \quad R_{iA} = \frac{16\delta^2 r_\omega^2 A_{i-1}^2}{9\beta^2}. \quad i = 2, 3, \cdots, n. \tag{44}$$

r_ω is given by Eq. (35) and r_Ω is obtained from r_ω by replacing ω by Ω. Equations (39) and (40) can be viewed as cubic equations for the variables μ_i^2 and A_i^2, respectively. Analytical expressions for the roots of the cubic equation of the form (Eq. 39) are given in [66]. We determine μ_i, A_i and then $Q_i = A_i/f$ by solving the Eqs. (39) and (40). We use $Q_{200} > Q_1$ as the criterion for undamped and enhanced signal propagation in the coupled oscillators.

We check the validity of the theoretical approach. In Fig. 13a we plot both the theoretically calculated Q_i and the numerically computed Q_i as a function of i for three values of g with $L = 2$, $\alpha = 1$ and $\delta = 2.5$. We observe a very good agreement of the theoretical Q_i with the numerical Q_i.

7.2 *Undamped and Damped Signal Propagations*

In Fig. 13a for each fixed value of g, for sufficiently large i, Q_i attains a saturation value. The variation of Q_i with i displays a kink-like dependence. That is, there is a critical number of oscillators for obtaining the maximum response and this number depends on the control parameters. An interesting observation in Fig. 13a is that

$Q_i > Q_1$ for $i > 1$, even in the absence of a high-frequency force. This implies that a coupling alone is able to give rise to an enhanced undamped signal propagation in the coupled oscillators. Figure 13b shows the influence of the number of time-delay terms in the coupling on Q_i, where $g = 175, \alpha = 1$ and $\delta = 5$. For $L = 2$ and 3 an undamped signal propagation occurs while for $L = 4$ a damped signal propagation takes place.

We call the limiting or saturation value of Q_i as $Q_{\rm L}$. Interestingly, we can determine $\mu_{\rm L}$, $A_{\rm L}$ and hence $Q_{\rm L} = A_{\rm L}/f$ from Eqs. (39) and (40), respectively. Substituting $\mu_i = \mu_{i-1} = \mu_{\rm L}$ and $A_i = A_{i-1} = A_{\rm L}$ for sufficiently large i in Eqs. (39) and (40) we obtain

$$\mu_{\rm L} = 0, \quad \left\{ \frac{4}{3\beta} \left[\Omega^2 - \omega_0^2 \pm \sqrt{\delta^2 r_\Omega^2 - d^2\Omega^2} \right] \right\}^{1/2} \tag{45}$$

and

$$A_{\rm L} = 0, \quad \left\{ \frac{4}{3\beta} \left[\omega^2 - \omega_{0\rm L}^2 \pm \sqrt{\delta^2 r_\omega^2 - d^2\omega^2} \right] \right\}^{1/2}, \tag{46}$$

where $\omega_{0\rm L}^2 = \omega_0^2 + \frac{3}{2}\beta\mu_{\rm L}^2$. $A_{\rm L} = 0$ and $\neq 0$ correspond to a damped and an undamped signal propagation, respectively. It is also possible to find out the condition on δ for undamped signal propagation. In Fig. 13 in all the examples of undamped signal propagation $Q_2 > Q_1$. This is further confirmed for a large set of parametric values. Therefore, we assume that if $Q_2 > Q_1$, then

$$Q_i \ge Q_{i-1} \ge \cdots > Q_3 > Q_2 > Q_1. \tag{47}$$

For Q_1 and Q_2 very much satisfactory analytical expressions are given by Eq. (27) with $i = 2$. The condition for $Q_2 > Q_1$ is $P_2 > 1$, where

$$P_2 = \frac{\delta\, r_\omega}{\sqrt{\left(\omega_{\rm r2}^2 - \omega^2\right)^2 + d^2\omega^2}}, \quad \omega_{\rm r2}^2 = \omega_0^2 + \frac{3\beta g^2\delta^2 r_\Omega^2}{2\Omega^8}. \tag{48}$$

In this equation, $\omega_{\rm r2}^2 \approx \omega_0^2$ because of $1/\Omega^8$ in the second term of $\omega_{\rm r2}^2$. Then $P_2 > 1$ gives

$$\delta > \delta_{\rm u} = \frac{\sqrt{\left(\omega_0^2 - \omega^2\right)^2 + d^2\omega^2}}{r_\omega}. \tag{49}$$

Undamped signal propagation takes place for $\delta > \delta_{\rm u}$. A very interesting result is that both $Q_{\rm L}$ and $\delta_{\rm u}$ are independent of the amplitude g of the high-frequency periodic force. We can confirm this in Fig. 13a where numerically computed Q_i for different

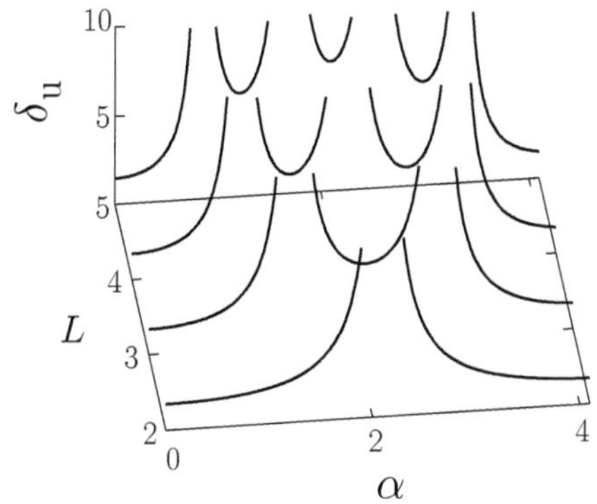

Fig. 14 Dependence of δ_u on the time-delay α and the number of time-delayed terms L. $\delta_u < 10$ are alone shown in this plot

values of g approach the same limiting value. Figure 14 describes the dependence of δ_u on the number of time-delayed terms L and the time-delay α. In this figure, for clarity, only the values $\delta_u < 10$ are shown. When $\delta_u > 10$ is also considered then the δ_u curve has L peaks for a given value of L. For δ values above the threshold curve an undamped signal propagation occurs. For $L = 2$ and $\alpha = 1, 1.5$ and 3 the theoretical and the numerically computed values of δ_u are (1.99, 1.96), (3.38, 3.31) and (2.32, 2.30), respectively. Figure 15 presents Q_L versus δ and α for four fixed values of L. In this figure we can clearly see the effect of the number of time-delayed terms L and the time-delay α on Q_L. The dependence of Q_L on α is nonmonotonic. Even for large values of δ there are intervals of α in which $Q_L = 0$ (damped signal propagation).

8 Conclusions

The role of the amplitude g of the high-frequency periodic force and the delay-time feedback parameters γ and α on VR is explored in the systems (Eqs. 1 and 2) through a theoretical approach. The theoretical treatment used in the present analysis allows us to predict the values of the control parameters at which resonance occurs, number of resonances, the maximum value of the response amplitude Q and explains the mechanism of resonance. The theoretical predictions of Q, g_{VR} and α_{VR} are in very good agreement with the numerical simulations. The presence of time-delay feedback is found to enrich the VR phenomenon. Particularly, the time-delay parameter α gives rise to a periodic or quasiperiodic pattern of VR profile. This feature of VR allows us to select different values (small or large) for the delay-time α to enhance the quality of the weak signal and it can be highly useful in optimizing the operation of multistable systems for the detection and regeneration of signals in a variety of experimental systems.

In the single oscillator, when the amplitude g of the high-frequency periodic force is varied, a single or a double resonance occurs depending upon the values

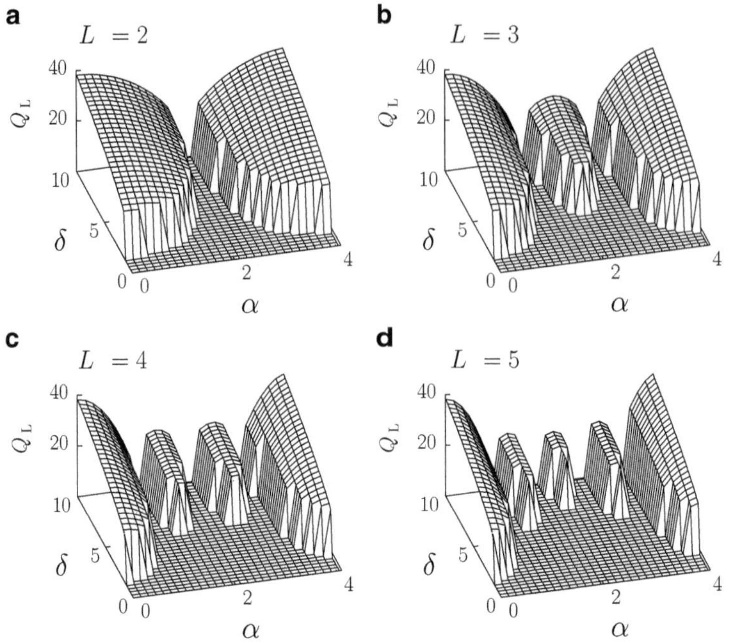

Fig. 15 Three-dimensional plot of Q_L versus δ and α for four fixed values of the number of time-delayed coupling terms L. Q_L is independent of the parameter g

of the parameters ω_0^2, ω, α, L and γ and is independent of the parameters d, f, β and Ω. In the γ (the strength of feedback term)—α (time-delay) parameter space the regions with $Q_{\max}(\gamma) > Q_{\max}(\gamma = 0)$ have L bands where L is the number of time-delayed feedback terms. The maximum value of response amplitude is found to decrease when the number of feedback terms increases. The response amplitude Q depends on all the parameters except f (the analysis performed in the present work is valid only for $|f| \ll 1$) while its value at resonance depends on the parameters $d, \omega, \gamma, \alpha$ and L.

More importantly, the theoretical approach is able to determine and explain the various features of signal propagation in coupled oscillators. One notable prediction is that in coupled oscillators the response amplitude as well as the dynamics is independent of the time-delay parameter α when the number of coupling terms is only one ($L = 1$). The system exhibits undamped signal propagation for appropriate choices of the parameters and these choices of parameters can be determined from the theoretical approach. We wish to stress that in the coupled oscillators system (2), even though only the first oscillator is driven by the high-frequency periodic force, fascinating results on signal propagation are obtained by the action of the unidirectional coupling with multiple time-delayed terms.

Acknowledgements MAFS acknowledges the financial support from the Spanish Ministry of Science and Innovation under Project No. FIS2009-09898.

References

1. Gammaitoni L, Hanggi P, Jung P, Marchesoni F (1998) Rev Mod Phys 70:223
2. McDonnell MD, Stocks NG, Pearce CEM, Abbott D (2008) Stochastic resonance. Cambridge University Press, Cambridge
3. Zambrano S, Casado JM, Sanjuan MAF (2007) Phys Lett A 366:428
4. Pikovsky AS, Kurths J (1997) Phys Rev Lett 78:775
5. Landa PS, McClintock PVE (2000) J Phys Math Gen 33:L433
6. Blekhman II, Landa PS (2004) Int J Non Lin Mech 39:421
7. Chizhevsky VN (2008) Int J Bifurcat Chaos 18:1767
8. Jeyakumari S, Chinnathambi V, Rajasekar S, Sanjuan MAF (2009) Phys Rev E 80:046608
9. Baltanas JP, Lopez L, Blekhman II, Landa PS, Zaikin A, Kurths J, Sanjuan MAF (2003) Phys Rev E 67:066119
10. Rajasekar S, Jeyakumari S, Chinnathambi V, Sanjuan MAF (2010) J Phys Math Theor 43:465101
11. Jeyakumari S, Chinnathambi V, Rajasekar S, Sanjuan MAF (2009) Chaos 19:043128
12. Rajasekar S, Abirami K, Sanjuan MAF (2011) Chaos 21:033106
13. Ullner E, Zaikin A, Garcia-Ojalvo J, Bascones R, Kurths J (2003) Phys Lett A 312:348
14. Yang JH, Zhu H (2012) Chaos 22:013112
15. Rajasekar S, Used J, Wagemakers A, Sanjuan MAF (2012) Comm Nonlinear Sci Numer Simulat 17:3435
16. Deng B, Wang J, Wei X, Tsang KM, Chan WL (2010) Chaos 20:013113
17. Jeevarathinam C, Rajasekar S, Sanjuan MAF (2013) Ecol Complex 15:33
18. Yao C, Zhan M (2010) Phys Rev E 81:061129
19. Yang JH, Liu XB (2011) Phys Scripta 83:065008
20. Chizhevsky VN, Giacomelli G (2008) Phys Rev E 77:051126
21. Atay FM (2010) Complex time-delay systems: theory and applications. Springer, Berlin
22. Lakshmanan M, Senthilkumar DV (2010) Dynamics of nonlinear time-delay systems. Springer, Berlin
23. Choe CU, Dahms T, Hovel P, Scholl E (2010) Phys Rev E 81:025205(R)
24. Fischer I, Vicente R, Buldu JM, Peil M, Mirasso CR, Torrent MC, Garcia-Ojalvo J (2006) Phys Rev Lett 97:123902
25. Vincente R, Gollo LL, Mirasso CR, Fischer I, Pipa G (2008) Proc Natl Acad Sci USA 105:17157
26. Flunkert V, D'Huys O, Danckaert J, Fischer I, Scholl E (2009) Phys Rev E 79:065201(R)
27. Ramana Reddy DV, Sen A, Johnston GL (2000) Phys Rev Lett 85:3381
28. Gassel M, Glatt E, Kaiser F (2007) Fluct Noise Lett 7:L225
29. Bonnin M, Corinto F, Gilli M (2007) Int J Bifurcat Chaos 17:4033
30. Dahlem MA (2009) Phil Trans Roy Soc Lond 367:1079
31. Yang JH, Liu XB (2010) J Phys A Math Theor 43:122001
32. Yang JH, Liu XB (2010) Chaos 20:033124
33. Jeevarathinam C, Rajasekar S, Sanjuan MAF (2011) Phys Rev E 83:066205
34. Hu D, Yang JH, Liu XB (2012) Comm Nonlinear Sci Numer Simulat 17:1031
35. Daza A, Wagemakers A, Rajasekar S, Sanjuan MAF (2013) Comm Nonlinear Sci Numer Simulat 18:411
36. Yang JH, Liu XB (2010) Phys Scripta 82:025006
37. Ahlborn A, Parlitz U (2005) Phys Rev E 72:016206
38. Shahverdiev EM, Shore KA (2008) Phys Rev E 77:057201

39. Gakkhar S, Negi K, Sahani SK (2009) Comm Nonlinear Sci Numer Simulat 14:850
40. Englert A, Kinzel W, Aviad Y, Butkovski M, Reidler I, Zigzag M, Kanter I, Rosenblum M (2010) Phys Rev Lett 104:114102
41. Englert A, Heiligenthal S, Kinzel W, Kanter I (2011) Phys Rev E 83:046222
42. Lee WS, Restrepo JG, Ott E, Antonsen TM (2011) Chaos 21:023122
43. Gakkhar S, Singh A (2012) Comm Nonlinear Sci Numer Simulat 17:914
44. Martinenghi R, Rybalko S, Jacquot M, Chembo YK, Larger L (2012) Phys Rev Lett 108:244101
45. Cushing JM (1977) Integro-differential equations and delay models in population dynamics. Springer, Berlin
46. Maza D, Mancini H, Boccaletti S, Genesio R, Arecchi FT (1998) Int J Bifurcat Chaos 8:1843
47. Khrustova N, Veser G, Mikhailov A, Imbihl R (1995) Phys Rev Lett 75:3564
48. Schanz M, Pelster A (2003) Phys Rev E 67:056205
49. Larger L, Goldgebuer J, Erheux T (2004) Phys Rev E 69:036210
50. Erneux T, Glorieux P (2010) Laser dynamics. Cambridge University Press, Cambridge
51. Argyris A, Syvridis D, Larger L, Annovazzi-Lodi V, Colet P, Fischer I, Garcia-Ojalvo J, Mirasso CR, Pesquera L, Shore KA (2005) Nature 438:343
52. Uchida A, Amano K, Inoue M, Hirano K, Naito S, Someya H, Oowada I, Kurashige T, Shiki M, Yoshimori S, Yoshimura K, Davis P (2008) Nat Photon 2:728
53. MacDonald N (1989) Biological delay systems: linear stability theory. Cambridge University Press, Cambridge
54. Hashemi M, Valizadeh A, Azizi Y (2012) Phys Rev E 85:021917
55. Suel GM, Garcia-Ojalvo J, Liberman LM, Elowitz MB (2006) Nature 440:545
56. Mergenthaler K, Engbert R (2007) Phys Rev Lett 98:138104
57. Cabrera JL, Milton JG (2002) Phys Rev Lett 89:158702
58. Bauer S, Brox O, Kreissl J, Sartorius B, Radziunas M, Sieber J, Wunsche HJ, Henneberger F (2004) Phys Rev E 69:016206
59. Gerstner W, van Hemmen JL (1993) Phys Rev Lett 71:312
60. Bacci A, Huguenard JR, Prince DA (2003) J Neurosci 23:859; (2006) Neuron 49:119
61. Diez-Martinez O, Segundo JP (1983) Biol Cybern 47:33
62. Pakdaman K, Vibert JF, Boussard E, Azmy N (1996) Neural Network 9:797
63. Kane DM, Shore KA (2005) Unlocking dynamical diversity: optical feedback effects on semiconductor lasers. Wiley, New York
64. Shahverdiev EM, Shore KA (2008) Phys Rev E 77:057201
65. Jeevarathinam C, Rajasekar S, Sanjuan MAF (2013) Chaos 23:013136
66. Pipes LA, Harvill LR Applied mathematics for engineers and physicists, 3rd edn. McGraw-Hill, New York

Experimental Studies of Noise Effects in Nonlinear Oscillators

Vadim S. Anishchenko, Tatjana E. Vadivasova, Alexey V. Feoktistov, Vladimir V. Semenov, and Galina I. Strelkova

Abstract In the paper the noisy behavior of nonlinear oscillators is explored experimentally. Two types of excitable stochastic oscillators are considered and compared, i.e., the FitzHugh–Nagumo system and the Van der Pol oscillator with a subcritical Andronov–Hopf bifurcation. In the presence of noise and at certain parameter values both systems can demonstrate the same type of stochastic behavior with effects of coherence resonance and stochastic synchronization. Thus, the excitable oscillators of both types can be classified as stochastic self-sustained oscillators. Besides, the noise influence on a supercritical Andronov–Hopf bifurcation is studied. Experimentally measured joint probability distributions enable to analyze the phenomenological stochastic bifurcations corresponding to the boundary of the noisy limit cycle regime. The experimental results are supported by numerical simulations.

1 Introduction

Since any real systems are subjected to random excitations, the influence of external noise on dynamical systems becomes an important research topic from both fundamental and applied viewpoints. A series of scientific monographs, among of which [1–9], and the majority of research papers are devoted to this problem.

V.S. Anishchenko • T.E. Vadivasova • A.V. Feoktistov • V.V. Semenov • G.I. Strelkova (✉)
Saratov State University, Astrakhanskaya str., 83, Saratov 410012, Russia
e-mail: wadim@info.sgu.ru; vadivasovate@yandex.ru; feoktistov_a_v@mail.ru; semenov_v_v@list.ru; strelkovagi@info.sgu.ru

V. Afraimovich et al. (eds.), *Nonlinear Dynamics and Complexity*, Nonlinear Systems and Complexity 8, DOI 10.1007/978-3-319-02353-3_10,

Nonlinear dynamical systems that possess stochastic oscillations arising from random excitation (noise) can be referred to a separate class. Without external excitation, a system is in its stable equilibrium state. Such systems are called stochastic oscillators and demonstrate the number of fundamental effects, such as stochastic resonance (SR) [10–13], coherence resonance (CR) [14, 15], and stochastic synchronization (SS) [12, 13, 16–18]. At certain conditions nonlinear stochastic oscillators possess some features of self-sustained oscillatory systems. This fact enables to call these systems as stochastic self-sustained oscillators [19]. One of the fundamental properties of self-sustained oscillations that can demonstrate stochastic self-sustained oscillators is their ability to synchronization. As noted above, this feature is also inherent in bistable stochastic oscillators [13, 16, 17] and excitable oscillators [13, 20–22]. Additionally, stochastic synchronization for excitable systems is completely similar to synchronization of a noisy self-sustained oscillator. This analogy is related with the fact that the spectrum of stochastic oscillations of excitable systems exhibits a peak at a certain characteristic nonzero frequency, which is sufficiently narrow in the CR regime.

The effect of coherence resonance is typical not only for excitable systems but it can also be observed in isochronous self-sustained oscillators with hard excitation in the presence of additive noise [23, 24]. With this, the CR effect has been established in a "subthreshold" region, i.e., in the region where there are no limit cycles in a deterministic system. In the presence of noise such a system can be considered as a stochastic oscillator.

An important problem in studying noisy dynamical systems is the analysis of stochastic bifurcations (i.e., bifurcations in the presence of noise). In [24–32] the authors explore bifurcations of noisy systems (stochastic bifurcations) and noise-induced transitions (new types of behavior that can initiate in the presence of noise only). When studying noisy systems, first of all the concept of "bifurcation" changes. For deterministic systems bifurcations are connected with qualitative changes in the stability or structure of limit sets in the phase space (stable and unstable equilibrium points, limit cycles, separatrix loops, etc.). In the presence of noise limit sets being typical for deterministic systems can no longer be distinguished and thus bifurcations of stochastic systems are often associated with qualitative changes in the probability distribution [6, 8, 9]. In many cases the theoretical analysis of stochastic bifurcations (except for systems with phase space dimension $N = 1$) is a complex problem and can be carried out only approximately. Therefore, experimental approaches gain in importance.

The present work is devoted to the experimental study of the noise influence on nonlinear oscillators. We consider effects of CR and SS in an electronic model of the excitable FitzHugh–Nagumo oscillator and in the Van der Pol oscillator with hard excitation, both subjected to additive broadband noise. Our goal is to identify the analogy in behavior of these two systems and thus to show that in the subthreshold regime and in the presence of additive noise the oscillator with hard excitation exhibits the properties of an excitable stochastic oscillator. And, as well as excitable oscillators like the FitzHugh–Nagumo oscillator, it has the features of a stochastic self-sustained oscillator and demonstrates the property of synchronization in the CR regime. The stochastic supercritical Andronov–Hopf bifurcation is also explored by using analog simulation. The effect of noise is analyzed in terms of probability distribution modifications, i.e., P-bifurcations and their dependence on the parameter of generation and noise level are established.

2 Excitable Oscillator as an Example of a Stochastic Self-sustained Oscillator: Properties of Stochastic Self-sustained Oscillations

One of the actual problems of modern nonlinear dynamics is to study the properties of the so-called excitable systems. Excitable stochastic oscillators represent two-state systems, one of which is a stable equilibrium and the other one (excitation state) is a transient process of return to the equilibrium. The peculiarity of excitable systems consists in the fact that the relaxation to an equilibrium takes place along phase trajectories in the form of a nearly closed loop, i.e., they are similar to the motion on a limit cycle. A system is in its equilibrium state without external forces. An external driving (kick) can pass the system to its excitation state, afterwards the system returns itself to its equilibrium. In the presence of relatively weak noise, such systems can demonstrate excitation and relaxation processes resulted in the appearance of undamped stochastic oscillations. These oscillations have a high degree of coherence (regularity) for a certain (optimal) noise intensity. This effect is called *coherence resonance* (CR) [14, 15]. A considerable amount of publications is devoted to the study of excitable oscillators, ensembles of such oscillators, and excitable media [20–22, 33, 34]. Excitable oscillatory regimes are typical for neuron activity [35]; therefore, their investigation is very important from the viewpoint of understanding of how the nerve system of living organisms operates. Properties of a sequence of excitation impulses (they are called spikes in biophysics) are mostly controlled by noise presented in a system. Besides, the variation of noise parameters can serve as an ordering factor that can make the system behavior more ordered. The analysis of models of stochastic excitable systems testifies an important role that noise can play in the wildlife.

One of the classical examples of excitable systems is the FitzHugh–Nagumo oscillator (FHN) [36, 37]. It is a simplified model of the Hodgkin–Huxley neuron describing the spike generation in axons of a large squid. The important fact has been established: stochastic oscillations in the CR regime can demonstrate the effect of phase-frequency synchronization [20–22]. It is known that synchronization is a characteristic feature of self-sustained oscillatory systems. The principal question arises: Could noise-induced oscillations of excitable systems be considered as a special type of self-sustained oscillations, namely, as stochastic self-sustained oscillations? To answer to this question, it is necessary to find out what is the difference between stochastic oscillations in excitable systems and stochastic oscillations arising from transformations of a random external force by nonlinear systems.

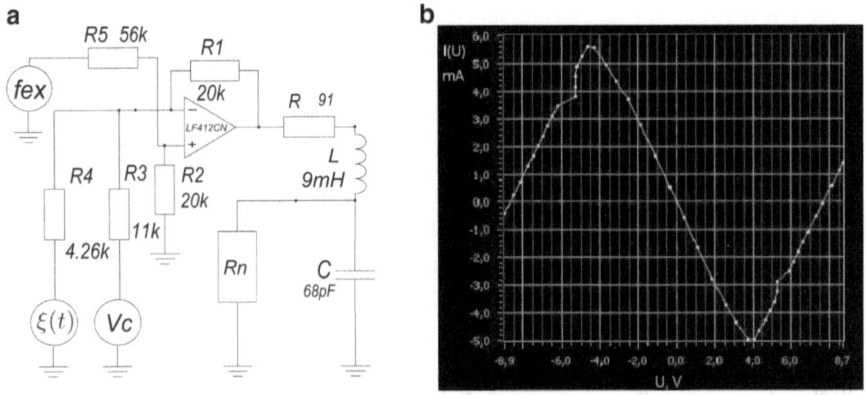

Fig. 1 Schematic diagram of the experimental setup with noise source $\xi(t)$ and external harmonic force $f_{ex}(t) = F(t)$ (**a**) and the experimental voltage-current characteristic of the nonlinear element (**b**). The nominal values of the circuit elements are: $R = 100$ Ohm; $L = 6.8 \cdot 10^{-3}$ H; $C = 6.8 \cdot 10^{-11}$ F; $V_c = 7.2$ V

For our experiments we choose the original electronic model of an excitable system that was suggested by FitzHugh and Nagumo. In this model a tunnel diode is used as a nonlinear element with an N-shaped voltage-current characteristic. We have slightly changed the original circuit to provide a more stable operation of the model [38]. Our circuit diagram is presented in Fig. 1. It only differs by the block that models the N-type nonlinearity in a different way (Fig. 1b) [39]. The voltage-current characteristic of the nonlinear element is qualitatively similar to the characteristic of the tunnel diode, can provide a stable functioning of the scheme and, that is also quite important, provides the simplest approximation in the polynomial form $I_N(U) = g_2U^3 - g_1U$ for mathematical simulation of the dynamical system.

2.1 Model Equations

Using Kirchhoff's laws we can write equations describing the system dynamics in the presence of noise source. The equations read

$$\begin{aligned} \frac{dU}{dt} &= \frac{1}{C}(g_1U - g_2U^3 - I), \\ \frac{dI}{dt} &= \frac{1}{L}(V_c + U - IR) + \frac{1}{L}\xi(t), \end{aligned} \tag{1}$$

where U is the voltage across the capacitor. The nominal values of the circuit elements, supply voltage and parameters of the nonlinear characteristic are indicated in Fig. 1 caption. $\xi(t)$ is the voltage of a broadband noisy generator with Gaussian

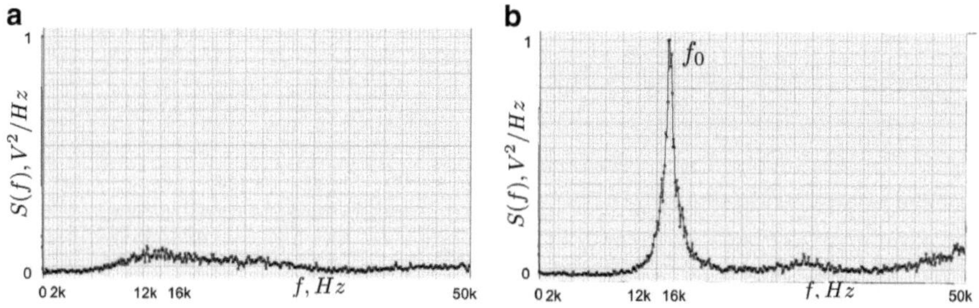

Fig. 2 Power spectra of oscillations for different values of the noise intensity: (**a**) near the threshold of oscillation origin ($\sigma_\xi = 900$ mV) and (**b**) in the regime of coherence resonance ($\sigma_\xi = 1,300$ mV)

distribution. Its mean square $\sigma_\xi = \sqrt{\langle \xi^2(t) \rangle}$ is regulated and is the characteristic of the noise intensity. Equation (1) can be reduced to a dimensionless form by renormalizing the variables and time

$$\begin{aligned} \varepsilon \dot{x} &= x - \alpha x^3 - y, \\ \dot{y} &= \gamma x - y + b + \sqrt{2D} n(\tau). \end{aligned} \tag{2}$$

The following notations are used:

$$U = U_0 x, \quad I = U_0 g_1 y, \quad \tau = \frac{R}{L} t,$$

$$\varepsilon = \frac{CR}{Lg_1}, \quad s = \frac{I_0}{U_0 g_1}, \quad \alpha = \frac{g_2}{g_1} U_0^2, \quad \gamma \frac{1}{Rg_1}, \quad b = \frac{V_c}{Rg_1 U_0}.$$

The points denote dimensionless time τ derivatives and U_0 is some constant potential (for example, $U_0 = V_c$). The broadband Gaussian noise source in the second equation (Eq. 2) is replaced by the normalized Gaussian white noise $n(\tau)$ ($\langle n(\tau) n(\tau + \Delta\tau) \rangle = \delta(\Delta\tau)$, $\delta(\Delta\tau)$ is Dirac's function) with the constant intensity

$$D = \frac{1}{Rg_1 U_0} \int_0^\infty \langle \xi(t) \xi((t + \theta)) \rangle d\theta. \tag{3}$$

Consider the behavior of the experimental system excited by noise and without the external regular force (signal $F(t)$ is absent). We record the signal realization, digitize it by using the ADC NI PCI-6133, and then process on a computer. As the noise intensity σ_ξ increases, stochastic oscillations are initially induced with a sufficiently wide power spectrum (Fig. 2a). Then, a relatively narrow spectral line is formed with the maximum at a certain characteristic frequency f_0 (Fig. 2b) that corresponds to the mean frequency of spike movement. This line has a minimal width at an optimal noise intensity. When the noise intensity grows further, the spectral line width increases again. The spectral line width at the half-power level and the power spectrum density at the maximum frequency are experimentally measured depending on the noise intensity. The obtained results are presented in

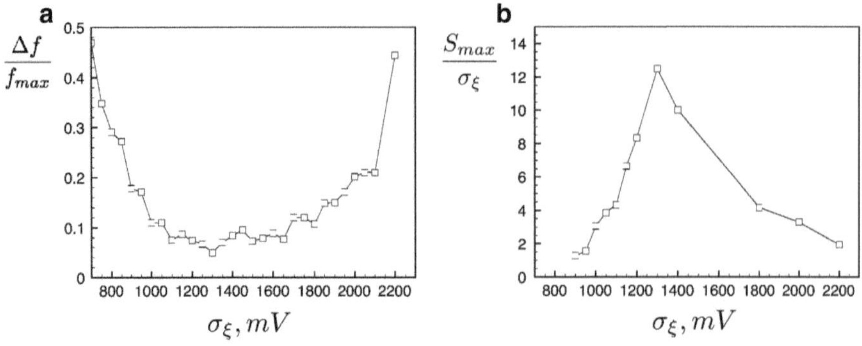

Fig. 3 Dependences of the relative spectrum width (**a**) and the normalized power spectrum density at the maximum (**b**) on the noise level σ_ξ

Fig. 3a, b. The experimental data clearly indicate the effect of coherence resonance when the spectrum width is minimal and hence the highest degree of oscillation coherence is achieved.

2.2 *External Synchronization of the FHN Oscillator by the Harmonic Signal in the Regime of coherence Resonance*

Now we study experimentally synchronization of stochastic oscillations in the FHN system. For this purpose the external harmonic signal $F(t) = A_{\text{ex}} \sin(2\pi f_{\text{ex}} t + \phi_0)$ is added to the circuit. The noise level corresponds to the regime of coherence resonance. To reveal the synchronization effect we fix the external amplitude $A_{\text{ex}} = 510\,\text{mV}$, vary the external frequency f_{ex}, and measure the power spectrum of oscillations (Fig. 4). The observed effect of locking is also illustrated in Fig. 5 where the frequency ratio $\Theta = f_{\text{ex}}/f_0$ is plotted as a function of the external frequency f_{ex} for the fixed external amplitude. It is seen from the figure that there is a finite region of synchronization $f_{\text{ex}} : f_0 = 1$ (at the basic tone).

The experimental results on external and mutual synchronization of stochastic oscillations in the FHN oscillator are described in more detail in the paper [40].

2.3 *Substantiation of Self-sustained Oscillatory Character of Stochastic Oscillations in the FHN System*

The experimental data presented above can testify that in spite of the fact that oscillations in the FHN system can be induced and sustained only in the presence of external noise, they are characterized by a complete set of the properties peculiar to self-sustained oscillatory processes. Let us discuss this in more detail.

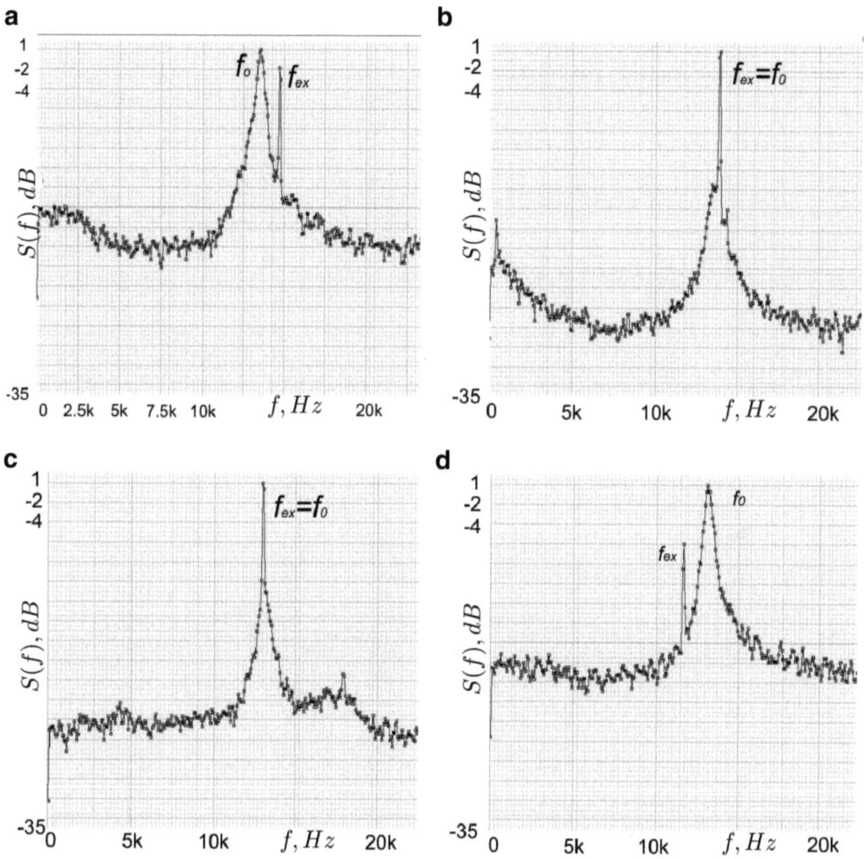

Fig. 4 Effect of frequency locking in the FHN system for different values of the external frequency: (**a**) no frequency locking ($f_{ex} = 14,800$ Hz), (**b**) frequency locking when the spectral line of system oscillations is shifted to the right and coincides with f_{ex} ($f_{ex} = 13,950$ Hz); (**c**) frequency locking inside the synchronization region when the system frequency follows f_{ex} ($f_{ex} = 13,050$ Hz); (**d**) exit from the synchronization region ($f_{ex} = 11,700$ Hz)

We first consider the question: Which attractor corresponds to oscillations in the regime of coherence resonance? The attractor definition of a non-autonomous system was introduced in [19], where it was shown that in this case the attractor is determined as a limit set in a functional (Hilbert) space for all possible solutions $\mathbf{x} = f(\mathbf{x}_0, \tau)$. Unfortunately, such a limit set cannot be pictured geometrically. However, if there is an attractor in the functional space, any trajectory of the FHN system tends to the same limit set of points in the phase plane. These points have a cumulative probability close to 1. Figure 6 presents several phase trajectories obtained numerically for system (2) with different initial conditions. After some relaxation time, the trajectories fall on the same limit set that resembles a slightly

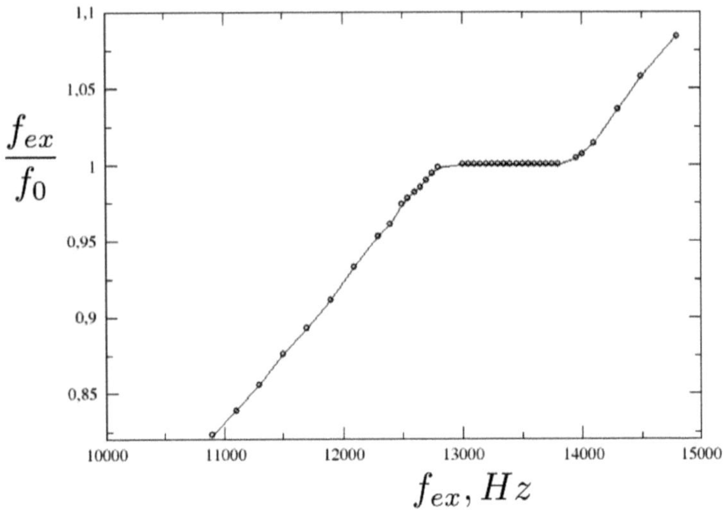

Fig. 5 Experimental dependence of the frequency ratio f_{ex}/f_0 on the external signal frequency f_{ex} for the fixed external amplitude $A_{ex} = 510$ mV

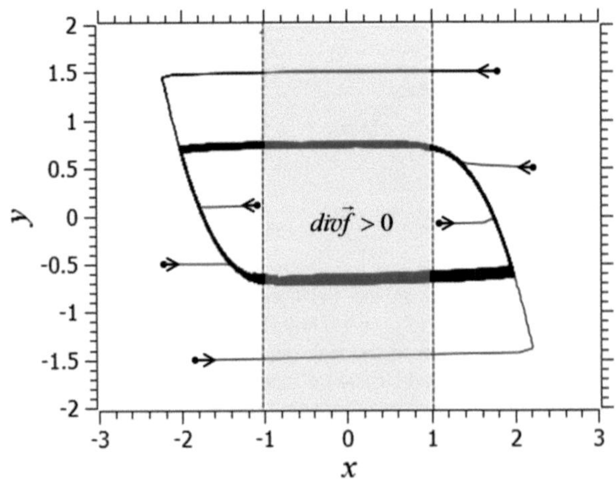

Fig. 6 Phase trajectories of the FHN system obtained numerically with different initial conditions for time $t = 500$. The *gray region* of the phase plane corresponds to the positive vector field divergence of system (Eq. 2) and negative dissipation. The parameters are: $\varepsilon = 0.01$, $\alpha = 1/3$, $\gamma = 1$, $b = 0.5$, $D = 0.0025$

noisy limit cycle. Thus, Fig. 6 illustrates the fact that the limit set of phase trajectories is independent of the initial conditions. The FHN system has no other limit sets.

Based on the numerical data, one can conclude that the FHN system possesses an attractor. Its presence does not yet prove that the system demonstrates a self-sustained oscillatory process. To prove that we must discuss how the energy is added to the system. Model (Eq. 2) can be written in the following oscillatory form:

$$\varepsilon\ddot{x} + (3\alpha x^2 - 1 + \varepsilon)\dot{x} + (1-\gamma)x - \alpha x^3 - b = \sqrt{2D}n(t). \tag{4}$$

We obtain the oscillator with the dissipation coefficient

$$\delta = \frac{3\alpha x^2 - 1}{\varepsilon} + 1, \tag{5}$$

that depends on the coordinate x and can have both positive and negative values. The vector field divergence div**f** of system (Eq. 4) is equal to the coefficient δ taken with an opposite sign and, consequently, also changes its sign depending on the x value. In the range $|x| < \sqrt{\frac{1-\varepsilon}{3\alpha}}$ the dissipation coefficient is negative and the divergence is positive. Thus, within a certain region of states the energy comes to the system, and the excitable oscillator behaves as a self-sustained oscillator. From the physical point of view, the conditions for energy supply are fulfilled when the voltage U of oscillations corresponds to the values in the negative part of the nonlinear element G_N characteristic. In this part the system is characterized by a negative resistance and the source energy increases the energy of oscillations.

These arguments indicate that in the presence of noise the FHN system sustains the oscillatory regime by implementing a synchronous nonlinear energy pumping from the source. The calculations and experimental measurements have confirmed the important fact: the power of the oscillatory process, that is proportional to $x^2(t)$, significantly exceeds the power of the noise source.

3 coherence Resonance and Stochastic Synchronization of Oscillations in an Oscillator with Hard Excitation

Now we consider an oscillator with hard excitation subjected to additive noise. It is known that stochastic oscillations in this oscillator can demonstrate the effect of CR when the noise intensity is varied [23, 24]. We explore experimentally the CR effect and the possibility of forced synchronization of stochastic oscillations with a minimal spectral line width.

3.1 Model Equations

The system being studied is the Van der Pol oscillator with a subcritical Andronov–Hopf bifurcation and is modelled by the following equation in dimensionless variables and with additive noise source:

$$\ddot{x} - (\lambda + \mu x^2 - x^4)\dot{x} + x = \sqrt{2D}n(t), (\mu > 0), \tag{6}$$

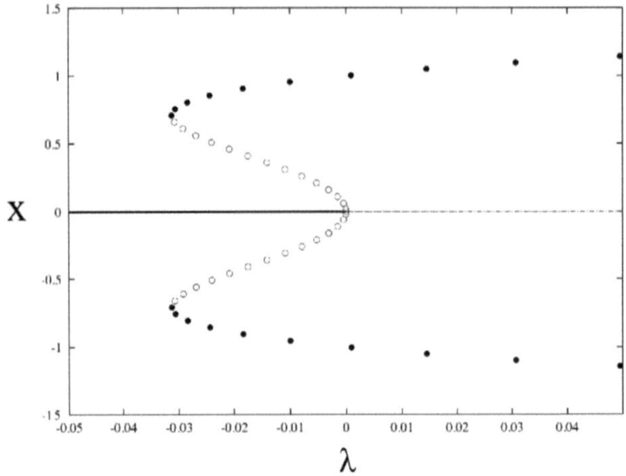

Fig. 7 Phase-parametric diagram of system (Eq. 6) for $\mu = 0.5$. The *solid circles* denote the coordinate of the stable limit cycle, the *empty circles*—the coordinate of the unstable limit cycle, the *thick solid line* denotes the stable equilibrium, and the *thin dashed line*—the unstable equilibrium

where $n(t)$ is a normalized Gaussian white noise source: $\langle n(t)\rangle \equiv 0$; $\langle n(t)n(t+\tau)\rangle = \delta(\tau)$. The brackets $\langle\ldots\rangle$ denote the statistical mean, $\delta(\tau)$ is Dirac's function, and D is the noise intensity. The parameter λ characterizes energy pumping to the system, and μ is responsible for energy dissipation. Without noise ($D = 0$) system (Eq. 6) has an equilibrium point at the origin of coordinates ($x = 0; y = \dot{x} = 0$), which is stable for $\lambda < 0$ and undergoes a subcritical (hard) Andronov–Hopf bifurcation at $\lambda = 0$. On the line $\lambda = -\frac{\mu^2}{8}$ a pair of limit cycles (stable and repeller ones) is born as a result of the tangent bifurcation. The phase-parametric diagram of system (Eq. 6) is shown in Fig. 7 for $D = 0$ and $\mu = 0.5$.

3.2 Experimental Setup

We explore the effects of CR and synchronization of noise-induced oscillations by using a specially constructed experimental setup of the oscillator with hard excitation. Its electronic circuit is shown in Fig. 8 and provides the possibility of connecting external periodic and noisy signals. The basic active elements are operational amplifiers $TL072$ and $LF412$, and multipliers $AD633$. The latter are used to form the nonlinearity of the second and fourth degrees. The other details include resistors and capacitors. The control parameter λ is given by the appropriate voltage from a voltage divider built on a multi-turn resistor. The parameter μ is set by another variable resistor (it is marked as $R7$ in Fig. 8). Component values are shown in the scheme. Noise voltage $\xi(t)$ is generated by a broadband noise

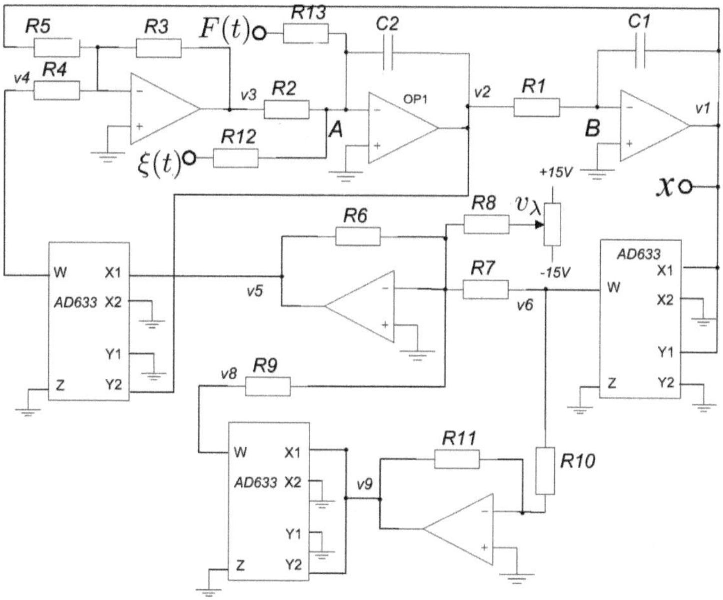

Fig. 8 Schematic diagram of the experimental setup

generator. Its distribution is Gaussian with zero mean $\langle\xi(t)\rangle = 0$. Mean square value $\sigma_\xi = \sqrt{\langle\xi^2(t)\rangle}$ is regulated and is a noise characteristic in our experiments. $F(t)$ represents the signal from a generator of harmonic oscillations: $F(t) = A_{\mathrm{ex}} \sin(2\pi f_{\mathrm{ex}} t + \phi_0)$. Amplitude A_{ex} and frequency f_{ex} can also be regulated.

Equations describing the circuit in Fig. 8 can be easily derived by using Kirchoff's equations. For nodes A and B indicated in the diagram we have

$$C_2 \frac{dv_2}{dt} + \frac{v_3}{R_2} + \frac{\xi(t)}{R_{12}} + \frac{F(t)}{R_{13}} = 0. \tag{7}$$

$$C_1 \frac{dv_1}{dt} + \frac{v_2}{R_1} = 0. \tag{8}$$

For voltage v_3 the following expression is valid

$$v_3 = \left(\frac{R_3 R_6}{10 R_4}\right)\left[\frac{\alpha}{10 R_7} v_1^2 + \frac{1}{R_8} v_\lambda - \frac{\beta}{1000 R_9}\left(\frac{R_{11}}{R_{10}}\right)^2 v_1^4\right] R_1 C_1 \frac{dv_1}{dt} - \frac{R_3}{R_5} v_1. \tag{9}$$

α and β are dimensional coefficients and their values are equal to unity ($\alpha = 1$ OhmV^{-2}, $\beta = 1$ V^{-2}). Choose the resistance values such that $R_1 = R_2 = R_4 = R_9 = R_{10} = R_{12} = R_{13} = R$, $C_1 = C_2 = C$, and $R_3 = R_6 = R_8 = R_{11} = R_5 = 10R$. Then from Eqs. (7)–(9) we obtain the equation describing the circuit

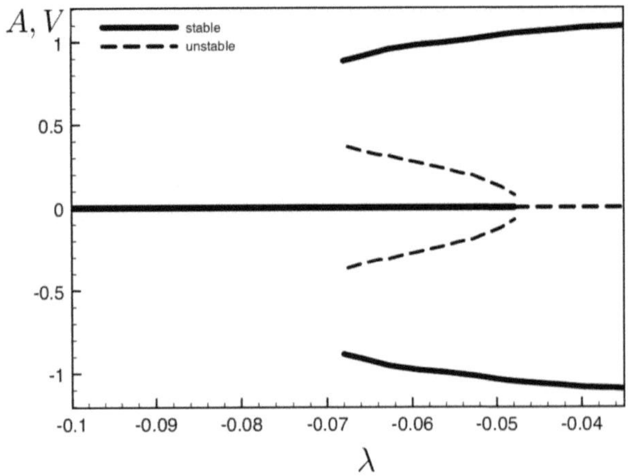

Fig. 9 Experimental phase-parametric diagram of the oscillator for $A_{ex} = 0$ and $\mu = 0.5$. The ordinates represent the value of x provided that $y = \dot{x} = 0$

$$\frac{1}{\omega_0^2}\frac{d^2x}{dt^2} - \frac{1}{\omega_0}\left[\lambda + \mu x^2 - \beta x^4\right]\frac{dx}{dt} + x = A_{ex}\sin(2\pi f_{ex}t + \phi_0) + \xi(t), \quad (10)$$

where the following notations are introduced

$$x = v_1, \ \omega_0 = \frac{1}{RC}, \ \lambda = v_\lambda, \ \mu = \alpha\frac{R}{R_7}. \quad (11)$$

Equation (10) is written with respect to the physical variable x in real time and includes dimensional parameters. However, it is easy to see that without harmonic force it can be reduced to (Eq. 6) by replacing $t_1 = \omega_0 t$. At the same time, as the noise generator used in the experiment is broadband, noise source $\xi(t)$ can be replaced by an equivalent source of white noise $\sqrt{2D}n(t_1)$, where D is the intensity of the equivalent white noise in the normalized time. This quantity is different in meaning and does not coincide quantitatively with the experimental characteristic of noise intensity $\xi(t)$.

Although Eq. (10) for $A_{ex} = 0$ corresponds exactly to dimensionless equation (Eq. 6), the circuit is described by using certain assumptions on operational amplifiers, which, strictly speaking, are approximate. Besides, all the radioelements used for developing the circuit have their own tolerances for the accuracy of ratings. Therefore, we plot an experimental phase-parametric diagram and compare it with the numerical one. This diagram is shown in Fig. 9.

Construction of the lines corresponding to the stable equilibrium and stable cycle is straightforward. The line for the unstable limit cycle is plotted as follows: the parameter λ is initially chosen so that the system has only the single equilibrium state—the stable focus. Then λ increases. As the initial condition is chosen in the

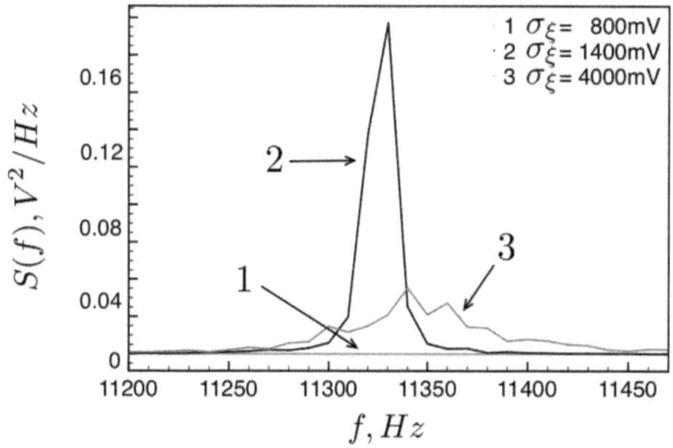

Fig. 10 Power spectra of noise-induced oscillations for $\lambda = -0.05775$ and $\mu = 0.5$

bistability region, the system is still in its equilibrium point. To get a point on the unstable cycle an external harmonic force of low frequency is applied to the system. Starting from a certain amplitude of the external force the phase point jumps from the stable focus to the stable limit cycle. This amplitude value is considered to be approximately equal to the amplitude of the unstable cycle in the system. The diagrams in Figs. 7 and 9 are qualitatively equivalent. Now we compare them quantitatively. In the mathematical model at $\mu = 0.5$ the stable and unstable limit cycles are born at $\lambda = -0.03125$, and the unstable limit cycle and stable focus merge for $\lambda = 0$. In the experiment the same bifurcations take place for $\lambda = -0.069$ and $\lambda = -0.0479$, respectively. Thus, the numerical and experimental values of λ are related by the formula:

$$\lambda_{\text{mm}} = 1.48(\lambda_{\text{exp}} + 0.0479), \tag{12}$$

where λ_{mm} is the parameter value of the mathematical model and λ_{exp} is the corresponding experimental value.

3.3 *Effect of coherence Resonance*

We choose the parameters λ and μ of the experimental scheme so that the oscillator regime corresponds to the bistability region. We apply the external noise of different intensity σ_ξ in the absence of harmonic force. Figure 10 shows power spectra of $x(t)$ oscillations for different noise levels.

It is seen that there is a certain optimal noise level when the spectral line width is minimal (Fig. 10, curve 2). Thus, the effect of coherence resonance takes place.

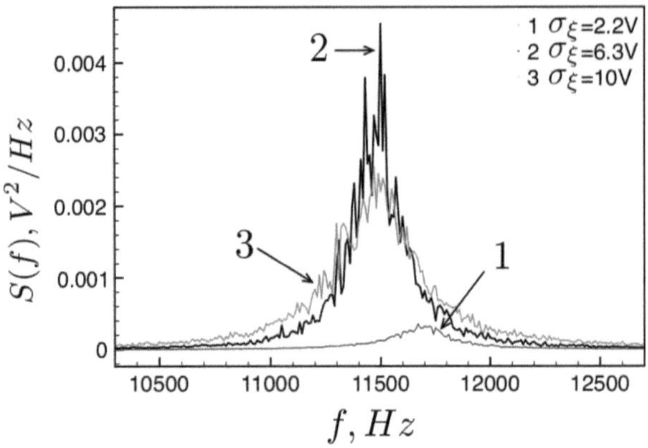

Fig. 11 Power spectra of noise-induced oscillations for $\lambda = -0.079$ and different values of σ_ξ

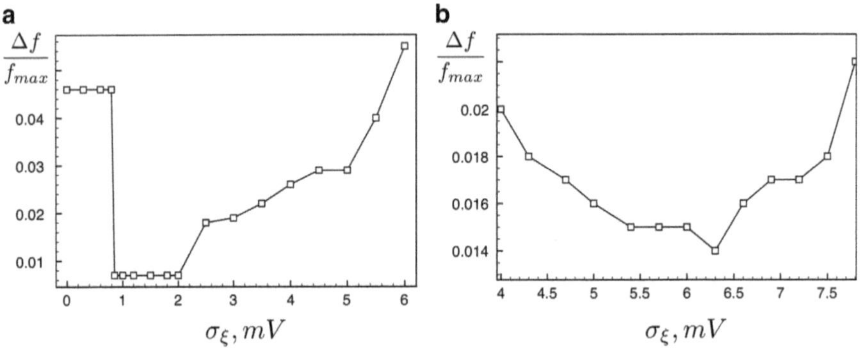

Fig. 12 Experimental dependence of Δf to f on the noise intensity σ_ξ for (**a**) $\lambda = -0.05775$, $\mu = 0.5$ and (**b**) $\lambda = -0.0792$, $\mu = 0.5$

Now we choose the parameters below the generation threshold, i.e., in the region where the stable focus is the unique attractor of the deterministic system. The corresponding evolution of the power spectrum with increasing the noise level is shown in Fig. 11. And in this case it is easy to see the CR effect. Similar results have been obtained numerically in [23, 24]. Experimental dependences of the relative width of the spectral line at the half-power level on the noise intensity σ_ξ are shown in Fig. 12 for the bistability region and in the subthreshold regime.

The performed experimental studies completely verify the numerical results presented in [24]. Thus, the isochronous oscillator with hard excitation both in the bistability region and in the subthreshold regime near the tangent bifurcation and in the presence of additive noise behaves like an excitable system demonstrating the CR effect. The experimental results obtained for the oscillator with hard excitation are very similar to the results of analogous experiments for the FitzHugh–Nagumo oscillator [40]. However, in this case the mechanism of CR is somewhat different than in excitable systems of the FitzHugh–Nagumo oscillator type or in systems with the tangent bifurcation of cycles on a torus. This has been described in [23,24] and is as follows. For weak noise a trajectory rotates mainly in the neighborhood of an equilibrium point. In this case the spectral line of stochastic oscillations has a maximum width defined by the parameter λ. As noise increases, the trajectory often leaves the vicinity of the equilibrium and falls in the region of a stable cycle. Rotations in this region correspond to the spectrum whose width is determined by the noise intensity. At a certain moderate noise level the spectrum width is minimal. The further growth of the noise intensity decreases the spectrum width.

It can be noted that the dependence of the spectrum width on the noise intensity is somewhat different in the two studied solutions. While in the subthreshold regime the spectrum width varies smoothly when changing the noise intensity and reaches a minimum at a separate point (Fig. 12b), in the bistability regime the spectrum width sharply decreases for low noise. Then, it is almost constant in a certain range of the noise intensity and starts growing only further (Fig. 12a).

3.4 *Synchronization of Stochastic Oscillations in the Oscillator with Hard Excitation*

To synchronize oscillations an external harmonic signal is applied to the circuit (Fig. 8) additionally to the noise source. As well as in the case of coherence resonance, we consider two cases. In the first case the circuit parameters ($\mu = 0.5$ and $\lambda = -0.05775$) correspond to the bistability regime of the noise-free oscillator. The noise intensity corresponds to the CR regime ($\sigma_\xi = 1.45$ V). We explore power spectra of oscillations when the external signal amplitude is varied and the frequency mismatch is constant $f_{\mathrm{ex}} = f_0 \pm \Delta$ ($\Delta = 450$ Hz). The obtained results are shown in Fig. 13a–c. It is clearly seen that for $f_{\mathrm{ex}} < f_0$ and as the external signal amplitude grows, the fundamental frequency f_0 of system oscillations is shifted to the external signal frequency f_{ex} and then locked at $A_{\mathrm{ex}} = 0.69$ V. Let the external signal frequency be $f_{\mathrm{ex}} = 11,800$ Hz. The frequency mismatch is the same as in the previous case $\Delta = 450$ Hz, but the spectral peak of the external signal is already located to the right of the fundamental frequency. We start increasing the external amplitude and follow the spectrum evolution. The results are shown in Fig. 13d–f. Contrary to expectations synchronization occurs through suppression. And however small the frequency mismatch, we cannot detect the locking region for $f_{\mathrm{ex}} > f_0$.

Now we consider the second case. The scheme parameters $\mu = 0.5$ and $\lambda = -0.0792$ are set so that the system phase space includes only a stable focus at the origin. The noise level corresponds to the CR regime. We set the external signal frequency to be less than the frequency of noise-induced oscillations ($f_{\mathrm{ex}} < f_0$) and increase the external amplitude. The evolution of power spectra is shown in Fig. 14a–c. As the amplitude grows, the fundamental frequency f_0 is shifted towards f_{ex} and then locked at $A_{\mathrm{ex}} = 0.30$ V. If $f_{\mathrm{ex}} > f_0$, the suppression of oscillations is observed similarly to the above case in Fig. 13. The corresponding evolution of power spectra is presented in Fig. 14d–f.

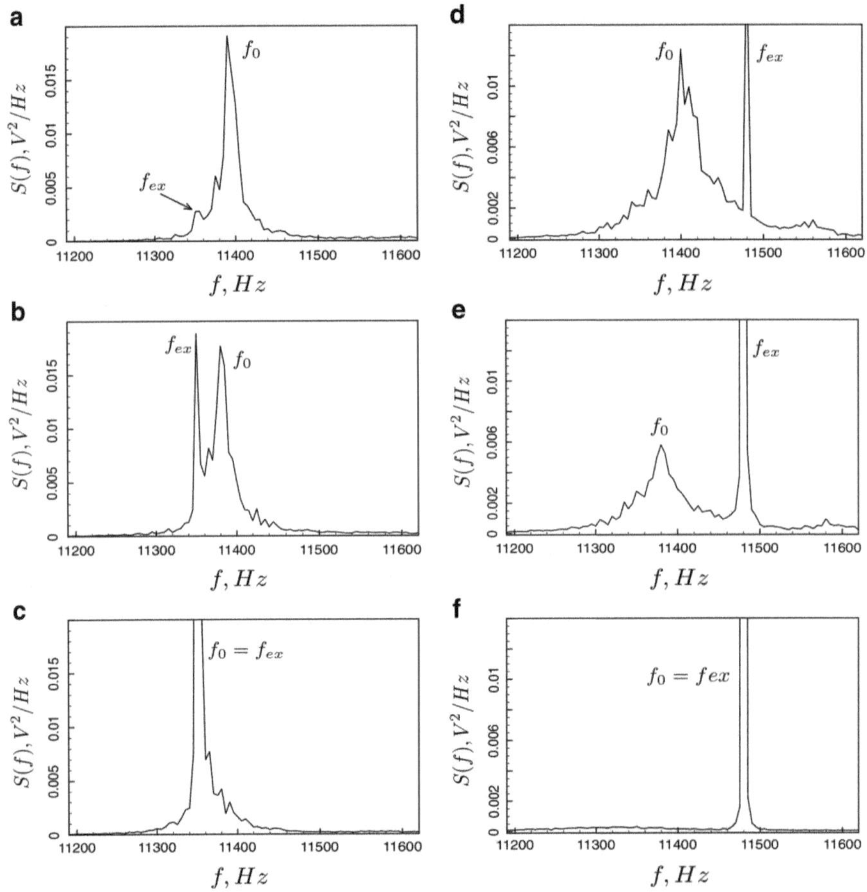

Fig. 13 Power spectra of synchronization of system oscillations in the regime of coherence resonance ($\sigma_\xi = 1,450$ mV). For external signal frequency $f_{\mathrm{ex}} = 11,350$ Hz and different values of external amplitude A_{ex}: (**a**) $A_{\mathrm{ex}} = 0.01$ V, (**b**) $A_{\mathrm{ex}} = 0.02$ V, (**c**) $A_{\mathrm{ex}} = 0.03$ V. For $f_{\mathrm{ex}} = 11,480$ Hz and different values of A_{ex}: (**d**) $A_{\mathrm{ex}} = 0.05$ V, (**e**) $A_{\mathrm{ex}} = 0.09$ V, (**f**) $A_{\mathrm{ex}} = 0.2$ V. The maximum at f_0 corresponds to stochastic oscillations in the system. The oscillator parameters are $\mu = 0.5$ and $\lambda = -0.05775$

Figure 15 shows experimentally plotted synchronization regions in the parameter plane "amplitude-frequency" of the external signal in the bistability regime (Fig. 15a) and in the subthreshold region (Fig. 15b).

In both cases the synchronization region has a certain peculiarity: its left boundary corresponds to the frequency f_0 locking and its right boundary—to the suppression of oscillations at f_0. It is known that the same feature is peculiar to the synchronization region of self-sustained oscillations of a deterministic system in the bistability regime [41].

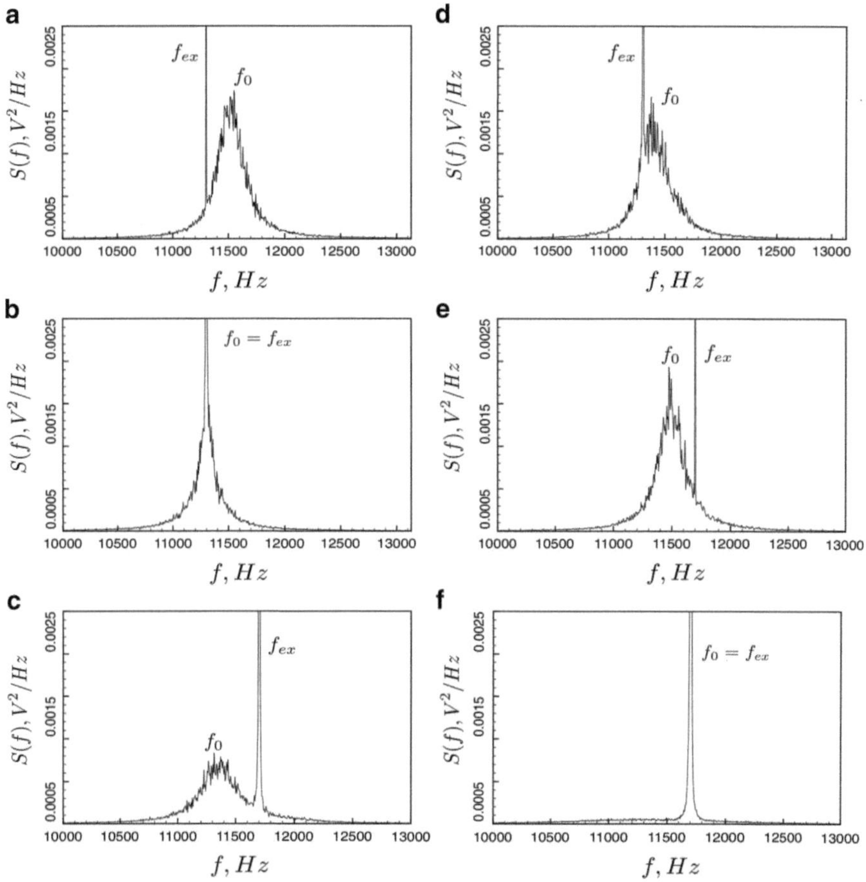

Fig. 14 Power spectra of oscillation synchronization in the CR regime ($\sigma_\xi = 6,300$ mV). For the external signal frequency $f_{ex} = 11,300$ Hz and different values of external amplitude A_{ex}: (**a**) $A_{ex} = 0.05$ V, (**b**) $A_{ex} = 0.20$ V, (**c**) $A_{ex} = 0.30$ V. For $f_{ex} = 11,700$ Hz and different values of A_{ex}: (**d**) $A_{ex} = 0.05$ V, (**e**) $A_{ex} = 0.45$ V, (f) $A_{ex} = 1.20$ V. The maximum at f_0 corresponds to stochastic oscillations in the system. The oscillator parameters are $\mu = 0.5$ and $\lambda = -0.0792$

4 Experimental Study of a Supercritical Andronov–Hopf Bifurcation in Noisy Self-sustained Oscillators

One of the fundamental bifurcations in dynamical systems is the Andronov–Hopf bifurcation that is resulted in the generation of oscillations. In a deterministic case, the appearance of self-sustained oscillations is related to the birth of a stable limit cycle from a focus-type equilibrium point. A stochastic Andronov–Hopf bifurcation leads to the formation of a probability distribution being typical for noisy self-sustained oscillations and having the form of a closed crater. A local maximum of the probability density is observed at points of a deterministic cycle and a

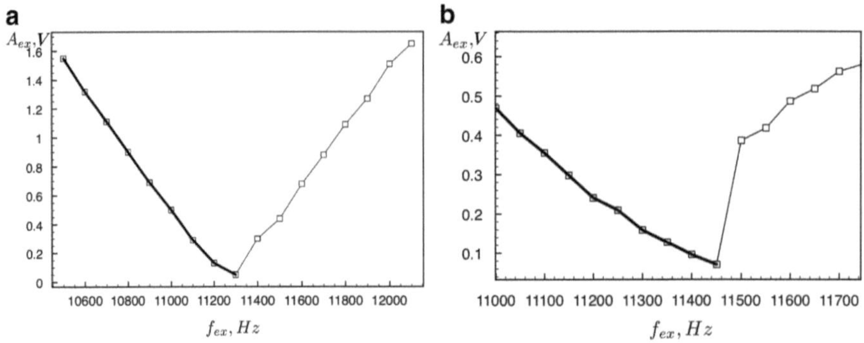

Fig. 15 Experimental regions of synchronization for the oscillator with hard excitation in the CR regime for (**a**) $\mu = 0.5$ and $\lambda = -0.05775$ and (**b**) $\mu = 0.5$ and $\lambda = -0.0792$. The *bold line* denotes the transition to the synchronization region through locking, and the *thin line*—through suppression

local minimum—at an unstable point. The stochastic Andronov–Hopf bifurcation has been explored in a number of research works for different noise statistics [24–26, 28, 29, 32, 42]. In particular, in [26] it has been shown analytically and numerically that the Andronov–Hopf bifurcation in the noisy Van der Pol oscillator occurs not in a single point (as follows from the quasi-harmonic approximation) but when passing through a so-called bifurcation interval corresponding to a gradual rebuilding of the probability distribution. The width of this interval increases when increasing the noise intensity. Thus, at a fixed value of the control parameter, which corresponds to the generation mode, the growth of the noise intensity leads to the transition into the bifurcation interval. At the same time, the crater-like form of the probability distribution is destroyed. The existence of the bifurcation interval, as well as the delayed character of the Andronov–Hopf bifurcation under a colored parametric noise is noted in theoretical studies [25, 29, 32].

Near bifurcations where a system is structurally unstable as well as in the presence of large noise both approximate analytical methods and numerical simulation methods can lead to significant errors. In this situation full-scale experiments are particularly important. However, there is a small number of publications in the scientific literature, which are devoted to experimental studies of stochastic bifurcations. As far as we know, the existence of the bifurcation interval for the stochastic Andronov–Hopf bifurcation has been experimentally verified only in [42] for an analog model of the brusselator with a low-frequency parametric noise.

In our work we aim to explore experimentally the peculiarities of the supercritical Andronov–Hopf bifurcation in various systems and for different noise statistics. Our studies include both full-scale and numerical experiments whose results are compared. We consider the Van der Pol oscillator as a classical model of a periodic self-sustained oscillator as well as Anishchenko–Astakhov's oscillator with inertial nonlinearity that is a more complex system demonstrating the supercritical Andronov–Hopf bifurcation.

4.1 Probability Distribution Evolution in Self-sustained Oscillators with Additive Noise

As a basic model for studying the appearance of self-sustained oscillations, we consider the Van der Pol oscillator. Stochastic equations of the noisy Van der Pol oscillator have the following form:

$$\begin{aligned} \dot{x} &= y, \\ \dot{y} &= (\varepsilon - x^2)y - x + \sqrt{2D}n(t). \end{aligned} \tag{13}$$

The variable x and time t are dimensionless quantities, $n(t)$ is a normalized Gaussian white noise source, and D is the noise intensity. Equation (13) describes many of self-sustained oscillatory systems of various nature, including a radio engineering oscillator. The parameter ε governs the oscillator regime. Without noise and at $\varepsilon = 0$ a supercritical ("soft") Andronov–Hopf bifurcation takes places resulted in the birth of a stable limit cycle from a stable focus at the origin ($x = 0, y = 0$). The analysis of Eq. (13) in the quasi-harmonic approximation shows that the character of the stochastic bifurcation and the bifurcation parameter value $\varepsilon = 0$ are independent of the noise intensity D. But the quasi-harmonic approximation is correct when $D \ll \varepsilon^2$ and may lead to erroneous results near the value $\varepsilon = 0$ and for a finite noise intensity. In [26] based on the analytical solution of the Fokker–Planck–Kolmogorov equation, the expression for the probability distribution in the self-sustained oscillator has been derived that is valid when $\varepsilon^2 \gg D$:

$$p(x, y) = C \exp\left\{-\frac{1}{32D}((x^2 + y^2)^2 - 8\varepsilon(x^2 + y^2)) - \frac{3}{4}xy\right\}. \tag{14}$$

The analysis of expression (Eq. 14) gives the following picture of the stochastic bifurcation. For $\varepsilon < -3D/2$ the distribution $p(x, y)$ has a single maximum at the origin, which corresponds to a noisy stable equilibrium at $x = 0,\ y = 0$. The range $-3D/2 < \varepsilon < 3D/2$ is the bifurcation interval. On its left boundary $\varepsilon = -3D/2$ at points $x_{1,2} = \pm\sqrt{(2\varepsilon + 3D)}$, $y_{1,2} = \pm\sqrt{(2\varepsilon + 3D)}$ there appear two maxima which lead to the destruction of the radial symmetry of the distribution. Within the bifurcation interval the distribution shape evolves smoothly and at $\varepsilon = 3D/2$ a closed crater is completed to form around the distribution minimum at the origin. Thus, the distribution is formed whose shape is typical for the self-sustained oscillatory regime in a noisy system. The bifurcation diagram corresponding to rebuildings of distribution (Eq. 14) is shown in Fig. 16.

We have carried out preliminary computer experiments by integrating numerically stochastic equations (Eq. 13) and plotted the probability distribution $p(x, y)$ by means of statistical processing of data obtained at a sufficiently large time interval. Our calculations show that there is a bifurcation interval corresponding to the stochastic Andronov–Hopf bifurcation. The probability density evolution when ε and D are varied, is, in general, in good agreement with the conclusions of the theory presented in [26].

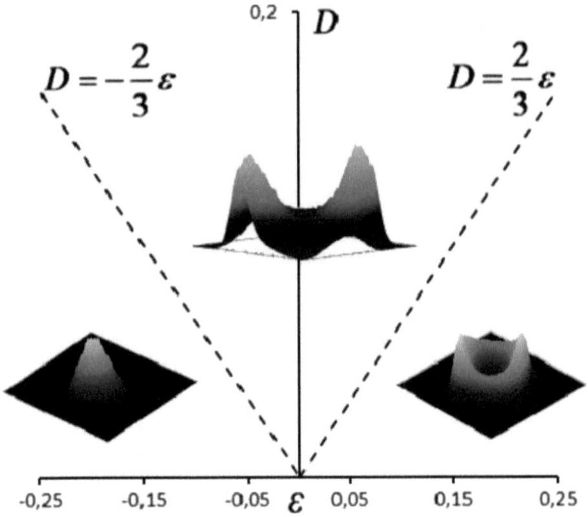

Fig. 16 Bifurcation diagram of system (Eq. 13) according to [26] (the *dashed lines* denote the bifurcation interval boundaries) and qualitative probability distributions in different regions of the diagram (numerical simulation)

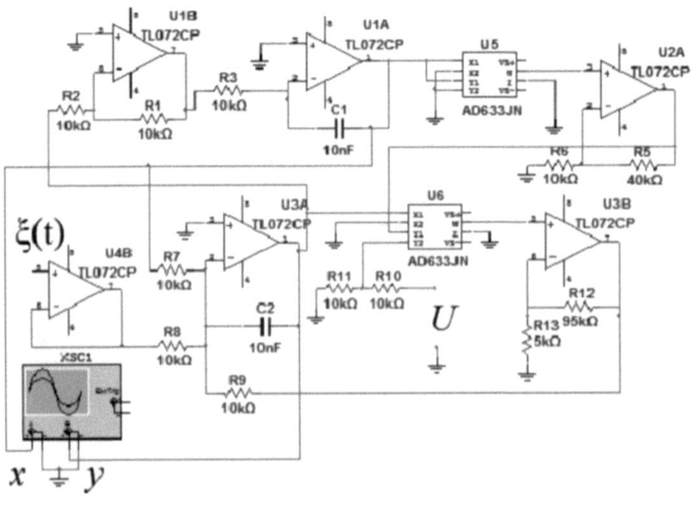

Fig. 17 Schematic diagram of the experimental setup (analog model of the Van der Pol oscillator)

We have also performed a series of full-scale experiments to study the effect of additive noise on the generation regime. To do this we have developed an experimental setup representing an analog model of system (Eq. 13). The diagram is shown in Fig. 17 and described by the following equations in physical variables:

$$R_0 C\dot{x} = y,$$
$$R_0 C\dot{y} = (U - x^2)y - x + \xi(t), \tag{15}$$

where x, y are voltages taken off at the corresponding outputs, U is the voltage that sets the parameter ε ($\varepsilon = U$). The scheme parameters are chosen so that the dimensionless variables x and $y = \dot{x}$ in Eq. (13) coincide quantitatively with the corresponding quantities in Eq. (15). Thus, Eqs. (13) and (15) differ only in scale over time, which is defined in Eq. (15) by parameters $R_0 = 10$ kOhm and $C = 10$ nF.[1] Equation (15) also include the random term $\xi(t)$ that describes the voltage produced by a noise generator. In our experiments we use a broadband Gaussian noise generator whose spectral density is almost constant in the frequency range 0–100 kHz. Within this frequency range we consider $\xi(t) = \sqrt{2\tilde{D}}n(t)$, where $n(t)$ is a normalized source with the unit power spectrum density. The quantity $2\tilde{D}$ is the power spectrum density of the applied noisy signal that can be controlled in the experiment. To compare the obtained results the normalized noise intensity D is found that corresponds to model (Eq. 13). Taking into account the time scale we have $D = \tilde{D}/R_0 C$. During our full-scale experiments $x(t)$ and $y(t)$ realizations are recorded, digitized by using the ADC NI PCI-6133, and then processed on a computer. This procedure enables to obtain the probability density $p(x, y)$ and observe its changes when ε and D are varied.

The carried out experiments have verified the existence of the bifurcation interval and the theoretical dependence of the distribution $p(x, y)$ form on the additive noise intensity, established in [26]. In our experiments when the noise intensity D grows the crater-like form of the distribution is destroyed for $\varepsilon > 0$. Figure 18a shows the experimentally obtained diagram of oscillator (Eq. 15) regimes in the parameter plane $(D, \varepsilon > 0)$. The dashed line marks the right boundary of the bifurcation interval for the Van der Pol oscillator $\varepsilon = 3D/2$. The evolution of distribution density $p(x, y)$ observed experimentally when increasing the noise intensity is illustrated in Fig. 18b. The obtained experimental results are in well qualitative agreement with the numerical results and theoretical conclusions.

It can be assumed that the specific character of probability distribution rebuilding when increasing the noise level depends on the properties of a dynamical system: on its dimension and the form of nonlinearity. However, in any case a closed crater corresponding to noisy self-sustained oscillations must be destroyed as noise increases. In order to verify this assumption, in addition to the Van der Pol oscillator with the two-dimensional phase space, we explore experimentally a

[1] All the terms in Eq. (15), regardless of the degree, are voltages taken off at different points of the scheme and measured in volts.

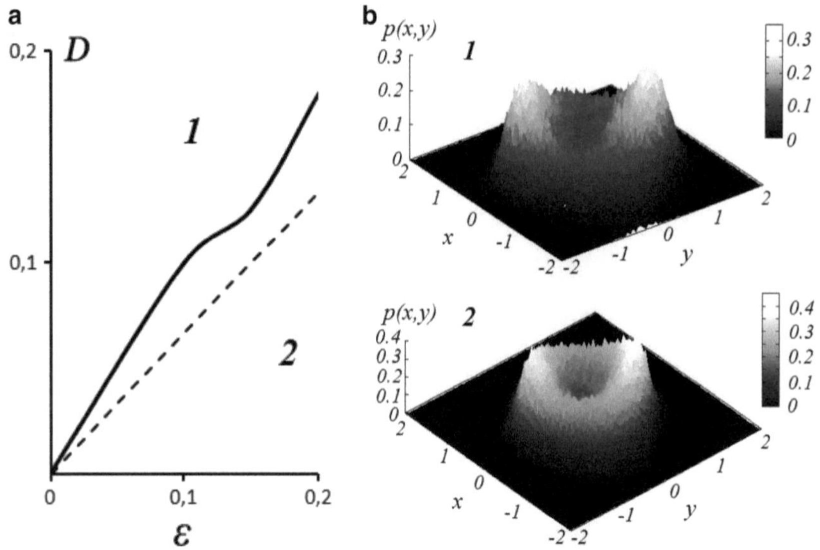

Fig. 18 (**a**) Experimentally obtained diagram of regimes for model (Eq. 15) (the theoretical boundary of the bifurcation interval according to [26] is marked by the *dashed line*) and (**b**) probability distribution densities typical for each region of the diagram

model of Anishchenko–Astakhov's oscillator with inertial nonlinearity with a three-dimensional phase space [43]. This oscillator is described by the following system of equations for dimensionless variables

$$
\begin{aligned}
\dot{x} &= mx + y - xz + \sqrt{2D}n(t), \\
\dot{y} &= -x, \\
\dot{z} &= -g(z - F(x)), \\
F(x) &= \begin{cases} x^2, & x \geq 0, \\ 0, & x < 0, \end{cases}
\end{aligned}
\tag{16}
$$

where m, g are control parameters of the system, $n(t)$ is a normalized white noise source, and D is the noise intensity. For $D = 0$ system (Eq. 16) has a single equilibrium point at the origin, that is stable for $m < 0$ and undergoes a supercritical Andronov–Hopf bifurcation at $m = 0$. When studying a stochastic bifurcation in three-dimensional system (Eq. 16) we consider a joint distribution of only two variables, x and y, as phase trajectories rotate around an unstable equilibrium in the plane of these variables. To implement full-scale experiments, an experimental setup is created which is an analog model of system (Eq. 16). The diagram is shown in Fig. 19 and described by the following equations in physical variables

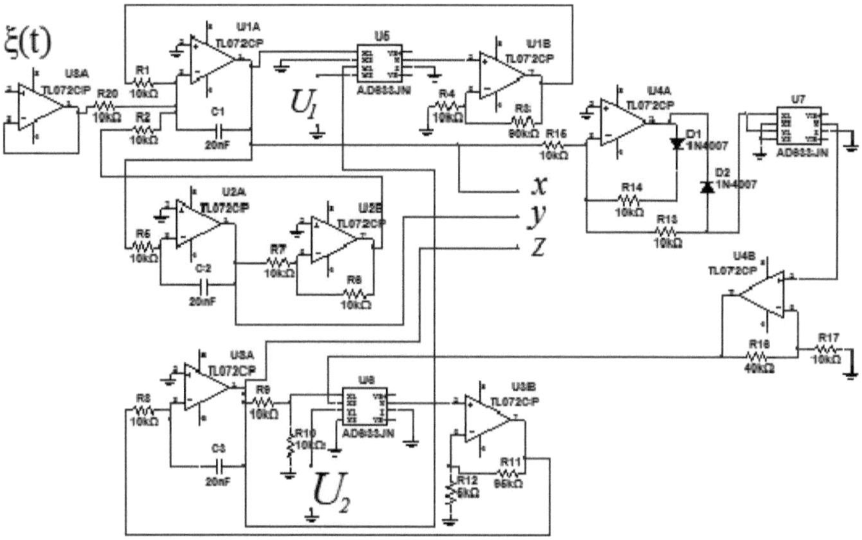

Fig. 19 Schematical diagram of the experimental setup (analog model of Anishchenko–Astakhov's oscillator)

$$
\begin{aligned}
R_0C\dot{x} &= U_1x + y - xz + \xi(t),\\
R_0C\dot{y} &= -x,\\
R_0C\dot{z} &= -U_2(z - F(x)),\\
F(x) &= \begin{cases} x^2, & x \geq 0,\\ 0, & x < 0, \end{cases}
\end{aligned}
\tag{17}
$$

where x, y, z are voltages from the corresponding outputs of the circuit, $U_{1,2}$ are voltages that set the values of the control parameters ($m = U_1, g = U_2$). As for the Van der Pol oscillator, the circuit parameters are chosen so that the dimensionless variables x and $y = \dot{x}$ in Eq. (16) are equal to the corresponding quantities in Eq. (17). The time scale is given in Eq. (17) by parameters $R_0 = 10$ kOhm and $C = 10$ nF. The random term $\xi(t)$ describes the voltage produced by a noise generator (as in the case of the Van der Pol oscillator): $\xi(t) = \sqrt{2\tilde{D}}n(t)$, $D = \tilde{D}/R_0C$.

Our numerical experiments have shown that in Eq. (17) a crater-like form of the distribution is also destroyed as the noise intensity D increases. Moreover, the crater wall is destroyed on one side only. This effect is well marked in Fig. 20a, b. The performed full-scale experiments have confirmed the same character of probability distribution evolution, which has been found in numerical simulation. Although analog model (Eq. 17) does not strictly quantitatively correspond to Eq. (16) (there is a minor shift of the bifurcation parameter values), the experimentally obtained evolution of the probability density with increasing the noise level (Fig. 20c, d) is in complete qualitative agreement with the numerical data.

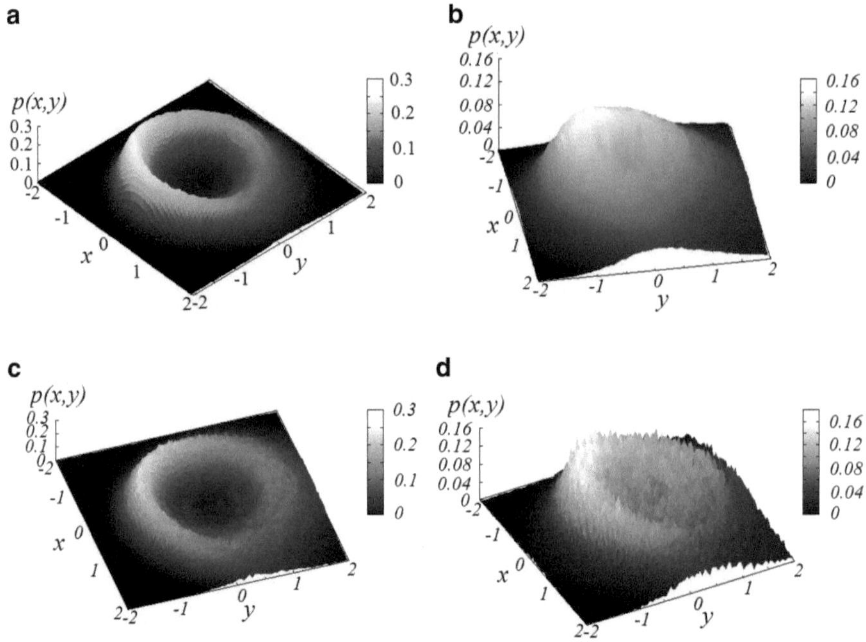

Fig. 20 Probability distributions in Anishchenko–Astakhov's oscillator for $m = 0.4, g = 0.25$ and for different values of the noise intensity in numerical simulation: (**a**) $D = 0.027$; (**b**) $D = 0.11$, and in full-scale experiments: (**c**) $D = 0.11$; (**d**) $D = 0.41$

4.2 *Probability Distribution Evolution in Self-sustained Oscillators with Multiplicative Noise*

It has been shown in the previous section that additive noise can lead to the qualitative rebuildings of the probability distribution. The destruction of the crater-like form of the distribution is observed for two qualitatively different oscillator models. This fact enables to suggest that this effect is general for a wide range of self-sustained oscillatory systems. The question remains open: How general this effect is with respect to the character of noise effect. One of the possible and widespread types of noise in radio devices is a parametric noise representing a random modulation of any parameters of a system. For example, in the Van der Pol oscillator the excitation parameter or the fundamental frequency may fluctuate randomly. In both cases the parametric noise is multiplicative as its intensity in the equations depends on one of the dynamical variables, x or y. Unfortunately, in the case of parametric noise we could not derive the expression for the probability density in the Van der Pol oscillator, similar to Eq. (14). In the theoretical work [25] dedicated to colored parametric noise, the conclusion is made concerning the existence of bifurcation interval and the delayed character of Andronov–Hopf bifurcation in the presence of noise with a large correlation time. At the same time it is indicated that for a small noise correlation the bifurcation can be advanced, i.e., it can occur earlier than in a deterministic system. The last statement contradicts the results presented in [29, 32, 42]. Thus, the nature of parametric noise effect on the supercritical Andronov–Hopf bifurcation remains unclear.

Now we study numerically and experimentally the stochastic supercritical Andronov–Hopf bifurcation in the same oscillators considered in the previous section but in the presence of parametric noise. We start with considering the Van der Pol oscillator. The dimensionless stochastic equations of the oscillator for the chosen type of parametric noise are as follows:

$$\begin{aligned} \dot{x} &= y, \\ \dot{y} &= (\varepsilon + \sqrt{2D}n(t) - x^2)y - x, \end{aligned} \tag{18}$$

where ε is the control parameter, $n(t)$ is a normalized Gaussian white noise source, and D is the noise intensity. Hence, the noise modulates the excitation parameter ε.

Full-scale experiments are carried out by using the analog model of the Van der Pol oscillator shown in Fig. 17. Instead of additive noise, a broadband random signal $\xi(t)$ is added to voltage U_0 which sets the control parameter ε. It should be noted that due to unavoidable natural fluctuations, the additive noise component still presents in the experimental setup, but it is very small compared with the external parametric noise produced by a special generator.

The experimental setup operation is described by the following equations:

$$\begin{aligned} R_0C\dot{x} &= y, \\ R_0C\dot{y} &= ((U_0 + \xi(t)) - x^2)y - x. \end{aligned} \tag{19}$$

As before, we assume $\xi(t) = \sqrt{2\tilde{D}}n(t)$, $\tilde{D} = R_0CD$.

Our numerical calculations have shown that the effect of parametric noise on the probability distribution has an essentially different character than the additive noise influence. The crater contracts, i.e., its diameter decreases, and this leads to the distribution formation with a single sharp peak. Such a distribution is typical for a noisy state of a stable equilibrium (Fig. 21a–c). At a fixed value of the control parameter ε corresponding the generation regime and as the noise intensity D increases, a qualitative rebuilding of the distribution takes place at a certain point. This situation can be treated as an inverse stochastic Andronov–Hopf bifurcation. Thus, the presence of parametric noise causes the bifurcation parameter value to increase as compared with the bifurcation point in a deterministic system. When the noise intensity increases, the bifurcation is more shifted that is consistent with the results presented in [29, 32, 42]. However, unlike [29, 42], we have not fixed the existence of the bifurcation interval: the transition from the distribution with the single maximum at the origin to a crater-like distribution (or vice-versa) when one of the control parameters is varied takes place at the same point.

Our full-scale experiments have completely confirmed the character of the distribution evolution, obtained numerically (Fig. 21d,e). The distribution evolution established for the Van der Pol oscillator in numerical and full-scale experiments is very similar to the theoretical results for the Hopf oscillator [32], which is strictly quasi-harmonic and for which, thus, the bifurcation interval cannot occur.

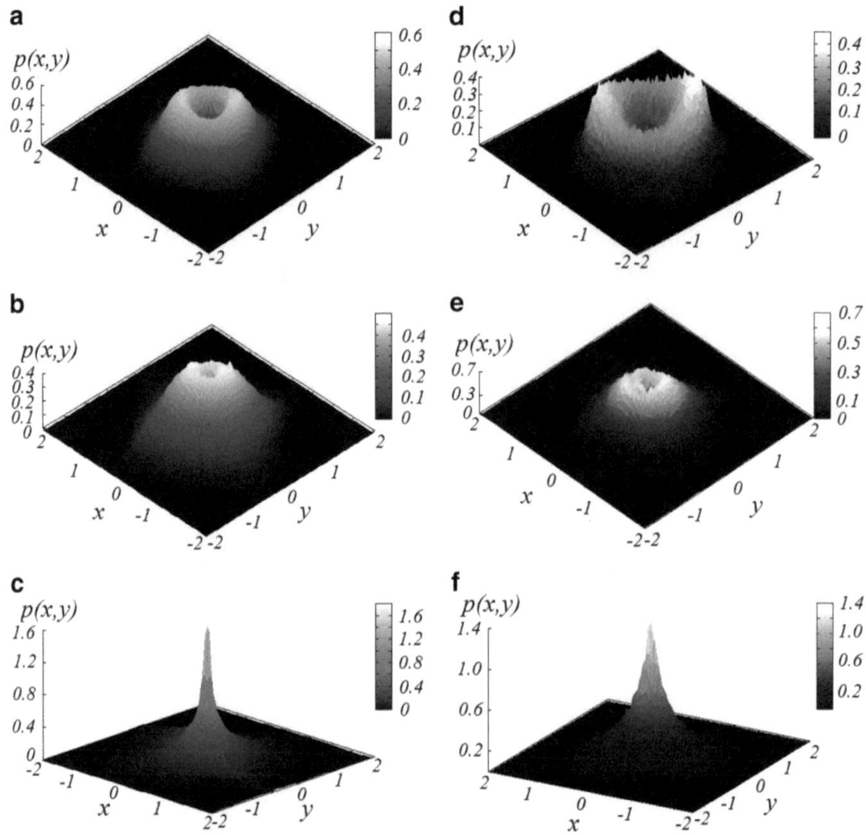

Fig. 21 Probability distributions for the Van der Pol oscillator at $\varepsilon = 0.10$ and for different values of the parametric noise intensity in numerical simulation: (**a**) $D = 0.05$; (**b**) $D = 0.09$; (**c**) $D = 0.19$, and in full-scale experiment: (**d**) $D = 0.055$; (**e**) $D = 0.12$; (**f**) $D = 0.19$

The experimentally obtained diagram of system (Eq. 19) regimes for $\varepsilon > 0$ is plotted in Fig. 22. The line shown in the parameter plane corresponds to the stochastic Andronov–Hopf bifurcation.

We have also conducted numerical and full-scale experiments for Anishchenko–Astakhov's oscillator in the presence of parametric noise modulating the parameter m. The following system is studied numerically:

$$\begin{aligned} \dot{x} &= (m + \sqrt{2D}n(t))x + y - xz, \\ \dot{y} &= -x, \\ \dot{z} &= -g(z - F(x)), \\ F(x) &= \begin{cases} x^2, x \geq 0, \\ 0, x < 0, \end{cases} \end{aligned} \tag{20}$$

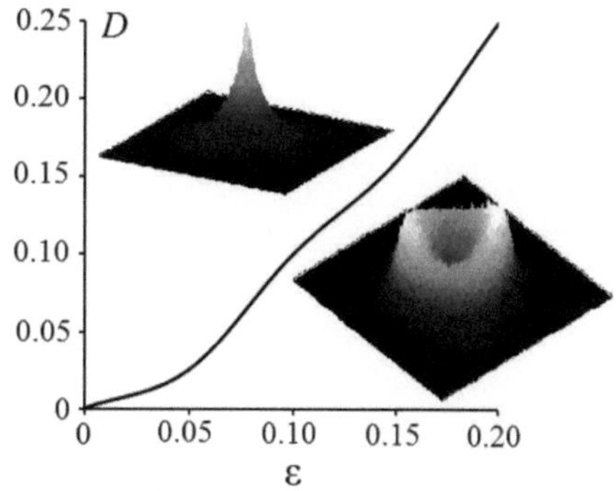

Fig. 22 Bifurcation diagram of system (Eq. 19)

where m, g are the control parameters of the system, $n(t)$ is a normalized Gaussian white noise source, and D is the noise intensity.

Full-scale experiments are carried out by using the analog model of Anishchenko–Astakhov's oscillator shown in Fig. 20. The additive noise is replaced by a random signal $\xi(t)$ added to voltage U_1 that sets the control parameter m. Voltage U_2 that gives the control parameter $g = U_2$ is fixed ($g = U_2 = 0.25$).

The probability density evolution is shown in Fig. 23 when the parametric noise intensity increases. Qualitatively the same evolution is observed both numerically (Fig. 23a–c) and experimentally (Fig. 23d–f). When increasing the noise intensity a rebuilding of the probability distribution takes place that consists in the decrease of the crater walls and in the simultaneous formation of a peak in the crater center.

Thus, the probability distribution evolution in Anishchenko–Astakhov's oscillator in the presence of parametric noise qualitatively differs from that observed in the Van der Pol oscillator (Fig. 21). One can assume that the system dimension plays a special role in this case. The equilibrium point in Anishchenko–Astakhov's oscillator becomes a saddle-focus after the bifurcation and besides an unstable two-dimensional manifold, has a stable one-dimensional manifold directed along the z axis. The parametric noise can lead to the appearance of phase trajectories tending to the equilibrium along the stable manifold. This can be resulted in that the distribution maximum at the origin (at the equilibrium) arises earlier than the crater corresponding to a limit cycle is destroyed.

5 Summary

We have experimentally studied two types of excitable oscillators: the classical model of an excitable oscillator—the FitzHugh–Nagumo system, and the Van der Pol oscillator with a subcritical Andronov–Hopf bifurcation. The carried out experiments have verified that in spite of the existing differences in mechanisms of stochastic oscillation initiation, both systems possess a particular similarity: they demonstrate the effect of "true" coherence resonance, i.e., the spectral line width

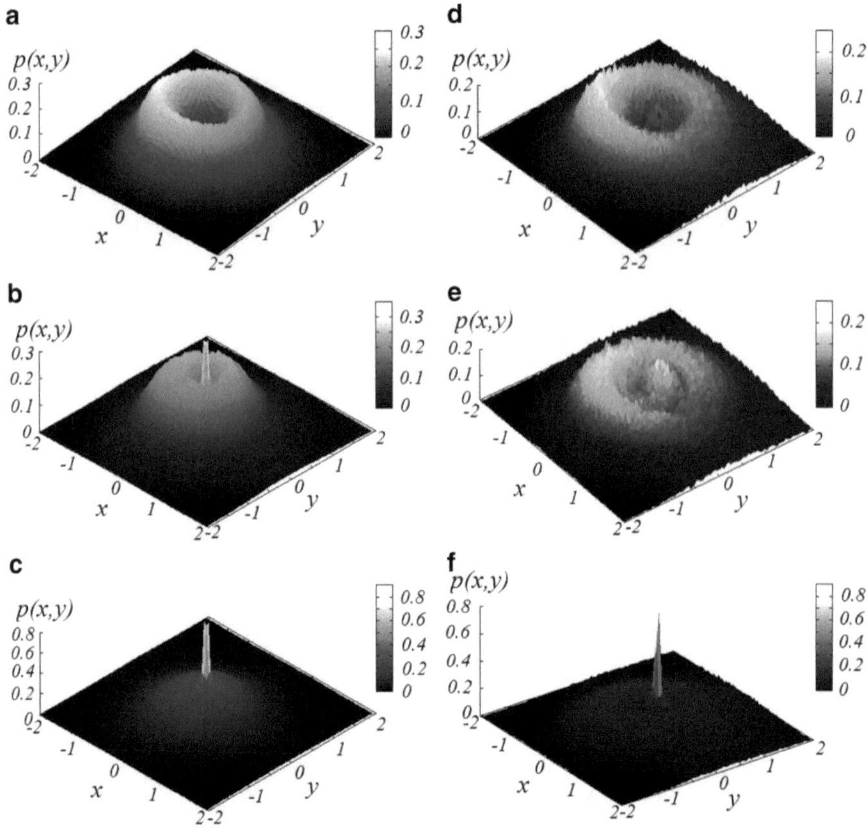

Fig. 23 Probability distribution densities in Anishchenko–Astakhov's oscillator for $m = 0.3$, $g = 0.25$ and for different values of the noise intensity in numerical simulation: (**a**) $D = 0.05$; (**b**) $D = 0.08$; (**c**) $D = 0.10$, and in full-scale experiment: (**d**) $D = 0.10$; (**e**) $D = 0.12$; (**f**) $D = 0.13$

decreases at a certain optimal noise level. For the oscillator with hard excitation the CR effect is observed not only in the bistability region but also before oscillations arise in a deterministic system. The last circumstance enables to treat the oscillator with hard excitation as a special type of an excitable system. Moreover, as a result of the performed experiments it has been first established that the effect of stochastic synchronization can be observed in the oscillator with hard excitation in the subthreshold regime as well as for the FitzHugh–Nagumo oscillator. Thus, the presence of synchronization effect is the main reason that enables to attribute stochastic oscillations of excitable systems of both types to self-sustained oscillatory regimes. A noise source presented in a system removes it from its stable equilibrium to the phase space region where a nonlinear process of energy pumping is switched on that leads to the origin of self-sustained oscillations.

We have also experimentally explored the effect of noise on the supercritical Andronov–Hopf bifurcation for two different self-sustained oscillators. Our experiments have shown that in the Van der Pol oscillator with additive noise there is a bifurcation interval theoretically established in [26]. When the excitation parameter of increases, the upper boundary of the bifurcation interval corresponds to the transition from a "two-maximum" distribution $p(x, y)$ to a distribution in the form of a closed crater, being characteristic for the regime of a noisy limit cycle. However, in Anishchenko–Astakhov's oscillator with 1.5 freedom degrees the nature of distribution $p(x, y)$ rebuildings is different. As noise increases, only one crater wall is destroyed resulted in the appearance of a one-maximum distribution. Correspondingly, in this case the noise simply leads to the bifurcation delay, which is shifted to higher values of the parameter of generation. The delay of stochastic bifurcation has also been observed in our experiments with parametric noise, which is consistent with the theoretical results presented in [29, 32]. With this, as in the case of additive noise, the character of distribution evolution for the Van der Pol oscillator and Anishchenko–Astakhov's oscillator is essentially different that can be related to a different phase space dimension of the systems. In all the considered cases the growth of the noise level in a system being in the generation regime leads to the disappearance (destruction or "constriction") of a closed crater in the distribution corresponding to a noisy limit cycle. In this sense one can speak of the noise-induced destruction (or suppression) of self-sustained oscillations.

Acknowledgments This work is supported by the Russian Ministry of Education and Sciences in the framework of the state contract N 14.B37.21.0751.

References

1. Stratonovich RL (1961) Selected problems of the fluctuation theory in radiotechnique. Sov. Radio, Moscow (in Russian)
2. Freidlin MI, Wentzell AD (1984) Random perturbations in dynamical systems. Springer, New York
3. Gardiner CW (1982) Handbook of stochastic methods. Springer series in synergetics, vol 13. Springer, Berlin
4. Risken H (1984) The Fokker–Planck equation. Springer series in synergetics, vol 18. Springer, Berlin
5. Van Kampen NG (1992) Stochastic processes in physics and chemistry, 2nd edn. North Holland, Amsterdam
6. Horsthemke W, Lefever R (1983) Noise induced transitions: theory and applications in physics, chemistry and biology. Springer series in synergetics, vol 15. Springer, Berlin
7. Graham R (1990) Macroscopic potentials, bifurcations and noise in dissipative systems. In: Moss F, McClintock PVE (eds) Theory of continuous Fokker–Planck systems. Noise in nonlinear dynamical systems, vol 1. Cambridge University Press, Cambridge
8. Arnold L (2003) Random dynamical systems. Springer, Berlin
9. Garcia-Ojalvo J, Sancho JM (1999) Noise in spatially extended systems. Springer, New York
10. Benzi R, Sutera A, Vulpiani A (1981) J Phys Math Gen 14:L453
11. Gammaitoni L, Marchesoni F, Menichella-Saetta E, Santucci S (1989) Phys Rev Lett 62:349

12. Anishchenko VS, Neiman AB, Moss F, Schimansky-Geier L (1999) Phys Usp 42:7
13. Anishchenko V et al (2002) Nonlinear dynamics of chaotic and stochastic systems: Tutorial and modern development. Berlin, Springer
14. Pikovsky A, Kurths J (1997) Phys Rev Lett 78:775
15. Lindner B, Schimansky-Geier L (1999) Phys Rev E 60(6):7270
16. Neiman AB (1994) Phys Rev E 49:3484–3488
17. Shulgin B, Neiman A, Anishchenko V (1995) Phys Rev Lett 75(23):4157
18. Anishchenko V, Neiman A (1997) Stochastic synchronization. In: Schimansky-Geier L, Pöschel T (eds) Stochastic dynamics. Springer, Berlin, p 155
19. Anishchenko VS, Vadivasova TE, Strelkova GI (2010) Eur Phys J Spec Top 187:109–125
20. Han SK, Yim TG, Postnov DE, Sosnovtseva OV (1999) Phys Rev Lett 83(9):1771
21. Neiman A, Schimansky-Geier L, Cornell-Bell A, Moss F (1999) Phys Rev Lett 83(23):4896
22. Hu B, Zhou Ch (2000) Phys Rev E 61(2):R1001
23. Ushakov OV, Wùnsche H-J, Henneberger F, Khovanov IA, Schimansky-Geier L, Zaks MA (2005) Phys Rev Lett 95:123903(4)
24. Zakharova A, Vadivasova T, Anishchenko V, Koseska A, Kurths J (2010) Phys Rev E 81:011106(1–6)
25. Lefever R, Turner J (1986) Phys Rev Lett 56:1631
26. Ebeling W, Herzel H, Richert W, Schimansky-Geier L (1986) Z Angew Math Mech 66:141
27. Schimansky-Geier L, Herzel H (1993) J Stat Phys 70:141
28. Arnold L, Sri Namachshivaya N, Schenk-Yoppe JR (19996) Int J Bifurcat Chaos 6:1947
29. Olarrea J, de la Rubia FJ (1996) Phys Rev E 53(1):268
30. Landa PS, Zaikin AA (1996) Phys Rev E 54(4):3535
31. Crauel H, Flandol F (1998) J Dynam Differ Equat 10:259
32. Bashkirtseva I, Ryashko L, Schurz H (2009) Chaos Solit Fract 39:7
33. Longtin A (1993) J Stat Phys 70:309
34. Baltanas JP, Casado JM (1998) Phys D 122:231
35. Izhikevich EM (2007) Dynamical systems in neuroscience: The geometry of excitability and bursting. Cambridge, MIT Press, Cambridge, MA
36. FitzHugh R (1955) Bull Math Biophys 17:257
37. Scott AC (1975) Rev Mod Phys 47:487
38. Anishchenko VS, Vadivasova TE, Feoktistov AV, Strelkova GI (2013) Stochastic oscillators. In: Rubio RG, Ryazantsev YS, Starov VM, Huang G-X, Chetverikov AP, Arena P, Nepomnyashchy AA, Ferrus A, Morozov EG (eds) Without bounds: A scientific canvas of nonlinearity and complex dynamics. Springer, Berlin, p 20
39. Makarov VA, del Rio E, Ebeling W, Velarde MG (2001) Phys Rev E 64:036601
40. Feoktistov AV, Astakhov SV, Anishchenko VS (2010) Izv. VUZ. Appl Nonlinear Dynam 18:33 (in Russian)
41. Kuznetsov AP, Milovanov SV (2003) Izv. VUZ. Appl Nonlinear Dynam 11:16 (in Russian)
42. Franzoni L, Mannella R, McClintock PVE, Moss F (1987) Phys Rev E 36:834
43. Anishchenko VS (2009) Complex oscillations in simple systems. URSS, Moscow (in Russian)

Index

V. Afraimovich et al. (eds.), *Nonlinear Dynamics and Complexity*,
Nonlinear Systems and Complexity 8, DOI 10.1007/978-3-319-02353-3,

MIX
Papier aus verantwortungsvollen Quellen
Paper from responsible sources
FSC® C105338
FSC
www.fsc.org

If you have any concerns about our products,
you can contact us on
ProductSafety@springernature.com

In case Publisher is established outside the EU,
the EU authorized representative is:
Springer Nature Customer Service Center GmbH
Europaplatz 3, 69115 Heidelberg, Germany

Printed by Libri Plureos GmbH
in Hamburg, Germany